Chemometrics in Food Chemistry

Chemometrics in Food Chemistry

Contributors

Riccardo Guidetti, Roberto Beghi et al.

AURIS
Reference

www.aurisreference.com

Chemometrics in Food Chemistry

Contributors: Riccardo Guidetti, Roberto Beghi et al.

Published by Auris Reference Limited

www.aurisreference.com

United Kingdom

Chemometrics in Food Chemistry

ISBN: 978-1-78154-872-1

British Library Cataloguing in Publication Data
A CIP record for this book is available from the British Library

Printed in the United Kingdom

Exclusively distributed by CBS Publishers & Distributors Pvt. Ltd.

Sales & Distribution Rights only for India, Pakistan, Bangladesh, Sri Lanka, Nepal and Bhutan. This book is not to be sold outside these territories.

Contents

List of Abbreviations .... *vii*

List of Contributors... ....*ix*

Preface... ....*xv*

Chapter 1   **Chemometrics in Food Technology** 1

Riccardo Guidetti, Roberto Beghi and Valentina Giovenzana

Chapter 2   **Chemometric Analysis of the Amino Acid Requirements of Antioxidant Food Protein Hydrolysates**... 45

Chibuike C. Udenigwe and Rotimi E. Aluko

Chapter 3   **Profiling of Fatty Acids Composition in Suet Oil Based on Gc–Ei-Qms And Chemometrics Analysis**.. 61

Jun Jiang and Xiaobin Jia

Chapter 4   **Comprehensive and Comparative Metabolomic Profiling of Wheat, Barley, Oat and Rye Using Gas Chromatography-Mass Spectrometry and Advanced Chemometrics** 75

Bekzod Khakimov, Birthe Møller Jespersen and Søren Balling Engelsen

Chapter 5   **Quantitative Analysis of Total Amino Acid in Barley Leaves Under Herbicide Stress Using Spectroscopic Technology and Chemometrics**.. 93

Yidan Bao, Wenwen Kong, Yong He, Fei Liu, Tian Tian and Weijun Zhou

Chapter 6   **Concomitant Use of Fourier Transform Infrared Attenuated Total Reflectance Spectroscopy and Chemometrics for Quantification of Multiple Adulterants in Roasted and Ground Coffee** 105

Nádia Reis, Adriana S. Franca and Leandro S. Oliveira

Chapter 7   **Identification of Imitation Cheese and Imitation Ice Cream Based on Vegetable Fat Using NMR Spectroscopy and Chemometrics**.. 119

Yulia B. Monakhova, Rolf Godelmann, Claudia Andlauer, Thomas Kuballa, and Dirk W.

Chapter 8   The Role of Visible and Infrared Spectroscopy Combined With
             Chemometrics to Measure Phenolic Compounds in Grape
             and Wine Sampless 137
             Cozzolino Daniel

Chapter 9   Exploratory Data Analysis With Latent Subspace Models 153
             José Camacho

Chapter 10   Metabolomics and Chemometrics as Tools for Chemo(Bio)
             Diversity Analysis - Maize Landraces and Propolis 187
             Marcelo Maraschin et al

Chapter 11   Parafac Analysis for Temperature-Dependent NMR Spectra of
             Poly(Lactic Acid) Nanocomposite 207
             Hideyuki Shinzawa, Masakazu Nishida, Toshiyuki Tanaka,
             Kenzi Suzuki and Wataru Kanematsu

Chapter 12   Using Principal Component Scores and Artificial Neural
             Networks In Predicting Water Quality Index 225
             Rashid Atta Khan, Sharifuddin M. Zain, Hafizan Juahir,
             Mohd Kamil Yusoff and Tg Hanidza T.I.

Chapter 13   Application of Chemometrics to the Interpretation of Analytical
             Separations Data.. 243
             James J. Harynuk, A. Paulina de la Mata and Nikolai A. Sinkov

             Citations 271

             Index.. 275

List of Abbreviations

ADF	Acid detergent fiber
ANN	Artificial Neural Networks
ATR	attenuated total reflectance
BP	Back propagation
BOD	biochemical oxygen demand
BBD	Box–Behnken design
CE	Capillary electrophoresis
COD	chemical oxygen demand
PARAFAC	concept of Parallel factor
CT	condensed tannins
CPMAS	Cross-polarization magic-angle
DSC	differential scanning calorimetry
DS	dummy scans
EWs	effective wavelengths
ED	Eigendecomposition
RPLC	example reverse-phase
EDA	Exploratory Data Analysis
FA	Fatty acid
FOV	field-of-view
FT	Fourier transform
GC	Gas chromatography
HILIC	hydrophilic interaction
IR	infrared
DRIFTS	Infrared Spectroscopy
IS	Internal standard
LV	latent variables
LOQ	limits of quantification
LDA	linear discriminant analysis
LDL	low-density lipoprotein
MMT	Montmorillonite
MSC	Multiple scatter correction
MCR	multivariate curve resolution
NIR	Near infrared
NDF	neutral detergent fiber
NPS	non-point source
NMR	nuclear magnetic resonance
PLS	Partial Least Square
PS	point source
PCA	Principal component analysis
RPD	ratio deviation in prediction
RG	Receiver gain

NIR	Region of near infrared
RC	Regression coefficients
RSD	relative standard deviations
SEC	size exclusion
FTIR	spectroscopy
SEP	standard error in prediction
SNV	standard normal variate

List of Contributors

Riccardo Guidetti
Department of Agricultural Engineering, Università degli Studi di Milano, Milano, Italy

Roberto Beghi
Department of Agricultural Engineering, Università degli Studi di Milano, Milano, Italy

Valentina Giovenzana
Department of Agricultural Engineering, Università degli Studi di Milano, Milano, Italy

Chibuike C. Udenigwe
The Department of Human Nutritional Sciences and the Richardson Centre for Functional Foods and Nutraceuticals, University of Manitoba, Winnipeg, MB R3T 2N2, Canada

Rotimi E. Aluko
The Department of Human Nutritional Sciences and the Richardson Centre for Functional Foods and Nutraceuticals, University of Manitoba, Winnipeg, MB R3T 2N2, Canada

Jun Jiang
Affiliated Hospital on Integration of Chinese and Western Medicine, Nanjing University of Chinese Medicine, Xianlin Avenue 138#, Xianlin University City, Nanjing 210023, China
Key Laboratory of New Drug Delivery System of Chinese Meteria Medica, Jiangsu Provincial Academy of Chinese Medicine, 100# Shizi Road, Nanjing 210028, China

Xiaobin Jia
Affiliated Hospital on Integration of Chinese and Western Medicine, Nanjing University of Chinese Medicine, Xianlin Avenue 138#, Xianlin University City, Nanjing 210023, China
Key Laboratory of New Drug Delivery System of Chinese Meteria Medica, Jiangsu Provincial Academy of Chinese Medicine, 100# Shizi Road, Nanjing 210028, China

Bekzod Khakimov
Department of Food Science, Faculty of Science, University of Copenhagen, Rolighedsvej 30, Frederiksberg C, 1958 Copenhagen, Denmark

Birthe Møller Jespersen
Department of Food Science, Faculty of Science, University of Copenhagen, Rolighedsvej 30, Frederiksberg C, 1958 Copenhagen, Denmark

Søren Balling Engelsen
Department of Food Science, Faculty of Science, University of Copenhagen, Rolighedsvej 30, Frederiksberg C, 1958 Copenhagen, Denmark

Yidan Bao
College of Biosystems Engineering and Food Science, Zhejiang University, Hangzhou 310058, China

Wenwen Kong
College of Biosystems Engineering and Food Science, Zhejiang University, Hangzhou 310058, China

Yong He
College of Biosystems Engineering and Food Science, Zhejiang University, Hangzhou 310058, China
Cyrus Tang Center for Sensor Materials and Applications, Zhejiang University, Hangzhou 310058, China

Fei Liu
College of Biosystems Engineering and Food Science, Zhejiang University, Hangzhou 310058, China

Tian Tian
College of Agriculture and Biotechnology, Zhejiang University, Hangzhou 310058, China

Weijun Zhou
College of Agriculture and Biotechnology, Zhejiang University, Hangzhou 310058, China

Nádia Reis
PPGCA, Universidade Federal de Minas Gerais, Avenida Antônio Carlos 6627, 31270-901 Belo Horizonte, MG, Brazil
DEMEC, Universidade Federal de Minas Gerais, Avenida Antônio Carlos 6627, 31270-901 Belo Horizonte, MG, Brazil

Adriana S. Franca
PPGCA, Universidade Federal de Minas Gerais, Avenida Antônio Carlos 6627, 31270-901 Belo Horizonte, MG, Brazil
DEMEC, Universidade Federal de Minas Gerais, Avenida Antônio Carlos 6627, 31270-901 Belo Horizonte, MG, Brazil

Leandro S. Oliveira
PPGCA, Universidade Federal de Minas Gerais, Avenida Antônio Carlos 6627, 31270-901 Belo Horizonte, MG, Brazil
DEMEC, Universidade Federal de Minas Gerais, Avenida Antônio Carlos 6627, 31270-901 Belo Horizonte, MG, Brazil

Yulia B. Monakhova
Chemisches und Veterinäruntersuchungsamt (CVUA) Karlsruhe, Weissenburger Strasse 3, 76187 Karlsruhe, Germany
Department of Chemistry, Saratov State University, Astrakhanskaya Street 83, 410012 Saratov, Russia
Bruker Biospin GmbH, Silbersteifen, 76287 Rheinstetten, Germany

Rolf Godelmann
Chemisches und Veterinäruntersuchungsamt (CVUA) Karlsruhe, Weissenburger Strasse 3, 76187 Karlsruhe, Germany

Claudia Andlauer
Chemisches und Veterinäruntersuchungsamt (CVUA) Karlsruhe, Weissenburger Strasse 3, 76187 Karlsruhe, Germany

Thomas Kuballa
Chemisches und Veterinäruntersuchungsamt (CVUA) Karlsruhe, Weissenburger Strasse 3, 76187 Karlsruhe, Germany

Dirk W. Lachenmeier
Chemisches und Veterinäruntersuchungsamt (CVUA) Karlsruhe, Weissenburger Strasse 3, 76187 Karlsruhe, Germany
Ministry of Rural Affairs and Consumer Protection, Kernerplatz 10, 70182 Stuttgart, Germany

Cozzolino Daniel
School of Agriculture, Food and Wine, Faculty of Sciences, The University of Adelaide, Waite Campus, PMB 1 Glen Osmond, Adelaide, SA 5064, Australia

José Camacho
Department of Signal Theory, Telematics and Communication, University of Granada, Granada Spain

Marcelo Maraschin et al
Plant Morphogenesis and Biochemistry Laboratory, Federal University of Santa Catarina, Florianopolis, SC, Brazil

Hideyuki Shinzawa
Research Institute of Instrumentation Frontier, Advanced Industrial Science and Technology (AIST)

Masakazu Nishida
Research Institute of Instrumentation Frontier, Advanced Industrial Science and Technology (AIST)

Toshiyuki Tanaka
Mikawa Textile Research Center, Aichi Industrial Technology Institute (AITEC)

Kenzi Suzuki
Department of Chemical Engineering, Graduate School of Engineering, Nagoya University Japan

Wataru Kanematsu
Research Institute of Instrumentation Frontier, Advanced Industrial Science and Technology (AIST)

Rashid Atta Khan
Chemistry Department, Faculty of Science, University of Malaya, Kuala Lumpur Malaysia

Sharifuddin M. Zain
Chemistry Department, Faculty of Science, University of Malaya, Kuala Lumpur Malaysia

Hafizan Juahir
Department of Environmental Science, Faculty of Environmental Study, University Putra Malaysia, Serdang

Mohd Kamil Yusoff
Department of Environmental Science, Faculty of Environmental Study, University Putra Malaysia, Serdang

Tg Hanidza T.I.
Department of Environmental Science, Faculty of Environmental Study, University Putra Malaysia, Serdang

James J. Harynuk
Department of Chemistry, University of Alberta Canada

A. Paulina de la Mata
Department of Chemistry, University of Alberta Canada

Nikolai A. Sinkov
Department of Chemistry, University of Alberta Canada

Preface

Chemometrics is the chemical discipline that uses mathematical and statistical methods to design or select optimal procedures and experiments, and to provide maximum chemical information by analyzing chemical data. Chemometrics is a necessary and powerful tool for the field of food analysis and control. For food science in general and food analysis and control in particular, there are several problems for which chemometrics are of utmost importance. Chemometrics in Food Chemistry include process control and monitoring, the possibility of using RGB or hyperspectral imaging techniques to nondestructively check food quality, calibration of multidimensional or hyphenated instruments etc. First chapter focuses on chemometrics in food technology. Second chapter focuses on chemometric analysis of the amino acid requirements of antioxidant food protein hydrolysates. Third chapter focuses on profiling of fatty acids composition in suet oil based on GC–EI-qMS and chemometrics analysis. Fourth chapter outlines a comprehensive protocol for GC-MS metabolomic profiling of phenolics and organic acids in grains, the performance of which is demonstrated through a comparison of the metabolite profiles of the main northern European cereal crops: wheat, barley, oat and rye. Fifth chapter focuses on the feasibility of developing a rapid and effective method for the quantification of TAA in barley leaves using NIR spectroscopy to provide a new monitoring method for herbicide injury. Sixth chapter proposes the joint use of Fourier Transform Infrared Attenuated Total Reflectance Spectroscopy (FTIR-ATR) and Partial Least Square (PLS) regression for the simultaneous qsuantification of four adulterants (coffee husks, spent coffee grounds, barley, and corn) in roasted and ground coffee. In seventh chapter, 400 MHz nuclear magnetic resonance (NMR) spectroscopy of the fat fraction of the products was used in the context of food surveillance to validate the labeling of milk-based products. The objective of eighth chapter is to present examples and to discuss different applications of visible (VIS), near infrared (NIR) and mid infrared (MIR) to assess and measure phenolic compounds in grape and wines. In ninth chapter, several standard tools for EDA with projection models, namely score plots, loading plots and biplots, are revised and their limitations are elucidated. Tenth chapter focuses on metabolomics and chemometrics as tools for chemo(bio)diversity analysis - maize landraces and propolis. Eleventh chapter provides a tutorial on the fundamental concept of Parallel factor (PARAFAC) analysis and a practical example of its application. Twelfth chapter focuses on principal component scores and artificial neural networks in predicting water quality index. Thirteenth chapter presents a brief overview of separations approaches, with a focus on the data that are derived from different methods and on phenomena in the separations approach that lead to challenges in data interpretation.

Chapter 1

CHEMOMETRICS IN FOOD TECHNOLOGY

Riccardo Guidetti, Roberto Beghi and Valentina Giovenzana

Department of Agricultural Engineering, Università degli Studi di Milano, Milano, Italy

INTRODUCTION

The food sector is one of the most important voices in the economic field as it fulfills one of the main needs of man. The changes in the society in recent years have radically modified the food industry by combining the concept of globalization with the revaluation of local production. Besides the production needs to be global, in fact, there are always strong forces that tend to re-evaluate the expression of the deep local production like social history and centuries-old tradition. The increase in productivity, in ever-expanding market, has prompted a reorganization of control systems to maximize product standardization, ensuring a high level of food security, promote greater compliance among all batches produced. The protection of large quantities of production, however, necessarily passes through systems to highlight possible fraud present throughout the production chain: from the raw materials (controlled by the producer) to the finished products (controlled by large sales organizations). The fraud also concern the protection of local productions: the products of guaranteed origin must be characterized in such a way to identify specific properties easily and detectable by objective means. The laboratories employ analytical techniques that are often inadequate because they require many samples, a long time to get the response, staff with high analytical ability. In a context where the speed is an imperative, technology solutions must require fewer samples or, at least no one (non-destructive techniques); they have to provide quick answers, if not immediate, in order to allow the operator to decide quickly about further steps to control or release the product to market; they must be easy to use, to promote their use throughout the production chain where it is not always possible to have analytical laboratories. The technologies must therefore be adapted to this new approach to production: the sensors and the necessary related data modeling, which allows the "measure", are evolving to meet the needs of the agri-food sector. The trial involves, often, Research Institutions on the side of Companies, a sign of a great interest and a high level of expectations. The manufacturers of technologies, often, provide devices that require calibration phases not always easy to perform, but that are often the subject of actual researches. These are particularly complex when the modeling approach must be based on chemometrics.

This chapter is essentially divided into two parts: the first part analyzes the theoretical principles of the most important technologies, currently used in the food industry, that used a chemometric approach for the analysis of data (spectrophotometry Vis/NIR (Visible and Near InfraRed) and NIR (Near InfraRed), Image Analysis with particular regard to Hyperspectral Image Analysis and Electronic Nose); the second part will present some case studies of particular interest related to the same technologies (fruit and vegetables, wine, meat, fish, dairy, olive, coffee, baked goods, etc.) (Frank & Todeschini, 1994; Massart et al., 1997 and 1998; Basilevsk, 1994; Jackson, 1991).

TECHNOLOGIES USED IN THE FOOD SECTOR COM-BINED WITH CHEMOMETRICS

NIR and Vis/NIR Spectroscopy

Among the non-destructive techniques has met a significant development in the last 20 years the optical analysis in the region of near infrared (NIR) and visible-near infrared (Vis/NIR), based on the use of information arising from the interaction between the structure of food and light.

Electromagnetic Radiation

Spectroscopic analysis is a group of techniques allowing to get information on the structure of matter through its interaction with electromagnetic radiation.

Radiation is characterized by (Fessenden & Fessenden, 1993):

- wavelength (λ), which is the distance between two adjacent maxima and is measured in nm;
- frequency (v), representing the number of oscillations described by the wave per unit of time and is measured in hertz (cycles/s);
- wave number (n), which represents the number of cycles per centimeter and is measured in cm^{-1}.

The entire electromagnetic spectrum is divided into several regions, each characterized by a range of wavelengths (Fig.1)

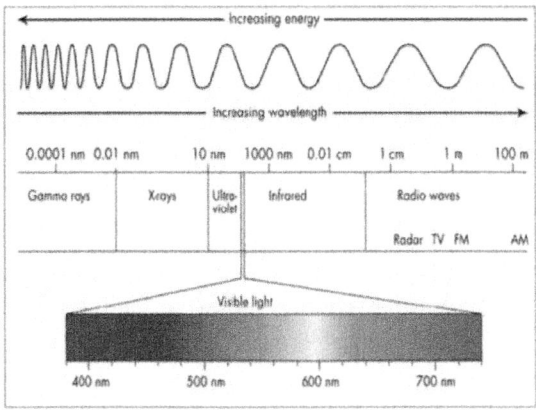

Figure. 1: The electromagnetic spectrum (Lunadei, 2008).

Transitions in the Near Infrared Region (NIR)

The radiation from the infrared region is able to promote transitions at vibrational level. The infrared spectroscopy is used to acquire information about the nature of the functional groups present in a molecule. The infrared region is conventionally divided into three subregions: near (750-2500 nm), medium (2500-50000 nm) and far infrared (50-1000 μm). Fundamental vibrational transitions, namely between the ground state and first excited state, take place in the mid-infrared, while in the region of near-infrared absorption bands are due to transitions between the ground state and the second or the third excited state. This type of transitions are called overtones and their absorption bands are generally very weak. The absorption bands associated with overtones can be identified and correlated to the corresponding absorption bands arising from the fundamental vibrational transitions because they fall at multiple wavelengths of these. Following the process of absorption of photons by molecules the intensity of the radiation undergoes a decrease. The law that governs the absorption process is known as the BeerLambert Law:

$$A = \log (I_0/I) = \log (1/T) = \varepsilon \cdot l \cdot c \tag{1}$$

where:

A = absorbance [log (incident light intensity/transmitted beam intensity)];
T = transmittance [beam intensity transmitted/incident light intensity];
I_0 = radiation intensity before interacting with the sample;
I = radiation intensity after interaction with the sample;
ε = molar extinction coefficient characteristic of each molecule ($l \bullet mol^{-1} \bullet cm^{-1}$);
l = optical path length crossed by radiation (cm);
c = sample concentration (mol/l).

The spectrum is a graph where in the abscissa is reported a magnitude related to the nature of radiation such as the wavelength (λ) or the wave number (n) and in the Y-axis a quantity related to the change in the intensity of radiation as absorbance (A) or transmittance (T).

Instruments

Since '70s producers developed analysis instruments specifically for NIR analysis trying to simplify them to fit also less skillful users, thanks to integrated statistical software and to partial automation of analysis. Instruments built in this period can be divided in three groups: desk instruments, compact portable instruments and on-line compatible devices. Devices evolved over the years also for the systems employed to select wavelength. First instruments used filter devices able to select only some wavelength (Fig. 2). These devices are efficient when specific wavelength are needed. Since the second half of '80s instruments capable to acquire simultaneously the sample spectrum in a specific interval of wavelength were introduced, recording the average spectrum of a single defined sample area (diode array systems and FT-NIR instruments) (Stark & Luchter, 2003). At the same time, chemometric data analysis growth helped to diffuse NIR analysis.

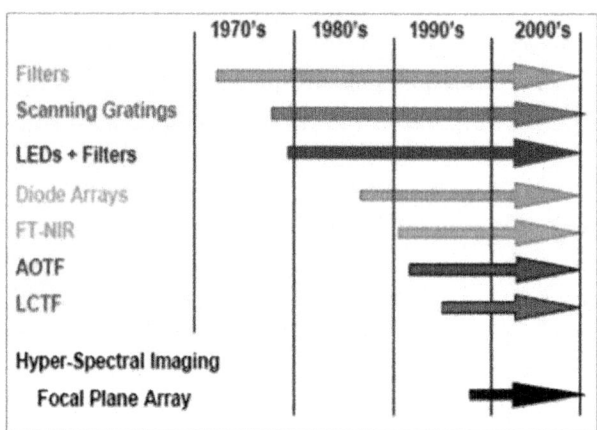

Figure. 2: Development of the different analysis technologies scheme (Stark & Luchter, 2003).

Particularly, food sector showed interest towards NIR and Vis/NIR instruments, both mobile and on-line. Devices based on diode array spectrophotometers and FT-NIR desk systems proved to be the best for this sector. Both in the case of portable and stationary instruments, the fundamental components of these systems are common and are four:

- Light source;
- Light radiation transport system;
- Sample compartment and measurement zone;
- Spectrophotometer and Personal Computers.

Light Source

Tungsten filament halogen lamps are chosen as the light source by most of the instruments. This is due to a good compromise between good performance and relatively low cost. This type of lamps are particularly suitable for use in low voltage. A little drawback may be represented by sensitivity to vibration of the filament. Halogen bulbs are filled with halogen gas to extend their lives by using the return of evaporated tungsten to the filament. The life of the lamp depends on the design of the filament and the temperature of use, on average ranges from a minimum of 50 hours and a maximum of 10000 hours at rated voltage. The lamp should be chosen according to the use conditions and the spectral region of interest. An increase in the voltage of the lamp may cause a shift of the peaks of the emission spectrum towards the visible region but can also lead to a reduction of 30% of its useful life. On the contrary, use of lower voltages can increase the lamp life together, however, with an intensity reduction of light radiation, especially in the visible region. Emission spectrum of the tungsten filament changes as a function of temperature and emissivity of the tungsten filament. The spectrum shows high intensity in the VNIR region (NIR region close to the area of the visible).

Even if less common, alternative light sources are available. For example, LED light sources and ad laser sources could be used. LED sources (light emitting diodes) are certainly interesting sources thank to their efficiency and their small size. They meet, however, a limited distribution due to limited availability of LEDs emitting at wavelengths in the NIR region. Technology to produce LEDs to cover most of the NIR region already exists, but demand for this type of light sources is currently too low and the development of commercial product of this type is still in an early stage. The use of laser sources guarantees very intense emission in a narrow band. But the reduced spectral range covered by each specific laser source can cause problems in some applications. In any case the complexity and high cost of these devices have limited very much their use so far, mostly restricted to the world of research.

Light Radiation Transport System

Light source must be very close to the sample to light it up with good intensity. This is not always possible, so systems able to convey light on the samples are needed. Thanks to optic fibers this problem was solved, allowing the development of different shapes devices. The use of fiber optics allows to separate the area of placement of the instrument from the measuring proper area. There are indeed numerous circumstances on products sorting line in which environmental conditions do not fulfill direct installation of measure instruments. For example, high temperature, excessive vibrations or lack of space are restricting factors to the use of on-line NIR devices. In all these situations optic fibers are the solution to the problem of conveying light. They transmit light from lamp to sample and from sample to spectrophotometer. They allow to have an immediate measure on a localized sample area, thanks to their small dimensions, reaching areas difficult to access. Furthermore, they are made of a dielectric material that protects from electric and electromagnetic interferences. 'Optic fibers' means fibers optically transparent, purposely studied to transmit light thanks to total internal reflection phenomenon. Internal reflection is said to be total because it is highly efficient, in fact more than 99,999% radiation energy is transmitted in every reflection. This means that radiation can be reflected thousands of times during the way without suffer an appreciable attenuation of intensity (Osborne et al., 1993). Optic fiber consists of an inner core, a covering zone and of an external protection cover. The core is usually made of pure silica, but can also be used plastics or special glasses. The cladding area consists of material with a lower refractive index, while the exterior is only to protect the fiber from mechanical, thermal and chemical stress. In figure 3 are shown the inner core and the cladding of an optical fiber. Index of refraction of inner core have to be bigger than cladding one. Each ray of light that penetrates inside the fiber with an angle $\leq \theta_{max}$ (acceptance angle) is totally reflected with high efficiency within the fiber.

Sample Compartment and Measurement Zone

Samples compartment and measurement zone are highly influenced by the technique of acquisition of spectra. Different techniques are employed, depending on type of samples, solid or liquid, small or large, to be measured in plan or in line, that influence the geometry of the measurement zone.

The techniques to acquire spectra are four: transmittance, reflectance, transflectance and interactance. They are different mainly for the different positioning of the light source and of the measurement sensor around the sample (Fig. 4).

Transmittance - The transmittance measurements are based on the acquisition of spectral information by measuring the light that goes through the whole sample (Lu & Ariana, 2002). The use of analysis in transmittance can explore much of the internal structure of the product. This showed that is a technique particularly well suited to detect internal defects. To achieve significant results with this technique is required a high intensity light source and a high sensitivity measuring device. This because intensity of light able to cross the product is often very low. The transmittance measurements generally require a particular geometry of the measuring chamber, which can greatly influence the design of the instrument.

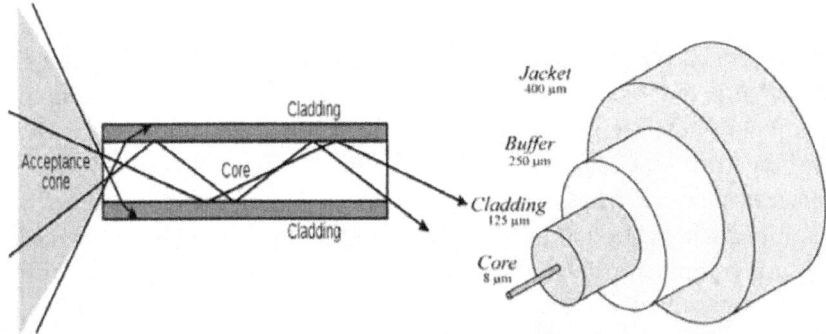

Figure. 3: Scheme of an optical fiber. The acceptance cone is determined by the critical angle for incoming light radiation. Buffer absorbs the radiation not reflected by the cladding. Jacket has a protective function.

Reflectance - This technique measures the component of radiation reflected from the sample. The radiation is not reflected on the surface but penetrates into the sample a few millimeters, radiation is partly absorbed and partly reflected back again. Measuring this component of reflected radiation after interacting with the sample is possible to establish a relationship of proportionality between reflectance and analyte concentration in the sample. The reflectance measurement technique is well suited to the analysis of solid matrices because the levels of intensity of light radiation after the interaction with the sample are high. This technique also allows to put in a limited space inside a tip the bundle of fibers that illuminate the sample and the fibers leading to the spectrophotometer the radiation after the interaction with the product. Therefore the use of this type of acquisition technique is particularly versatile and is suitable for compact, portable instruments, designed for use in field or on the process line. The major drawback using this technique is related to the possibility to investigate only the outer area of the sample without having the chance to go deep inside.

Transflectance – This technique is used in case it is preferable to have a single point of measurement, as in the case of acquisitions in reflectance. In this case, however, the incident light passes through the whole sample, is reflected by a special reflective surface, recross the sample and strikes the sensor located near the area of illumination. The incident light so makes a double passage through the sample. Obviously this type of technique can be used only in the case of samples very permeable to light radiation such as partially transparent fluid. It is therefore not applicable to solid samples.

Interactance - This technique is considered a hybrid between transmittance and reflectance, as it uses characteristics of both techniques previously seen. In this case the light source and sensor are located in areas near the sample but between them physically separated. So the radiation reaches the measurement sensor after interacting with part of internal structure of the sample. This technique is mainly used in the analysis of big solid samples, for example, a whole fruit. Interactance is thus a compromise between reflectance and transmittance and has good ability to detect internal defects of the product combined with a good intensity of light radiation. This analysis is widely used on static equipment where, through the use of special holders, is easily obtained the separation between the areas of incidence of light radiation and the area to which the sensor is placed. It is instead difficult to use this configuration on-line because is complicated to place a barrier between incident and returning light to the sensor directly on the process line.

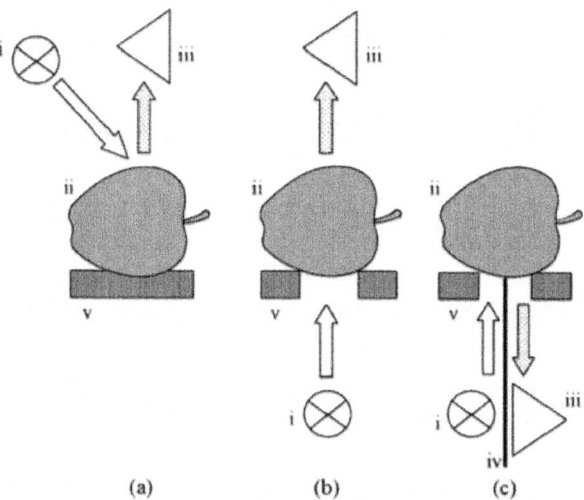

(a)   (b)   (c)

Figure. 4: Setup for the acquisition of (a) reflectance, (b) transmittance, and (c) inter-actance spectra, with (i) the light source, (ii) fruit, (iii) monochromator/detector, (iv) light barrier, and (v) support. In interactance mode, light due to specular reflection is physically prevented from entering the monochromator by means of a light barrier (Nicolai et al., 2007).

Spectrophotometer and Personal Computers

Spectrophotometer can be considered the heart of an instrument for NIR analysis. The employed technology for the wavelengths selection greatly influences the performance of the instrument. For example, the use of filters allows instruments to record the signal of a single wavelength at a time. Modern instruments (diode array instruments and interferometers) allow to record the spectrum of the entire wavelengths range. Instruments equipped with a diode array spectrophotometer are those who have met the increased use for portable and online applications in food sector. This is due to their compact size, versatility and robustness, thanks to the lack of moving parts during operation and also thanks to a relatively low cost. As seen before, fiber optic sensor collects the portion of the electromagnetic radiation after interaction with the internal structure of the sample and transfers it to the spectrophotometer. The optical fiber is connected to the optical bench of the instrument. The optical bench allows to decompose the electromagnetic radiation and recording the intensity at different wavelengths.

Optical bench of this type of instrument generally consists of five components:

Optical fiber connector: connects the fiber optic with the optical bench of the instrument.

First spherical mirror (collimating mirror), has the function of collimating the light and send it to the diffraction grating.

Diffraction grating: in this area of the instrument, the light is split into different wavelengths and sent to the second spherical mirror.

Second spherical mirror (focussing mirror), collects diffracted radiation from the grating and sends them to the CCD sensor.

Matrix CCD sensor (diode array): records the signal intensity at each wavelength.

High sensitivity of the CCD matrix sensor compensate the low intensity of light radiation input due to the reduced diameter of the optical fibers used. Sensors used are generally Sidiode array or InGaAs-diode array. The first ones are certainly the most common and cheap and allow the acquisition of the spectrum in the range between 400 and 1100 nm, so are used for Vis/NIR analysis. InGaAs sensors, more expensive, are used in applications requiring the acquisition of spectra at longer wavelengths, their use should range from 900 to 2300 nm. Recorded signal by the CCD sensor is digitized and acquired by a PC using the software management tool of the instruments. Software records and allows to display graphically the spectrum of the analyzed sample.

The management software also allows to interface with the spectrophotometer enabling to change some parameters during the acquisition of spectra.

Image Analysis

In the food industry, since some time, there is a growing interest in image analysis techniques, since the appearance of a food contains a variety of information directly related to the quality of the product itself and this characteristics are difficult to measure through use of classical methods of analysis. In addition, image analysis techniques: provide information much more accurate than human vision, are objective and continuous over time and offer the great advantage of being non-destructive. These features, enable vision systems to be used in real time on the process lines, allowing on-line control and automation of sorting and classification within the production cycle (Guanasekaran & Ding, 1994). The objective of the application of image analysis techniques, in the food sector, is the quantification of geometric and densitometric characteristic of image, acquired in a form that represents meaningful information (at macro and microscopic level) of appearance of an object (Diezak, 1988). The evolution of these techniques and their implementation in the vision machine in form of hardware and specialized software, allows a wide flexibility of applications, a high capacity of calculation and a rigorous statistical approach. The benefits of image analysis techniques (Brosnan & Sun, 2004) that rely on the use of machine vision systems can be summarized as follows:

are non-destructive techniques;

- techniques are use-friendly, rapid, precise, accurate and efficient;
- generate objective data that can be recorded for analysis deferred;
- allow a complete analysis of the lots and not just a single sample of the lot;
- reduce the involvement of human personnel in performing tedious tasks and allow the automation of various functions that would require intensive work shifts;
- are reasonably affordable cost.

These suggest the reason that drives scientific research, of the agro-food sector, to devote to the study and analysis of machine vision systems to analyze the internal and external quality characteristics of food, valued according to the optical properties of the products. With a suitable light source, it is possible to extract information about color, shape, size and texture. From these features, it is possible to know many objective aspects of the sample, to be able to correlate, through statistical analysis, the characteristics defined by quality parameters (degree of maturation, the presence of external or internal mechanical defects, class, etc.) (Du & Sun, 2006, Zheng et al., 2006). The image analysis may have several applications in the food industry: as a descriptor or as gastronomic and technologic parameter. Vision machine can also be used to know size, structure, color in order to quantify the macro and microscopic surface defects of a product or for the characterization and identification of foods or to monitor the shelf life (Riva, 1999). "Image analysis" is a wide designation that include, in addition to classical studies in grayscale and RGB images, the analysis of images collected by mean multiple spectral channels (multispectral) or, more recently, hyperspectral images, technique exploited for its full extension in the spectral direction. The hyperspectral image (Chemical and Spectroscopic Imaging) is an emerging technology, non-destructive, which complements the conventional imaging with spectroscopy in order to obtain, from an object, information, both spectral and spatial. The hyperspectral images are digital images in which each element (pixel) is not a single number or a set of three numbers, like the color pictures (RGB), but a whole spectrum associated with that point. They are three-dimensional blocks of data. Their main advantage is that they provide spatial information necessary for the study of non-homogeneous samples. The advantage of this technique is the ability to detect in a foodstuff even minor constituents, isolated spatially.

To support image analysis, chemometric techniques are necessary to process and to model, data sets, in order to extract the highest possible information content. Methods of classification, modeling, multivariate regression, similarity analysis, principal components analysis, methods of experimental design and optimization, must be applied on the basis of each different condition and needs.

The Vision System

The machine vision systems, appeared in the early sixties and then spread over time, in many fields of application, are composed of a lighting system, a data acquisition system connected to a computer, via a capture card, which digitizes (converts the signal into numerical form) and stores the analogic electrical signal, at the output from the camera sensor (Russ et al., 1988). The scanned image is then "converted" into a numerical matrix. Captured images are elaborated by appropriate processing softwares in order to acquire the useful information. In figure 5 shows an example of vision machine.

It is important to choose the localization of light source, but also the type of light source (incandescent, halogen, fluorescent, etc.) influences the performance of the analysis. In fact, although the light sources emitting electromagnetic radiation corresponding to the visible (VIS, 400-700 nm), ultraviolet (UV 10-400 nm) and near infrared (NIR, 700-1400 nm) are the most widely used, to create a digital image can also be used other types of light sources in order to emit different radiations, depending on the purpose of analysis. For example, to determine the internal structure of objects and/ or identify any internal defects, it's possible to use an X-ray source, even if, although this type of source gives good results, its application is much more widespread in the medical field than in the agro-food, this is due to high costs of equipment and low speed of operating.

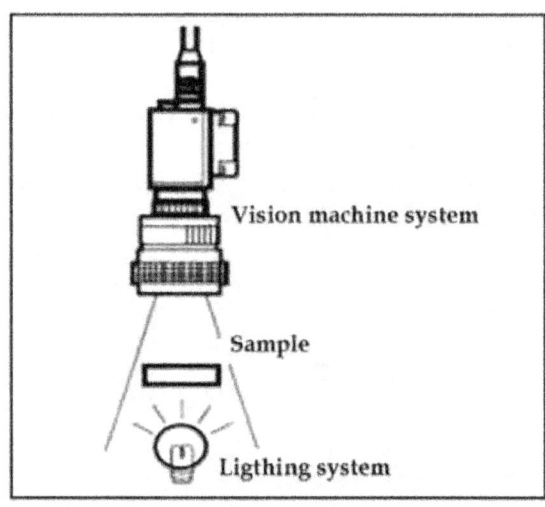

Figure. 5: Example of image acquisition of an inspected sample.

The Digital Image

A digital image is generated from the conversion of an analogic video signal produced by a digital camera into an electronic signal (scanning), then stored in the memory of a PC in the form of binary information. Any digital image can be considered as an array of points, the pixels, that make up the smallest element of an image. Each pixel contains a double set of information: its position in space, identified by the values (x, y) and the value of its intensity. Digital images can be represented using only two colors, typically black and white (binary image), or shades of gray (monochrome image) or a range of colors (multichannel image). The value of light intensity will be different depending on the type of image. The pixels, in the binary images, can have, as the intensity value, or 0 (equivalent to black) or 1 (white). The value of intensity in monochrome images, will be within a range, defined gray scale, from 0 to L, which usually corresponds to the interval from 0 (or 1) to 255 (or 256), where a value of 0 corresponds to black, a value of 255 corresponds to white, and intermediate values to the various shades of gray. Finally, in multi-channel images, the color of each pixel will be identified by three or four values, depending on the reference color model. For example, in RGB color space, each pixel will be characterized by a three values, each between 0 and 255, respectively, corresponding to the intensity in the red, green and blue. When all three values are 0, the color of object is black, and when all three have maximum value, the object will be white, while, when there are equal levels of R, G and B, the gray color is generated. The images of this type, in fact, may be considered as a three-dimensional matrix, consisting of three overlapping matrices having the same number of rows and columns, where the elements of the first matrix represent the pixel intensity in the red channel, those in the second matrix, the green channel and those of third matrix, in the blue channel.

Multispectral and Hyperspectral Images

RGB images, represented by three overlapping monochrome images, are the simplest example of multichannel images. In medical applications, in geotechnical, in the analysis of materials and of remote sensing, instead, are often used sensors capable of acquiring multispectral and hyperspectral images, two particular types of multi-channel images. The multispectral images are typically acquired in three/ten spectral bands including in the range of Vis, but also in the field of IR, fairly spaced (Aleixos et al., 2002). In this way it's possible to extract a larger number of information from the images respect those normally obtained from the analysis of RGB images. An example of bands normally used in this type of analysis, are the band of blue (430-490 nm), green (491-560 nm), red (620-700 nm), NIR (700-1400 nm), MIR (1400-1750 nm). The spectral combinations can be different depending on the purpose of analysis. The combination of NIR-RG (near infrared, red, green) is often used to identify green areas in satellite images, because the green color reflects a lot in the NIR wavelength. The combination of NIR-R-B (near infrared, red, blue) is very useful to verify the ripening stage of fruit, this is due to the chlorophyll that shows a peak of adsorption in the wavelength of the red. Finally, the combination of NIR-MIR-blue (NIR, MIR and blue) is useful to observe the sea depth, the green areas in remote sensing images. Hyperspectral imaging (HSI) combines spectroscopy and the traditional imaging to form a three-dimensional structure of multivariate data (hypercube). The hyperspectral images are consist of many spectral bands acquired in a narrow and contiguous way, allowing to analyze each pixel in the multiple wavelengths simultaneously and, therefore, to obtain a spectrum associated with a single pixels. The set of data constituting an hyperspectral image can be thought as a kind of data cube, with two spatial directions, ideally resting on the surface observed, and a spectral dimension. Extracting a horizontal plane from the cube it is possible to get a monochrome image, while the set of values, corresponding to a fixed position in the plane (x, y), is the spectrum of a pixel of the image (Fig. 6).

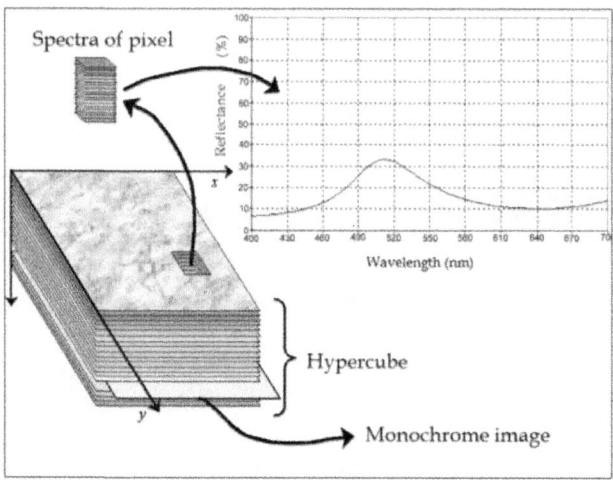

Figure. 6: Example of hyperspectral image.

With the hyperspectral imaging, you can acquire the spectra in reflectance, in transmission and fluorescence as a function of the different kind of sample to analysis, even if the most of the scientific works, present in the literature, using spectral images acquired in reflectance, transmission and emission. The significant time savings that can be made to the industrial production processes, encourage the use of this instrumentation. The hyperspectral image analysis has many advantages, but still has some defects. The advantages of using hyperspectral analysis for what concerns the agro-food sector can be summarized as follows:

- does not necessary to prepare the test sample;
- it is a non-invasive, non-destructive methodology, it avoids the sample loss that can be used for other purposes or analysis;
- can be regarded as an economic tool that it allows a saving of time, labor, reagents, and a strong cost-saving for the waste treatment;
- for each pixel of the analyzed sample is possible to have the full spectrum and not a only absorbance value for few wavelength;
- many constituents can be determined simultaneously within a sample, such as color and morphological characteristics;
- due to its high spectral resolution, it is possible to estimate both

qualitative than quantitative information;

- it is also possible to select a single region of interest of the sample, and save it in a spectral library.

As mentioned previously, one of the advantages HSI is the large volume of data available in each hypercube, with which to create the calibration and validation set. But, the information derived from the analysis, contain also redundant information. This abundance of data has two drawback, one due to the high computational load of heavy data size and the second is due to the long acquisition times, given the size of the data being collected (Firtha et al. 2008). Therefore, it is desirable to reduce the load to manageable levels, especially if the goal is the application of HSI techniques in real time, on-line on production lines. In fact, in many cases, the large amount of data acquired from the spectral image, is appropriately reduced (with chemometric processing) so as to select only those wavelengths interesting for the intended purpose. Once the spectral bands of interest were identified, a multispectral system, with only selected wavelengths, can be engineered a system for industrial application. Another negative aspect is that the spectral image analysis is an indirect method to which it is necessary to apply appropriate chemometric techniques and a procedure of data transfer. The spectral image is not suitable for liquid and homogeneous samples. In fact, the value of this type of image is evident when applied to heterogeneous samples, and many foods are an excellent heterogeneous matrix. Despite the novelty of applying HSI in the food sector, many jobs are already present in the literature. The traditional image analysis, based on a computer system, has had a strong development in the food sector with the aim of replacing the human eye on saving costs and improving efficiency, speed and accuracy. But the computer vision technology is not able to select between objects of similar colors, to make complex classifications, to predict quality characteristics (e.g. chemical composition) or detect internal defects. Since the quality of a food is not an individual attribute but it contains a number of inherent characteristics of the food itself, to measure the optical properties of food products has been one of the most studied non-destructive techniques for the simultaneous detection of different quality parameters. In fact, the light reflected from the food contains information about constituents near and at the surface of the foodstuff. Near-infrared spectroscopy technology (NIRS) is rapid, non-destructive, easy to apply on-line and off-line. With this technology, it is possible to obtain spectroscopic information about the components of analyzed sample, but it is not possible to know the position of the component. The only characteristic of appearance (color, shape, etc.) however, are easily detectable with conventional image analysis. The combination of image analysis technology and spectroscopy is the chemical imaging spectroscopy that allows to get spatial and spectral information for each pixel of the foodstuff. This technology allowing to know the location of each chemical component in the scanned image. Table 1 summarizes the main differences between the three analytical technologies: imaging, spectroscopy and hyperspectral imaging.

Electronic Nose (E-Nose)

"An instrument which comprises an array of electronic chemical sensors with partial specificity and appropriate pattern recognition system, capable of recognizing simple or complex odors" is the term of "electronic nose" coined in 1988 by Gardner and Bartlett (Gardner and Bartlett, 1994).

Table 1: Main differences among imaging, spectroscopy and hyperspectral imaging techniques (ElMarsy & Sun, 2010).

Features	Imaging	Spectroscopy	Hyperspectral Imaging
Spatial information	√	x	√
Spectral information	x	√	√
Multi-costituent information	x	√	√
Building chimica images	x	x	√
Flexibility of spectral information extraction	x	x	√

Scientific interest in the use of electronic noses was formalised, the first time, in a workshop on chemosensory information processing during a session of the North Atlantic Treaty Organization (NATO) that was entirely dedicated to the topic of artificial olfaction. Since 1991, interest in biological sensors technology has grown considerably as is evident by numerous scientific articles. Moreover, commercial efforts to improve sensor technologies and to develop tools of greater sophistication and improved capabilities, with diverse sensitivities, are with ever-expanding (Wilson & Baietto, 2009). Electronic noses are emerging as an innovative analytical-sensorial tool to characterize the sensory comparison of food in terms of freshness, determination of geographical origin, seasoning. The first electronic nose goes back to the '80s, when Persaud and Dodd of the University of Warwick (UK) tried to model and simulate the operation of the olfactory system of mammals with solid state sensors. Since then, artificial olfactory systems are designed closer to the natural one. The electronic nose is a technology that tends to replace/complement the human olfactory system. The tool does not analyze the chemical composition of the volatile fraction, but it identifies the olfactory fingerprint. Currently, these electronic devices are characterized by complex architecture, where it is possible to try to reproduce the functioning of the olfactory system of mammals. The tool is a biomimetic system that is designed to mimic the functioning of the olfactory systems that we find in nature, specifically human olfactory system. Typically, an electronic nose collects information through an array of sensors, able to respond in a selective mode and reversible to the presence of chemicals, generating electrical signals as a function of their concentration. Currently, the sensors that have reached the highest level of development are made from metal oxides semiconductor (MOS). The sensors are usually characterized by fast response, low energy consumption, small size, high sensitivity, reliability, stability and reproducibility. In addition to semiconductor of metal, the sensor can be made of transistors, plated with metal semiconductor (MOSFETs), or conductive polymers. The MOS sensors are inorganic, typically made of tin oxide, zinc oxide, titanium oxide, tungsten oxide. The absorption of gas by them change their conductivity. These sensors operate at high temperatures, between 200 and 500 °C and are relatively cheap. In figure 7 is represented the main parts of a typical sensor.

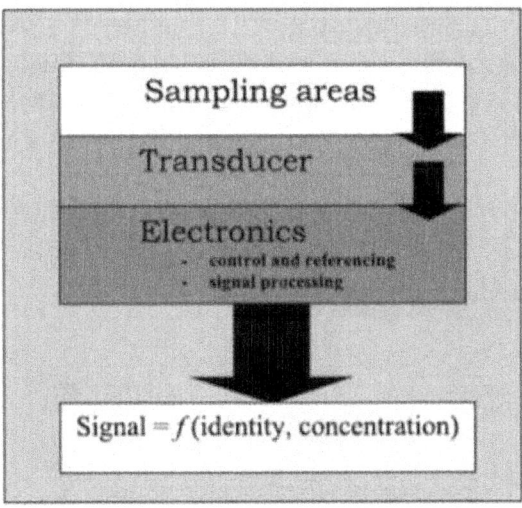

Figure. 7: The main parts of a typical sensor (Deisingh et al. 2004)

Below, ten possible MOS sensors for reading specific molecules (Table 2):

Table 2: Example of sensors in the electronic nose, with their categories of compounds that can to determine

Sensor	Molecules detected
W1C	Aromatic compounds
W5S	Oxides of nitrogen, low specificity
W3C	Ammonium compounds, aromatic
W6S	Hydrogen
W5C	Alkanes, aromatic compounds, less polar compounds
W1S	Methane, low specificity
W1W	Sulfur compounds, terpenes, limonene, pyrazines
W2S	Alcohol, partially aromatic compounds, low specificity
W2W	Aromatic compounds, organic sulfur compounds
W3S	Methane

The information is initially encoded as electrical unit, but are immediately captured and digitized in order to be numerically translated by a computer system. In practice, an odorant is described by the electronic nose, based on the responses of individual sensors, as a point or a region of a multidimensional space. Thanks to special algorithms, derived from the discipline called pattern recognition, the system is able to build an olfactory map in order to allow a qualitative and quantitative analysis, discriminating a foodstuff simply by its olfactory fingerprint. The architecture of an electronic nose (Fig. 8) is significantly dependent on the application for which it is designed. In general, the electronic nose, is characterized by the presence of a vacuum system, a large number of gas sensors, a subsystem of acquisition and digitization and by a processing subsystem able to implement appropriate algorithms for classification or regression.

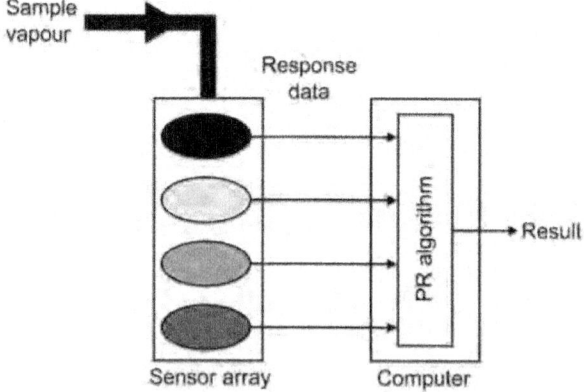

Figure. 8: A generalized structure of an electronic nose. (Deisingh et al. 2004)

The principle of working which operates the electronic nose is distinctly different from that of commonly used analytical instruments (e.g. gas chromatograph). The e-nose gives an overall assessment of the volatile fraction of the foodstuff that is, in large part responsible for the perception of the aroma of the investigated sample, without the need to separate and identify the various components. All the responses of the sensors resulted from the electronic nose creates a "map" of non-specific signals that constitute the profile of the food product, also called olfactory fingerprints. The goal is to find a relationship between the set of independent variables, resulted from the sensor, and the set of dependent variables, characteristics of the sample. Chemometrics software required for data set processing, in environmental and food sectors, allows to process the data by methods of multivariate analysis such as PCA (Principal Component Analysis), LDA (Linear Discriminant Analysis), PLS (Partial Least Square Analysis), DFA (Discriminant Function Analysis). As example the Principal Component Analysis (PCA) is a method for detecting patterns in data sets and express them in order to highlight their similarities and/or differences. Example of electronic nose applications, in the food sector, could be: to monitor of foodstuff shelf life, to check certified quality or the trademark DOP, to make microbiological tests, to check controlled atmosphere in the packaging, to control fermentation stage or to identify the presence of components of the packaging transferred in the product, or to verify the state of cleaning of kegs (on-line measures).

Chemometrics in fOod Sector

Chemometrics is an essential part of NIR and Vis/NIR spectroscopy in food sector. NIR and Vis/NIR instrumentation in fact must always be complemented with chemometric analysis to enable to extract useful information present in the spectra separating it both from not useful information to solve the problem and from spectral noise. Chemometric techniques most used are the principal component analysis (PCA) as a technique of qualitative analysis of the data and PLS regression analysis as a technique to obtain quantitative prediction of the parameters of interest (Naes et al., 2002; Wold et al., 2001; Nicolai et al., 2007; Cen & He, 2007).The developed models should be tested using independent samples as validation sets to verify model accuracy and robustness. To evaluate model accuracy, the statistics used were the coefficient of correlation in calibration (rcal), coefficient of correlation in prediction (rpred), root mean square error of calibration (RMSEC), and root mean square error of prediction (RMSEP).

Correlation coefficients (r_{cal} and r_{pred}):

$$r_{cal} \text{ or } r_{pred} = \sqrt{1 - \frac{\sum_{i=1}^{n}(y_i - \hat{y}_i)^2}{\sum_{i=1}^{n}(y_i - \bar{y})^2}}$$

(2)

where yi are the reference values, ŷi are the values predicted by the PLS model, and ÿ is the averaged reference value. Standard errors of calibration and prediction (RMSEC and RMSEP):

$$RMSEC \text{ or } RMSEP = \sqrt{\frac{\sum_{i=1}^{n}(y_i - \hat{y}_i)^2}{n}}$$

(3)

where n is the number of validated objects, and ŷi and yi are the predicted and measured values of the ith observation in the calibration or validation set, respectively. This value gives the average uncertainty that can be expected for predictions of future samples. The optimum calibrations should be selected based on minimizing the RMSEP. Percent errors (RMSEC% and RMSEP%) could be also calculated as: RMSEC (%) = RMSEC/averaged reference values of each parameter. Prediction capacity of a model can be evaluated with the ratio performance deviation (RPD) (Williams & Sobering, 1996). The RPD is defined as the ratio of the standard deviation of the response variable to the RMSEP. RPD value > 2.5 means that the model has good prediction accuracy.

APPLICATIONS

NIR and Vis/NIR Spectroscopy

During the last 50 years, there has been a lot of emphasis on the quality and safety of the food products, of the production processes, and the relationship between the two (Burns and Ciurczak, 2001). Near infrared (NIR) spectroscopy has proved to be one of the most efficient and advanced tools for monitoring and controlling of process and product quality in food industry. A lot of work has been done in this area. This review focuses on the use of NIR spectroscopy for the analysis of foods such as meat, fruit, grain, dairy products, oil, honey, wine and other areas, and looks at the literature published in the last 10 years.

Fruit and Vegetables

Water is the most important chemical constituent of fruits and vegetables and water highly absorbs NIR radiation, so the NIR spectrum of such materials is dominated by water. Further, the NIR spectrum is essentially composed of a large set of overtones and combination bands. This, in combination with the complex chemical composition of a typical fruit or vegetable causes the NIR spectrum to be highly convoluted. Multivariate statistical techniques are required to extract the information about quality attributes which is buried in the NIR spectrum. Developments in multivariate statistical techniques such as partial least squares (PLS) regression and principal component analysis (PCA) are then applied to extract the required information from such convoluted spectra (Cozzolino et al., 2006b; McClure, 2003; Naes et al., 2004; Nicolai et al., 2007;). The availability of low cost miniaturised spectrophotometers has opened up the possibility of portable devices which can be used directly on field for monitoring the maturity of fruit. Guidetti et al. (2008) tested a portable Vis/NIR device (450-980 nm) for the prediction of ripening indexes (soluble solids content and firmness) and presence of compounds with functional properties (total anthocyanins, total flavonoids, total polyphenols and ascorbic acid) of of blueberries ('Brigitta' and 'Duke' varieties). Good predictive statistics were obtained with correlation coefficients (r) between 0.80 and 0.92 for the regression models built for fresh berries (Table 3). Similar results were obtained for the regression models for homogenized samples with r > 0.8 for all the indexes. Results showed that Vis/NIR spectroscopy is an interesting and rapid tool for assessing blueberry ripeness.

Table 3: Results of PLS models for fresh 'Duke' berry samples (r = coefficient of correlation; RMSEC = root mean square of the standard error in calibration; RMSECV = root mean square of the standard error in cross-validation; LV = latent variables). All data were preprocessed by second derivative of reduced and smoothed data.

Dependent variable	LV	Calibration		Cross validation	
		r_{cal}	RMSEC	r_{cv}	RMSECV
TSS (°Brix)	4	0.86	0.78	0.85	0.79
Young's Module (MPa)	3	0.87	0.65	0.87	0.66
Total anthocyanins (mg/g f. w.)	4	0.87	0.31	0.87	0.31
Total flavonoids (mg cat/g)	4	0.87	0.37	0.86	0.37
Total polyphenols (mg cat/g f. w.)	11	0.82	0.20	0.81	0.20
Ascorbic acid (mg/100 g f. w.)	4	0.84	1.01	0.83	1.02

Meat

In literature there are numerous applications of NIR spectroscopy for the analysis of meat quality. One of the most important aim is to monitor the freshness of meat products. Sinelli et al. in 2010 investigated the ability of Near Infrared spectroscopy to follow meat freshness decay. PCA was applied by authors to the data and was able to discriminate samples on the basis of storage time and temperature. The modelling of PC scores versus time allowed the setting of the time of initial freshness decay for the samples (6–7 days at 4.3 °C, 2–3 days at 8.1 °C and less than 1 day at 15.5 °C). Authors reported that results showed the feasibility of NIR for estimating quality decay of fresh minced beef during marketing. Sierra et al. in 2007 conducted a study for the rapid prediction of the fatty acid (FA) profile of ground using near infrared transmittance spectroscopy (NIT). The samples were scanned in transmittance mode from 850 to 1050 nm. NIT spectra were able to accurately predict saturated R2=0,837, branched R2=0,701 and monounsaturated R2=0,852 FAs. Results were considered interesting because intramuscular fat content and composition influence consumer selection of meat products. Andrés et al. in 2007 implemented a study to evaluate the potential of visible and near infrared reflectance (NIR) spectroscopy to predict sensory characteristics related to the eating quality of lamb meat samples. A total of 232 muscle samples from Texel and Scottish Blackface lambs was analyzed by chemical procedures and scored by assessors in a taste panel and these parameters were predicted from Vis/NIR spectra. The results obtained by authors suggested that the more important regions of the spectra to estimate the sensory characteristics are related to the absorbance of intramuscular fat and water content in meat samples. Even in the meat industry have been tried online applications of NIR spectroscopy. A study was conducted by Prieto et al. in 2009a to assess the on-line implementation of visible and near infrared reflectance (Vis/NIR) spectroscopy as an early predictor of beef quality traits, by direct application of a fibre-optic probe to the muscle immediately after exposing the meat surface in the abattoir. Authors reported good correlation results only for prediction of colour parameters while less good results were achieved for sensory parameters. NIR spectroscopy could be used for the detection of beef contamination from harmful pathogens and the protection of consumer safety. Amamcharla et al. in 2010 investigated the potential of Fourier transform infrared spectroscopy (FTIR) to discriminate the Salmonella contaminated packed beef. Principal component analysis was performed on the entire spectrum (4000–500 cm^{-1}). Authors obtained encouraging classification results with different techniques and confirmed that NIR could be used for non-destructive discrimination of Salmonella contaminated packed beef samples from uncontaminated ones. A review published by Prieto et al. in 2009b indicates that NIR showed high potential to predict chemical meat properties and to categorize meat into quality classes. But authors underlined also that NIR showed in different cases limited ability for estimating technological and sensory attributes, which may be mainly due to the heterogeneity of the meat samples and their preparation, the low precision of the reference methods and the subjectivity of assessors in taste panels.

Grains, Bread and Pasta

Grains including wheat, rice, and corn are main agricultural products in most countries. Grain quality is an important parameter not only for harvesting, but also for shipping (Burns and Ciurczak, 2001). In many countries, the price of grain is determined by its protein content, starch content, and/or hardness, often with substantial price increments between grades. Measurement of carotenoid content of maize by Vis/NIR spectroscopy was investigated by Brenna and Berardo (2004). They generated calibrations for several individual carotenoids and the total carotenoid content with good results (R2 about 0,9). Several applications can be found in literature regarding the use of NIR for the prediction of the main physical and rheological parameters of pasta and bread. De Temmerman et al. in 2007 proposed near-infrared (NIR) reflectance spectroscopy for in-line determination of moisture concentrations in semolina pasta immediately after the extrusion process. Several pasta samples with different moisture concentrations were extruded while the reflectance spectra between 308 and 1704 nm were measured. An adequate prediction model was developed based on the Partial Least Squares (PLS) method using leave-one-out crossvalidation. Good results were obtained with R2 = 0,956 and very low level of RMSECV. This creates opportunities for measuring the moisture content with a low-cost sensor. Zardetto & Dalla Rosa in 2006 studied the evaluation of the chemical and physical characteristics of fresh egg pasta samples obtained by using two different production methodologies: extrusion and lamination. Authors evaluated that it is possible to discriminate the two kinds of products by using FT-NIR spectroscopy. FT-NIR analysis results suggest the presence of a different matrix–water association, a diverse level of starch gelatinization and a distinct starch–gluten interaction in the two kinds of pasteurised samples. The feasibility of using near infrared spectroscopy for prediction of nutrients in a wide range of bread varieties mainly produced from wheat and rye was investigated by Sørensen in 2009. Very good results were reported for the prediction of total contents of carbohydrates and energy from NIR data with R2 values of 0.98 and 0.99 respectively. Finally, a quick, non-destructive method, based on Fourier transform near-infrared (FT-NIR) spectroscopy for egg content determination of dry pasta was presented by Fodor et al. (2011) with good results.

Wine

Quantification of phenolic compounds in wine and during key stages in wine production is therefore an important quality control goal for the industry and several reports describing the application of NIR spectroscopy to this problem have been published. Grape composition at harvest is one of the most important factors determining the future quality of wine. Measurement of grape characteristics that impact product quality is a requirement for vineyard improvement and for optimum production of wines (Carrara et al., 2008). Inspection of grapes upon arrival at the winery is a critical point in the wine production chain (Elbatawi & Ebaid, 2006).

An optical, portable, experimental system (Vis/NIR spectrophotometer) for nondestructive and quick prediction of ripening parameters of fresh berries and homogenized samples of grapes in the wavelength range 450-980 nm was built and tested by Guidetti et al. (2010) (Fig. 9). Calibrations for technological ripening and for anthocyanins had good correlation coefficients (rcv > 0.90). These models were extensively validated using independent sample sets. Good statistical parameters were obtained for soluble solids content (r > 0.8, SEP < 1.24 °Brix) and for titratable acidity (r > 0.8, SEP < 2.00 g tartaric acid L-1), showing the validity of the Vis/NIR spectrometer. Similarly, anthocyanins could be predicted accurately compared with the reference determination (Table 4). Finally, for qualitative analysis, spectral data on grapes were divided into two groups on the basis of grapes' soluble content and acidity in order to apply a classification analysis (PLS-DA). Good results were obtained with the Vis/ NIR device, with 89% of samples correctly classified for soluble content and 83% of samples correctly classified for acidity. Results indicate that the Vis/ NIR portable device could be an interesting and rapid tool for assessing grape ripeness directly in the field or upon receiving grapes in the wine industry.

Figure. 9: Images of spectral acquisition phases on fresh berries and on homogenized samples.

Table 4: Results of PLS models for homogenized samples

Parameter	Pretreat-ment[a]	LV	Calibration		Validation	
			r	RMSEC	r	RMSEP
TSS (°Brix)	MSC+d2	5	0.93	0.95	0.75	0.95
Titratable acidity (g tart. acid dm⁻³)	MSC+d2	6	0.95	1.16	0.85	1.12
pH	MSC+d2	5	0.85	0.08	0.80	0.13
PA (mg dm⁻³)	MSC+d2	5	0.95	80.90	0.78	129.00
EA (mg dm⁻³)	MSC+d2	3	0.93	57.70	0.84	77.70
TP (OD 280 nm)	MSC+d2	4	0.80	3.74	0.70	5.81

The application of some chemometric techniques directly to NIR spectral data with the aim of following the progress of conventional fermentation and maturation was investigated by Cozzolino et al. (2006b). The application of principal components analysis (PCA) allowed similar spectral changes in all samples to be followed over time. The PCA loading structure could be explained on the basis of absorptions from anthocyanins, tannins, phenolics, sugar and ethanol, the content of which changed according to different fermentation time points. This study demonstrated the possibility of applying NIR spectroscopy as a process analytical tool for the wine industry. Urbano-Cuadrado et al. (2004) analysed by Vis/NIR spectroscopy different parameters commonly monitored in wineries. Coefficients of determination obtained for the fifteen parameters were higher than 0.80 and in most cases higher than 0.90 while SECV values were close to those of the reference method. Authors said that these prediction accuracies were sufficient for screening purposes. Römisch et al. in 2009 presented a study on the characterization and determination of the geographical origin of wines. In this paper, three methods of discrimination and classification of multivariate data were considered and tested: the classification and regression trees (CART), the regularized discriminant analysis (RDA) and the partial least squares discriminant analysis (PLS-DA). PLS-DA analysis showed better classification results with percentage of correct classified samples from 88 to 100%. Finally, PLS and artificial neural networks (ANN) techniques were compared by Janik et al. in 2007 for the prediction of total anthocyanin content in redgrape homogenates.

Other Applications

The applications of Vis/NIR and NIR spectroscopy and their chemometric techniques are present in many other sectors of the food industry. In literature are reported works relating to the dairy, oil, coffee, honey and chocolate industry. In particular, interesting studies have been conducted by some authors for the application of NIR spectroscopy in detecting the geographical origin of raw materials and finished products, defending the protected designation of origin (PDO). Olivieri et al. (2011) worked out the exploration of three different class-modelling techniques to evaluate classification abilities based on geographical origin of two PDO food products: olive oil from Liguria and honey from Corsica. Authors developed the best models for both Ligurian olive oil and Corsican honey by a potential function technique (POTFUN) with values of correctly classified around 83%. González-Martín et al. in 2011 presented a work on the evaluation by near infrared reflectance (NIR) spectroscopy of different sensorial attributes of different type of cheeses, taking as reference data the evaluation of the sensorial properties obtained by a panel of eight trained experts. NIR spectra were collected with a remote reflectance fibre optic probe applying the probe directly to the cheese samples and the calibration equations were developed by using modified partial least-squares (MPLS) regression for 50 samples of cheese. Authors stated that obtained results can be considered good and acceptable for all the parameters analyzed (presence of holes, hardness, chewiness, creamy, salty, buttery flavour, rancid flavour, pungency and retronasal sensation).

The quality of coffee is related to the chemical constituents of the roasted beans, whose composition depends on the composition of green beans (i.e., un-roasted). Unroasted coffee beans contain different chemical compounds, which react amongst themselves during coffee roasting influencing the final product. For this reason, monitoring the row materials and the roasting process is very important. Ribeiro et al. in 2011 elaborated PLS models correlating coffee beverage sensory data and NIR spectra of 51 Arabica roasted coffee samples. Acidity, bitterness, flavour, cleanliness, body and overall quality of coffee beverage were considered. Results were good and authors confirmed that it is possible to estimate the quality of coffee using PLS regression models obtained by using NIR spectra of roasted Arabica coffees. Da Costa Filho in 2009 elaborated a rapid method to determine sucrose in chocolate mass using near infrared spectroscopy. Data were modelled using partial least squares (PLS) and multiple linear regression (MLR), achieving good results (correlation coefficient of 0.998 and 0.997 respectively for the two chemometric techniques). Results showed that NIR can be used as rapid method to determine sucrose in chocolate mass in chocolate factories.

Image Analysis

The chemical imaging spectroscopy is applied to various fields, from astronomy to agriculture (Baranowski et al., 2008, Monteiro et al., 2007, V.

Smail, 2006), from the pharmaceutical industry (Lyon et al. 2002, Roggo et al., 2005) to medicine (Ferris et al. 2001, Zheng et al., 2004). But in recent years, has also found use for quality control and safety in food (Gowen et al., 2007b). In general, classify or quantify the presence of compounds in a sample is the main purpose of the application in the food hyperspectral analysis. There already exist algorithms for classification and regression but, improved algorithms efficiency, could be a target, as well as create datasets, identify anomalies or objects with different spectral characteristics, compared hyperspectral image with those of data library. These goals can be achieved only if the experimental data are processed with chemometric methods. K-nearest neighbors and hierarchical clustering are examples of multivariate analysis that allow to get information from spectral and spatial data (Burger & Gowen, 2011). With the use of spectral image, which allows to obtain in a single determination, spectral and spatial information characterizing the sample, it is possible to identify which chemical species are present and how they are distributed in a matrix. Several chemometric techniques are available for the development of regression models (for example partial least squares regression, principal components regression, and linear regression) capable of estimating the concentrations of constituents in a sample, at the pixel level, allowing the spatial distribution or the mapping of a particular component in the sample analyzed. Moreover the hyperspectral image, combined with chemometric technique, is a powerful method to identify key wavelengths in order to develop of multispectral system, for on-line applications. Karoui & De Baerdemaeker (2006) wrote a review about the analytical methods coupled with chemometric tools for the determination of the quality and identity of dairy products. Spectroscopic techniques (NIR, MIR, FFFS front face fluorescence spectroscopy, etc.), coupled with chemometric tools have many potential advantages for the evaluation of the identity dairy products (milk, ice cream, yogurt, butter, cheese, etc).

In another review Sankaran et al. (2010), compared the benefits and limitations of advanced techniques and multivariate methods to detect plant diseases in order to assist in monitoring health in plants under field conditions. These technologies include evaluation of volatile profiling (Electronic Nose), spectroscopy (fluorescence, visible and infrared) and imaging (fluorescence and hyperspectral) techniques for disease detection. In literature it's possible find several examples of applications of spectroscopic image analysis. Hyperspectral imaging could be used as critical control points of food processing to inspect for potential contaminants, defects or lesions. Their absence is essential for ensuring food quality and safety. In some case the application on-line was achieved. Ariana & Lu (2010), evaluated the internal defect and surface color of whole pickles, in a commercial pickle processing. They used a prototype of on-line hyperspectral imaging system, operating in the wavelength range of 400–1000 nm. Color of the pickles was modeled using tristimulus values: there were no differences in chroma and hue angle of good and defective pickles. PCA was applied to the hyperspectral images: transmittance images at 675–1000 nm were much more effective for internal defect detection compared to reflectance images for the visible region of 500–675 nm. A defect classification accuracy was of 86% compared with 70% by the human inspectors. Mehl et al. (2002), used hyperspectral image analysis and PCA, like chemometrics technique, to reduce the information resulting from HIS and to identify three spectral bands capable of separating normal from contaminated apples. These spectral bands were implemented in a multispectral imaging system. On 153 samples, it's possible to get a good separation between normal and contaminated (scabs, fungal, soil contaminations, and bruises) apples was obtained for Gala (95%) and Golden Delicious (85%), separations were limited for Red Delicious (76%). HSI application for damage detection on the caps of white mushrooms (Agaricus bisporus) was investigated from Gowen et al. (2007a). They employed a pushbroom line-scanning HSI instrument (wavelength range: 400–1000 nm). They investigated two data reduction methods. In the first method, PCA was applied to the hypercube of each sample, and the second PC (PC 2) scores image was used for identification of bruise-damaged regions on the mushroom surface. In the second method PCA was applied to a dataset comprising of average spectra from regions normal and bruise-damaged tissue. The second method performed better than the first when applied to a set of independent mushroom samples. Further, they (Gowen et al., 2009) identified mushrooms subjected to freeze damage using hyperspectral imaging. In this case they used Standard Normal Variate (SNV) transformation to pretreat the data, then they applied a procedure based on PCA and LDA to classify spectra of mushrooms into undamaged and freeze-damaged groups. The undamaged mushrooms and freeze-damaged mushrooms could be classified with high accuracy (>95% correct classification) after only 45 min thawing (at 23 ± 2 °C) at that time freeze–thaw damage was not visibly evident. A study on fruits and vegetables (Cubero et al., 2010) used ultraviolet or near-infrared spectra to explore defects or features that the human eye is unable to see, with the aim of applying them for automatic inspection. This work present a summary of inspection systems for fruit and vegetables and the latest developments in the application of this technology to the inspection of internal and external quality of fruits and vegetables.

Li et al. (2011) detected common defects on oranges using hyperspectral (wavelength range: 400-1000) reflectance imaging. The disadvantage of studied algorithm is that it could not discriminate between different types of defects. Bhuvaneswari et al. (2011) compared three methods (electronic speck counter, acid hydrolysis and flotation and near-infrared hyperspectral imaging) to investigate the presence of insect fragments (Tribolium castaneum_ Coleoptera: Tenebrionidae) in the semolina (ingredient for pasta and couscous). NIR hyperspectral imaging is a rapid, nondestructive method, as electronic speck counter, but they showed different correlation between insect fragments in the semolina and detection of specks in the samples: R2 = 0.99 and 0.639-0.767 respectively. For NIR hyperspectral image technique, the prediction model were developed by PLS regression. The most important features in meat are tenderness, juiciness and flavour. Jackmana et al., (2011) wrote a review about recent advances in the use of computer vision technology in the quality assessment of fresh meats. The researcher support that the best opportunities for improving computer vision solutions is the application of hyperspectral imaging in combination with statistical modelling. This synergy can provide some additional information on meat composition and structure. However, in parallel, new image processing algorithms, developed in other scientific disciplines, should be carefully considered for potential application to meat images. Other applications concern the possibility of estimating a correlation between characteristics (physical or chemical) of the food and the spectra acquired with spectroscopic image. Moreover these techniques were able to locate and quantify the characteristic of interest within the image. In most cases the range of wavelength used in applications of hyperspectral images is 400- 1000 nm but Maftoonazad et al. (2010) used artificial neural network (ANN) modeling of hyperspectral radiometric (350-2500 nm) data for quality changes associated with avocados during storage. Respiration rate, total color difference, texture and weight loss of samples were measured as conventional quality parameters during storage. Hyperspectral imaging was used to evaluate spectral properties of avocados. Results indicated ANN models can predict the quality changes in avocado fruits better than the conventional regression models. While Mahesh et al. (2011) used near-infrared hyperspectral images (wavelength range: 960– 1700 nm), applied to a bulk samples, to classify the moisture levels (12, 14, 16, 18, and 20%) on the wheat. Principal components analysis (PCA) was used to identify the region (1260– 1360 nm) with more information. The linear and quadratic discriminant analyses (LDA) and quadratic discriminant analysis (QDA) could classify the sample based on moisture contents than also identifying specific moisture levels with a god levels of accuracy (61- 100% in several case). Spectral features at key wavelengths of 1060, 1090, 1340, and 1450 nm were ranked at top in classifying wheat classes with different moisture contents. Manley et al. (2011) used near infrared hyperspectral imaging combined with chemometrics techniques for tracking diffusion of conditioning water in single wheat kernels of different hardnesses. NIR analysers is a commonly, non-destructive, non-contact and fast solution for quality control, and a used tool to detect the moisture-content of carrot samples during storage but Firtha (2009) used hyperspectral system that is able to detect the spatial distribution of reflectance spectrum as well. Statistical analysis of the data has shown the optimal intensity function to describe moisture-content. The intent of Junkwon et al. (2009), was to develop a technique for weight and ripeness estimation of palm oil (Elaeis guieensis Jacq. var. tenera) bunches by hyperspectral and RGB color images. In the hyperspectral images, the total number of pixels in the bunch was also counted from an image composed of three wavelengths (560 nm, 680 nm, and 740 nm), while the total number of pixels of space between fruits was obtained at a wavelength of 910 nm. Weight-estimation equations were determined by linear regression (LR) or multiple linear regression (MLR). As a result, the coefficient of determination (R2) of actual weight and estimated weight were at a level of 0.989 and 0.992 for color and hyperspectral images, respectively. About the estimation of palm oil bunch ripeness the bunches was classified in 4 classes of ripeness (overripe, ripe, underripe, and unripe) (Fig. 10). Euclidean distances between the test sample and the standard 4 classes of ripeness were calculated, and the test sample was classified into the ripeness class. In the classification based on color image, (average RGB values of concealed and not-concealed areas), and by hyperspectral images (average intensity values of fruits pixels from the concealed area), the results of validation experiments with the developed estimation methods indicated acceptable estimation accuracy.

Figure. 10: Bunches of palm oil: a) unripe, b) underripe, c) ripe, and d) overripe (Junkwon et al., 2009)

Nguyen et al. (2011) illustrated the potential of combination of hyperspectral imaging chemometrics and image processing as a process monitoring tool for the potato processing industry. They predicted the optimal cooking time by hyperspectral imaging (wavelength range 400–1000 nm). By partial least squares discriminant analysis (PLS-DA), cooked and raw parts of boiled potatoes, were discriminated successfully. By modeling the evolution of the cooking front over time the optimal cooking time could be predicted with less than 10% relative error.

Yu H. & MacGregor J.F. (2003) applied multivariate image analysis and regression for prediction of coating content and distribution in the production of snack foods. Elaboration tools based on PCA and PLS was used for the extraction of features from RGB color images and for their use in predicting the average coating concentration and the coating distribution. On-line and off-line imaging were collected from several different snack food product lines and were used to develop and evaluate the methods. The better methods are now being used in the snack food industry for the on-line monitoring and control of product quality. Siripatrawan et al. (2011) have developed a rapid method for the detection of Escherichia coli contamination in packaged fresh spinach using hyperspectral imaging (400–1000 nm) and chemometrics techniques. The PCA was implemented to remove redundant information of the hyperspectral data and artificial neural network (ANN) to correlate spectra with number of E. coli and to construct a prediction map of all pixel spectra of an image to display the number of E. coli in the sample. In this study (Barbin et al. 2011) a hyperspectral imaging technique (range from 900 to 1700 nm) was developed to achieve fast, accurate, and objective determination of pork quality grades. The sample investigated were 75 pork cuts of longissimus dorsi muscle from three quality grades. Six significant wavelengths (960, 1074, 1124, 1147, 1207 and 1341 nm) that explain most of the variation among pork classes were identified from 2nd derivative spectra. PCA was carried out and the results indicated that pork classes could be precisely discriminated with overall accuracy of 96%. Algorithm was developed to produce classification maps of the investigated sample. Valous et al. (2010) communicated perspectives and aspects, relating to imaging, spectroscopic and colorimetric techniques on the quality evaluation and control of hams. These no-contact and no-destructive techniques, can provide useful information regarding ham quality. Hams are considered a heterogenic solid system: varying colour, irregular shape and spatial distribution of pores. Fat-connective tissue, water, protein contribute to the microstructural complexity. This review paying attention on applications of imaging and spectroscopy techniques, for measuring properties and extracting features that correlate with ham quality. In literature is present a review (Mathiassena et al., 2011) that focused the attention on application of imaging technologies (VIS/NIR imaging, VIS/NIR imaging spectroscopy, planar and computed tomography (CT) X-ray imaging, and magnetic resonance imaging) to inspection of fish and fish products. Nicolai et al. (2007) wrote a review about the applications of non-destructive measurement of fruit and vegetable quality. Measurement principles are compared, and novel techniques (hyperspectral imaging) are reviewed. Special attention is paid to recent developments in portable systems. The problem of calibration transfer from one spectrophotometer to another is introduced, as well as techniques for calibration transfer. Chemometrics is an essential part of spectroscopy and the choice, of corrected techniques, is primary (linear or nonlinear regression, such as kernel-based methods are discussed). The principal objective of spectroscopy system applications in fruit and vegetables sector have focused on the nondestructive measurement of soluble solids content, texture, dry matter, acidity or disorders of fruit and vegetables. (root mean square error of prediction want to be achieved).

Electronic Nose

The preservation of quality in post-harvest is the prerequisite for agri-food products in the final stages of commercialization. The fruit quality is related to the appearance (skin color, size, shape, integrity of the fruit), to the sensorial properties (hardness and crispness of the flesh, juicy, acid/sugars) and to safety (residues in fruit). The agri-food products contain a variety of information, directly related to their quality, traditionally measured by means of tedious, time consuming and destructive analysis. For this reason, there is a growing interest in easy to use, rapid and non-destructive techniques useful for quality assessment. Currently, electronic noses are mainly applied in the food industry to recognize the freshness of the products, the detection of fraud (source control, adulteration), the detection of contaminants. An essential step in the analysis with an electronic nose, is the high performance of statistical elaboration. The electronic nose provides multivariated results that need to be processed using chemometric techniques. Even if the best performing programs are sophisticated and, consequently, require the operation of skilled personnel, most companies have implemented user-friendly software for data treatment in commercially available electronic noses (Ampuero & Bosset, 2003). A commercial electronic nose, as a non-destructive tool, was used to characterise peach cultivars and to monitor their ripening stage during shelf-life (Benedetti et al. 2008). Principal component analysis (PCA) and linear discriminant analysis (LDA) were used to investigate whether the electronic nose was able to distinguish among four diverse cultivars. Classification and regression tree (CART) analysis was applied to characterise peach samples into the three classes of different ripening stages (unripe, ripe, over-ripe). Results classified samples in each respective group with a cross validation error rate of 4.87%. Regarding the fruit and vegetable sector Torri et al. (2010) investigated the applicability of a commercial electronic nose in monitoring freshness of packaged pineapple slices during storage. The obtained results showed that the electronic nose was able to discriminate between several samples and to monitor the changes in volatile compounds correlated with quality decay. The second derivative of the transition function, used to interpolate the PC1 score trend versus the storage time at each temperature, was calculated to estimate the stability time. Ampuero and Bosset (2003), presented a review about the application of electronic nose applied to dairy products. The present review deal with as examples the evaluation of the cheese ripening, the detection of mould in cheese, the classification of milk by trademark, by fat level and by preservation process, the classification and the quantification of off-flavours in milk, the evaluation of Maillard reactions during heating processes in block-milk, as well as the identification of single strains of disinfectant–resistant bacteria in mixed cultures in milk. For each application correspond the chemometric method to extrapolate the maximum information. PCA analysis was carried out in order to associate descriptors (chocolate, caramel, burnt and nutty), typical of volatiles generated by Maillard reactions during milk heating. In another case PCA showed a correctly classification of sample in function of the origin of off-flavours. In figure 11 is showed an example of result carried out by DFA (discriminant function analysis) statistical technique. In this case the aim of researcher was to classify samples of a given product by their place of production.

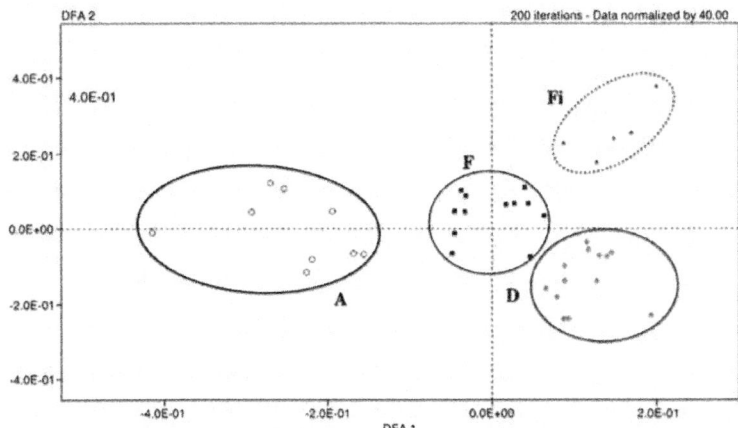

Figure. 11: Classification of Emmental cheese by the geographic origin performed with an electronic nose based on mass spectrometry. The graph shows DFA 1 vs. DFA

2 with 100% group classification based on five variables. No validation set was considered due to the limited number of samples. A: Austria, D: Germany, F: France, Fi: Finland (Pillonel et al, 2003)

The potential of electronic nose technique was investigated to monitoring storage time and the quality attribute of eggs by Yongwei et al. (2009). Using techniques of multivariate analysis was distinguished eggs under cool and room-temperature storage. Results showed that the E-nose could distinguish eggs of different storage time under cool and room-temperature storage by LDA, PCA, BPNN and GANN. Good distinction between eggs stored for different times were obtained by PCA and LDA (results by LDA were better than those obtained by PCA). By means BP neural network (BPNN) and the combination of a genetic algorithm and BP neural network (GANN) carried out good predictions for egg storage time (GANN demonstrated better correct classification rates than BPNN). The quadratic polynomial step regression (QPSR) algorithm established models that described the relationship between sensor signals and egg quality indices (Haugh unit and yolk factor). The QPST models showed an high predictive ability ($R2 = 0.91$-0.93). Guidetti et al. (2011) used electronic nose and infrared thermography to detect physiological disorders on apples (Golden Delicious and Stark Delicious). In particular the aim was to differentiate typical external apple diseases (in particular, physiological, pathological and entomological disorders). The applicability of the e-nose is based on the hypothesis that apples affected by physiopathology produce different volatile compounds from those produced by healthy fruits. The electronic nose data were elaborated by LDA in order to classify the apples into the four classes. Figure 12 shows how the first two LDA functions discriminate among classes. Considering Stark variety, function 1 seems to discriminate among the physiopathologies while function 2 discriminate the healthy apples from those with physiological disorders. The error rate and the cross validation error rate were of 2.6% and 26.3% respectively. In the case of Golden variety, along the first function there is the separation of Control samples from the apples affected by diseases, while in the vertical direction (function 2) there is an evident discrimination among the three physiopathologies. The error rate and the cross validation error rate were of 0.8% and 18% respectively.

Cerrato Oliveros et al. (2002) selected array of 12 metal oxide sensors to detected adulteration in virgin olive oils samples and to quantify the percentage of adulteration by electronic nose. Multivariate chemometric techniques such as PCA were applied to choose a set of optimally discriminant variables. Excellent results were obtained in the differentiation of adulterated and non-adulterated olive oils, by application of LDA, QDA. The models provide very satisfactory results, with prediction percentages >95%, and in some cases almost 100%.

The results with ANN are slightly worse, although the classification criterion used here was very strict. To determine the percentage of adulteration in olive oil samples multivariate calibration techniques based on partial least squares and ANN were employed. Not so good results were carried out, even if there are exceptions. Finally, classification techniques can be used to determine the amount of adulterant oil added with excellent results.

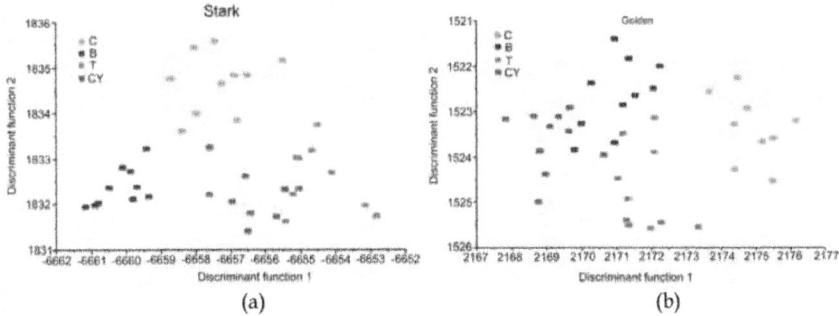

(a)   (b)

Figure. 12: a) Canonical discriminant functions of LDA for Stark variety; b) Canonical discriminant functions of LDA for Golden variety. C=control, B=bitter pit, T=scab and CY=Cydia Pomonella.

CONCLUSION

This work shows the principal non-destructive applications for analysis in food sector. They are rapid techniques used in combination with chemometrics analysis for qualitative and quantitative analysis. NIR spectroscopy is the technique that has been developed further in recent years. This success is because spectral measurement for one sample could be done in a few seconds. Numerous samples could be analyzed and multiindexes analysis can be carried out.

Compared with traditional methods, NIR and Vis/NIR are less expensive because of no demand of other materials such as chemical reagents except the electrical consumption. Many works are focused on the study of chemometrics. This is because an important challenge is to build robust calibration models, in fact it is important to apply chemometric methods able to select useful information from a great deal of spectral data. Moreover food researchers and analysts are looking for the sensitive wavelength in Vis/NIR region representing the characteristics of food products, with the aim of develop some simple and low-cost instruments (Cen & He, 2007). HSI is the new frontier for optical analysis of foods. The performance of HSI instrumentation has developed such that a full hypercube can now be acquired in just a few seconds. In tandem with these developments, advances in component design have led to reductions in the size and cost of HSI systems. This has led to increased interest in their online implementation for quality monitoring in major industries such as food and pharmaceutical (Burger & Gowen, 2011). In future, with further improvement, the HSI system could meet the need of a commercial plant setting. The equipment that the food industry has at its disposal is certainly complex and not easy to use. The chemometric approach has allowed, through different applicative researches, to arrive at algorithms that can support the analysis in the entire food chain from raw material producers to large retail organizations. Despite this, we are still faced with instrumentation with not easy usability and relatively high cost: the studies must move towards a more simplified instrumental approach through greater integration of hardware with software. The challenges are many: optimizing the information that you are able to extract from raw data and aimed at specific problems, simplify electronic components, increase the level of interaction tool/operator. In conclusion the only way of an interdisciplinary approach can lead to the solution of a system that can provide at different level more immediate response and more food safety and quality.

REFERENCES

1.   Aleixos, N.; Blasco, J.; Navarrón, F. & Moltó, E. (2002). Multispectral inspection of citrus in realtime using machine vision and digital signal processors. Computers and Electronics in Agriculture. Vol.33, N.2, pp. 121-137

2.   Amamcharla, J. K.; Panigrahi, S.; Logue, C. M.; Marchello, M. & Sherwood, J. S. (2010). Fourier transform infrared spectroscopy (FTIR) as a tool for discriminating Salmonella typhimurium contaminated beef. Sens. & Instrumen. Food Qual., Vol.4, pp. 1–12

3.   Ampuero, S. & Bosset, J.O. (2003). The electronic nose applied to dairy products: a review. Sensors and Actuators B Vol.94, pp. 1–12

4.   Andrés, S.; Murray, I.; Navajas, E.A.; Fisher, A.V.; Lambe, N.R. & Bünger, L. (2007). Prediction of sensory characteristics of lamb meat samples by near infrared reflectance spectroscopy. Meat Science Vol.76, pp. 509–516

5.   Ariana, D.P. & Lu, R.A. (2010). Evaluation of internal defect and surface color of whole pickles using hyperspectral imaging. Journal of Food Engineering Vol.96, pp. 583–590

6.   Baranowski, P.; Lipecki, J.; Mazurek, W. & Walczak, R.T. (2008). Detenction of watercore in 'Gloster' apples using thermography.

Postharvest Biology and Technology Vol.47, pp. 358

7. Barbin, D.; Elmasry, G.; Sun, D.W.; Allen, P. (2011). Near-infrared hyperspectral imaging forgrading and classification of pork. Meat Science. Article in press

8. Basilevsk, A. (1994) Statistical factor analysis and related methods: theory and applications. ileyInterscience Publication. ISBN 0-471-57082-6

9. Benedetti, S.; Buratti, S.; Spinardi, A.; Mannino, S. & Mignani, I. (2008). Electronic nose as a non-destructive tool to characterize peach cultivars and to monitor their ripening stage during shelf-life. Postharvest Biology and Technology Vol.47, pp. 181–188

10. Bhuvaneswari, K.; Fields, P. G.; White, N.D.G.; Sarkar, A. K.; Singh, C. B. & Jayas, D. S. (2011). Image analysis for detecting insect fragments in semolina. Journal of Stored Products Research Vol.47, pp. 20-24

11. Brenna, O.V. & Berardo, N. (2004). Application of near-infrared reflectance spectroscopy (NIRS) to

12. the evaluation of carotenoids content in maize. J. Agric. Food Chem. Vol.52, 5577

13. Brosnan, T. & Sun, D.W. (2004). Improving quality inspection of food products by computer vision: a review. Journal of Food Engineering. Vol.61, pp. 3-16

14. Burger, J. & Gowen, A. (2011). Data handling in hyperspectral image analysis. Chemometrics and intelligent Laboratory Systems Vol.108, pp. 13-22

15. Burns, D.A. & Ciurczak, E.W. (2001). Second ed. In: Handbook of Near-Infrared Analysis.., Marcel Dekker, New York. Vol.27, N.28, pp. 729–782

16. Carrara, M.; Catania, P.; Vallone, M. & Piraino, S. (2008). Mechanical harvest of grapes: Assessment of the physical-mechanical characteristics of the berry in order to improve the quality of wines. In Proc. Intl. Conf. on Agricultural Engineering: Agricultural and Biosystems Engineering for a Sustainable World (AgEng 2008). CIGR

17. Cen, H. & He, Y. (2007). Theory and application of near infrared reflectance spectroscopy in determination of food quality. Trends in Food Science & Technology.Vol.18, pp. 72-83

18. Cerrato Oliveros, M.C.; Pérez Pavón, J. L.; García Pinto, C.; Fernández Laespada, M. E.; Moreno Cordero, B. & Forina, M. (2002). Electronic nose based on metal oxide semiconductor sensors as a fast alternative for the detection of adulteration of virgin olive oils. Analytica Chimica Acta

Vol.459, pp. 219–228

19.   Cozzolino, D.; Cynkar, W.; Janik, L.; Dambergs, R. G. & Gishen, M. (2006a). Analysis of grape and wine by near infrared spectroscopy — A review. J Near Infrared Spectros, Vol.14,pp. 279–289

20.   Cozzolino, D.; Parker, M.; Dambergs, R.G.; Herderich, M. & Gishen, M. (2006b).

21.   Chemometrics and visible-near infrared spectroscopic monitoring of red wine fermentation in a pilot scale. Biotechnol. Bioeng. Vol. 95, pp. 1101

22.   Cubero, S.; Aleixos, N.; Moltó, E.; Gómez-Sanchis, J. & Blasco, J. (2010). Advances in Machine Vision Applications for Automatic Inspection and Quality Evaluation of Fruits and Vegetables. Food Bioprocess Technol Vol.4, pp.487–504

23.   Da Costa Filho, P. A. (2009). Rapid determination of sucrose in chocolate mass using near infrared spectroscopy. Analytica Chimica Acta Vol.631, pp. 206–211

24.   De Temmerman, J.; Saeys, W.; Nicolai, B. & Ramon, H. (2007). Near infrared reflectance

25.   spectroscopy as a tool for the in-line determination of the moisture concentration in

26.   extruded semolina pasta. Biosystems Engineering Vol.97, pp. 313–321

27.   Deisingh, A.K.; Stone, D.C. & Thompson, M. (2004). Applications of electronic noses and tongues in food analysis. International Journal of Food Science and Technology Vol.39, pp.587–604

28.   Diezak, J.D. (1988). Microscopy and imagine analysis for R&D, Special report. Food Technol. pp. 110-124

29.   Du, C.J. & Sun, D.W. (2006). Learning techniques used in computer vision for food quality evaluation: a review. Journal of food engineering. Vol.72, N.1, pp. 39–55

30.   Elbatawi, I. E. & Ebaid, M. T. (2006). A new technique for grape inspection and sorting classification. Arab Universities J. Agric. Sci. Vol.14, N.2, pp. 555-573

31.   ElMarsy, G. & Sun, D.W. (2010). Hyperspectral imaging for food quality, analysis and control. Book, N.1, pp. 3-43

32.   (http://elsevier.insidethecover.com/searchbook. jsp?isbn=9780123740854) Ferris, D.; Lawhead, R.; Dickman, E.; Holtzapple, N.; Miller, J.; Grogan, S.; et al. 2001.

33.   Multimodal hyperspectral imaging for the noninvasive diagnosis of cervical neoplasia. Journal of Lower Genital Tract Disease Vol.5, N.2,

pp. 65-72

34. Fessenden, R. J. & Fessenden, J. S. (1993). Chimica organica. Cap. 9: Spettroscopia I: Spettri infrarossi, Risonanza Magnetica Nucleare. Piccin Padova, Italy Firtha, F. (2009). Detecting moisture loss of carrot samples during storage by hyperspectral imaging system. Acta Alimentaria Vol.38, N.1, pp. 55-66

35. Firtha, F.; Fekete, A.; Kaszab, T.; Gillay, B.; Nogula-Nagy, M.; Kovács, Z. & Kantor, D.B. (2008). Methods for improving image quality and reducing data load of nir hyperspectral images. Sensors 2008, 8, 3287-3298

36. Fodor, M.; Woller, A.; Turza, S. & Szigedi, T. (2011). Development of a rapid, non-destructive method for egg content determination in dry pasta using FT-NIR technique. Journal of Food Engineering 107, 195–199

37. Frank, I.E. & Todeschini, R. (1994). The Data Analysis Handbook. Elsevier. ISBN 0-444-81659-3, included in series: Data Handling in Science and Technology Gardner, J.W. & Bartlett, P.N. (1994). A brief history of electronic noses. Sens. Actuat. B: Chem. Vol.18, pp. 211-220

38. González-Martín, M.I.; Severiano-Pérez, P.; Revilla, I.; Vivar-Quintana, A.M.; HernándezHierro, J.M.; González-Pérez, C. & Lobos-Ortega, I.A. (2011). Prediction of sensory attributes of cheese by near-infrared spectroscopy. Food Chemistry Vol.127, pp. 256–263

39. Gowen, A.A.; O'Donnell, C.P.; Cullen, P.J.; Downey, G. & Frias, J.M. (2007b). Hyperspectral imaging - an emerging process analytical tool for food quality and safety control. Trends in Food Science & Technology Vol.18, pp.590-598

40. Gowen, A.A.; O'Donnell, C.P.; Taghizadeh, M.; Cullen, P.J.; Frias, J.M. & Downey, G. (2007a). Hyperspectral imaging combined with principal component analysis for bruise damage detection on white mushrooms (Agaricus bisporus). J. of chemometric DOI:10.1002/cem.1127

41. Gowen, A.A.; Taghizadeh, M. & O'Donnell, C.P. (2009). Identification of mushrooms subjected to freeze damage using hyperspectral imaging. Journal of Food Engineering Vol.93, pp.7–12

42. Guanasekaran, S. & Ding, K. (1994). Using computer vision for food quality evaluation. Food Technol. Vol.15, pp. 1-54;

43. Guidetti, R.; Beghi, R. & Bodria, L. (2010). Evaluation of Grape Quality Parameters by a Simple Vis/NIR System. Transaction of the ASABE, Vol.53 N.2, pp. 477-484, ISSN: 2151-0032

44. Guidetti, R.; Beghi, R.; Bodria, L.; Spinardi, A.; Mignani, I. & Folini,

L. (2008). Prediction of blueberry (Vaccinium corymbosum) ripeness by a portable Vis-NIR device. Acta Horticulturae, n° 310, ISBN 978-90-66057-41-8, pp. 877-885

45. Guidetti, R.; Buratti, S. & Giovenzana, V. (2011). Application of Electronic Nose and Infrared Thermography to detect physiological disorders on apples (Golden Delicious and Stark Delicious). CIGR Section VI International Symposium on Towards a Sustainable

46. Food Chain Food Process, Bioprocessing and Food Quality Management. Nantes, France - April 18-20, 2011

47. Jackmana, P.; Sun D.W. & Allen P. (2011). Recent advances in the use of computer vision technology in the quality assessment of fresh meats. Trends in Food Science & Technology Vol.22, pp. 185-197

48. Jackson, J.E. (1991). A user's guide to principal components. Wiley-Interscience Publication. ISBN 0-471-62267-2

49. Janik, L.J.; Cozzolino, D.; Dambergs, R.; Cynkar W. & Gishen, M. (2007). The prediction of total anthocyanin concentration in red-grape homogenates using visiblenear- infrared spectroscopy and artificial neural networks. Anal. Chim. Acta pp. 594-107

50. Junkwon, P.; Takigawa, T.; Okamoto, H.; Hasegawa, H.; Koike, M.; Sakai, K.; Siruntawineti, J.; Chaeychomsri, W.; Sanevas, N.; Tittinuchanon, P. & Bahalayodhin, B. (2009).

51. Potential application of color and hyperspectral images for estimation of weight and ripeness of oil palm (elaeis guineensis jacq. var. tenera). Agricultural Information Research Vol.18, N.2, pp. 72–81

52. Karoui, R. & De Baerdemaeker, J. (2006). A review of the analytical methods coupled with chemometric tools for the determination of the quality and identity of dairy products. Food Chemistry Vol.102, pp. 621–640;

53. Li, J.; Rao, X. & Ying, Y. (2011). Detection of common defects on oranges using hyperspectral reflectance imaging. Computers and Electronics in Agriculture Vol.78, pp. 38–48

54. Lu, R. & Ariana, D. (2002). A Near-Infrared Sensing Technique for Measuring Internal Quality of Apple Fruit. Applied Engineering in Agricolture, Vol.18, N.5, pp. 585-590

55. Lunadei, L. (2008). Image analysis as a methodology to improve the selection of foodstuffs. PhD thesis in "Technological innovation for agro-food and environmental sciences", Department of Agricultural Engineering, Università degli Studi di Milano

56. Lyon, R. C.; Lester, D. S.; Lewis, E. N.; Lee, E.; Yu, L. X.; Jefferson, E. H.; et al. (2002). Nearinfrared spectral imaging for quality assurance of pharmaceutical products: analysis of tablets to assess powder blend homogeneity. AAPS PharmSciTech Vol.3, N.3, pp. 17

57. Maftoonazad, N.; Karimi, Y.; Ramaswamy, H.S. & Prasher, S.O. (2010). Artificial neural network modeling of hyperspectral radiometric data for quality changes associated with avocados during storage. Journal of Food Processing and Preservation ISSN 1745-4549

58. Mahesh, S.; Jayas, D. S.; Paliwal, J. & White, N. D. G. (2011). Identification of wheat classes at different moisture levels using near-infrared hyperspectral images of bulk samples. Sens. & Instrumen. Food Qual. Vol.5, pp. 1–9

59. Manley, M.; Du Toit, G. & Geladi, P., (2011). Tracking diffusion of conditioning water in single wheat kernels of different hardnesses by near infrared hyperspectral imaging. Analytica Chimica Acta Vol.686, pp. 64–75

60. Massart, D.L.; Buydens, L.M.C.; De Jong, S.; Lewi P.J. & Smeyers-Verbek, J. (1998). Handbook of Chemometrics and Qualimetrics: Part B. Edited by B.G.M. ISBN: 978-0-444-82853-8, included in series: Data Handling in Science and Technology

61. Massart, D.L.; Vandeginste, B.G.M.; Buydens, L.M.C.; De Jong, S.; Lewi P.J. & SmeyersVerbek, J. (1997). Handbook of Chemometrics and Qualimetrics: Part A. Elsevier, ISBN: 0-444-89724-0, included in series: Data Handling in Science and Technology

62. Mathiassena, J. R.; Misimib, E.; Bondøb, M.; Veliyulinb, E. & Ove Østvik, S. (2011). Trends in application of imaging technologies to inspection of fish and fish products. Trends in Food Science & Technology Vol.22, pp. 257-275

63. McClure, W. F. (2003). 204 years of near infrared technology: 1800 - 2003. Journal of Near Infrared Spectroscopy, Vol.11, pp. 487–518

64. Mehl, P. M.; Chao, K.; Kim, M.; Chen, Y. R. (2002). Detection of defects on selected apple cultivars using hyperspectral and multispectral image analysis. Applied Engineering in Agriculture Vol.18, N.2, pp. 219-226

65. Monteiro, S.; Minekawa, Y.; Kosugi, Y.; Akazawa, T. & Oda, K. (2007). Prediction of sweetness and amino acid content in soybean crops from hyperspectral imagery. ISPRS Journal of Photogrammetry and Remote Sensing Vol.62, N.1, pp. 2–12

66. Naes, T.; Isaksson, T.; Fearn, T. & Davies, T. (2002). A user-friendly guide to multivariate calibration and classification. Chichester, UK: NIR

Publications ISBN 0-9528666-2-5

67. Nguyen, D.; Trong, N.; Tsuta, M.; Nicolaï, B.M.; De Baerdemaeker, J. & Saeys, W. (2011). Prediction of optimal cooking time for boiled potatoes by hyperspectral imaging. Journal of Food Engineering Vol.105, pp. 617–624

68. Nicolai, B. M.; Beullens, K.; Bobelyn, E.; Peirs, A.; Saeys, W.; Theron & K. I., Lammertyna J. (2007). Non-destructive measurement of fruit and vegetable quality by means of NIR spectroscopy: A review. Postharvest Biology and Technology, Vol.46, pp. 99–118

69. Oliveri, P.; Di Egidio, V.; Woodcock, T. & Downey, G. (2011). Application of class-modelling techniques to near infrared data for food authentication purposes. Food Chemistry Vol.125, pp. 1450–1456

70. Osborne, B.G.; Fearn, T. & Hindle, P.H. (1993). Practical NIR Spectroscopy with Applications in Food and Beverage Analylis. Cap. 4: Fundamentals of near infrared instrumentation, pp. 73-76. Longman Scientific & Technical

71. Pillonel, L.; Ampuero, S.; Tabacchi, R. & Bosset, J.O. (2003). Analytical methods for the determination of the geographic origin of Emmental cheese, volatile compounds by GC–MS–FID and electronic nose, Eur. J. Food Res. Technol. Vol.216, pp. 179–183

72. Prieto, N.; Roehe, R.; Lavín, P.; Batten, G. & Andrés, S. (2009b). Application of near infrared reflectance spectroscopy to predict meat and meat products quality: A review. Meat Science Vol.83, pp. 175–186

73. Prieto, N.; Ross, D.W.; Navajas, E.A.; Nute, G.R.; Richardson, R.I.; Hyslop, J.J.; Simm, G. & Roehe, R. (2009a). On-line application of visible and near infrared reflectance spectroscopy to predict chemical–physical and sensory characteristics of beef quality. Meat Science Vol.83, pp. 96–103

74. Ribeiro, J.S.; Ferreira, M.M.C. & Salva, T.J.G. (2011). Chemometric models for the quantitative descriptive sensory analysis of Arabica coffee beverages using near infrared spectroscopy. Talanta Vol.83, pp.1352–1358

75. Riva, M. (1999). Introduzione alle tecniche di Image Analysis. http://www.distam.unimi.it/image_analysis/image0.htm

76. Roggo, Y.; Edmond, A.; Chalus, P. & Ulmschneider, M. (2005). Infrared hyperspectral imaging for qualitative analysis of pharmaceutical solid forms. Analytica Chimica Acta, Vol.535 N.1-2, pp. 79-87

77. Römisch, U.; Jäger, H.; Capron, X.; Lanteri, S.; Forina, M. & Smeyers-

Verbeke, J. (2009).

78. Characterization and determination of the geographical origin of wines. Part III: multivariate discrimination and classification methods. Eur. Food Res. Technol. Vol.230, pp. 31–45

79. Russ, J. C.; Stewart, W. & Russ, J. C., J. C. (1988). The measurement of macroscopic image. Food

80. Technol. Vol.42, pp. 94-102

81. Sankaran, S.; Mishra, A.; Ehsani, R. & Davis, C. (2010). A review of advanced techniques for detecting plant diseases. Computers and Electronics in Agriculture Vol.72, pp. 1–13

82. Sierra, V.; Aldai, N.; Castro, P.; Osoro, K.; Coto-Montes, A. & Oliva, M. (2007). Prediction of the fatty acid composition of beef by near infrared transmittance spectroscopy. Meat Science Vol.78, pp. 248–255

83. Sinelli, N.; Limbo, S.; Torri, L.; Di Egidio, V. & Casiraghi, E. (2010). Evaluation of freshness decay of minced beef stored in high-oxygen modified atmosphere packaged at different temperatures using NIR and MIR spectroscopy. Meat Science Vol.86, pp. 748–752

84. Siripatrawan, U.; Makino, Y.; Kawagoe, Y. & Oshita, S. (2011). Rapid detection of Escherichia coli contamination in packaged fresh spinach using hyperspectral imaging. Talanta Vol.85, pp. 276–281

85. Smail, V.; Fritz, A. & Wetzel, D. (2006). Chemical imaging of intact seeds with NIR focal plane array assists plant breeding. Vibrational Spectroscopy Vol.42, N.2, pp. 215-221

86. Sørensen, L.K. (2009). Application of reflectance near infrared spectroscopy for bread analyses. Food Chemistry Vol.113, pp. 1318–1322

87. Stark, E.K. & Luchter, K. (2003). Diversity in NIR Instrumentation, in Near Infrared Spectroscopy.: Proceeding oh the 11th International Conference. NIR Publication, Chichester, UK, pp. 55-66

88. Torri, L.; Sinelli, N. & Limbo S. (2010). Shelf life evaluation of fresh-cut pineapple by using an electronic nose. Postharvest Biology and Technology Vol.56, pp. 239–245

89. Urbano-Cuadrado, M.; de Castro, M.D.L.; Perez-Juan, P.M.; Garcia-Olmo, J. & Gomez-Nieto, M.A. (2004). Near infrared reflectance, spectroscopy and multivariate analysis in enology—Determination or screening of fifteen parameters in different types of wines.Anal. Chim. Acta Vol.527, pp. 81-88

90. Valous, N. A.; Mendoza, F. & Sun, D.W. (2010). Emerging non-contact imaging, spectroscopic and colorimetric technologies for quality

evaluation and control of hams: a review. Trends in Food Science & Technology Vol.21, pp. 26-43

91.   Williams, P. C. & Sobering, D. (1996). How do we do it: A brief summary of the methods we use in developing near infrared calibrations. In A. M. C. Davies & P. C. Williams (Eds.), Near infrared spectroscopy: the future waves (pp. 185–188). Chichester: NIR Publications

92.   Wilson, A. D. & Baietto, M. (2009). Applications and advances in electronic-nose technologies. Sensors Vol.9, pp. 5099-5148

93.   Wold, S.; Sjöström, M. & Eriksson, L. (2001). PLS-regression: a basic tool of chemometrics. Chemom. Intell. Lab. Syst. Vol.58, pp. 109–130

94.   Yongwei, W.; Wang, J.; Zhou, B. & Lu, Q. (2009). Monitoring storage time and quality attribute of egg based on electronic nose. Analytica Chimica Acta Vol.650, pp. 183–188

95.   Yu, H. & MacGregor, J.F., (2003). Multivariate image analysis and regression for prediction of coating content and distribution in the production of snack foods. Chemometrics and Intelligent Laboratory Systems Vol.67, pp. 125–144

96.   Zardetto, S. & Dalla Rosa, M. (2006). Study of the effect of lamination process on pasta by physical chemical determination and near infrared spectroscopy analysis. Journal of Food Engineering Vol.74, pp. 402–409

97.   Zheng, C.; Sun, D.W. & Zheng, L. (2006). Recent developments and applications of image features for food quality evaluation and inspection: a review. Trends in food Science &Technology. Vol.17, pp. 642-655

98.   Zheng, G.; Chen, Y.; Intes, X.; Chance, B. & Glickson, J. D. (2004). Contrast-enhanced nearinfrared (NIR) optical imaging for subsurface cancer detection. Journal of Porphyrins and Phthalocyanines Vol.8, N.9, pp. 1106-1117.

Chapter 2

CHEMOMETRIC ANALYSIS OF THE AMINO ACID REQUIREMENTS OF ANTIOXIDANT FOOD PROTEIN HYDROLYSATES

Chibuike C. Udenigwe and Rotimi E. Aluko

The Department of Human Nutritional Sciences and the Richardson Centre for Functional Foods and Nutraceuticals, University of Manitoba, Winnipeg, MB R3T 2N2, Canada

ABSTRACT

The contributions of individual amino acid residues or groups of amino acids to antioxidant activities of some food protein hydrolysates were investigated using partial least squares (PLS) regression method. PLS models were computed with amino acid composition and 3-$z$scale descriptors in the X-matrix and antioxidant activities of the samples in the Y-matrix; models were validated by cross-validation and permutation tests. Based on coefficients of the resulting models, it was observed that sulfur-containing (SCAA), acidic and hydrophobic amino acids had strong positive effects on scavenging of 2,2-diphenyl-1-picrylhydrazyl (DPPH) and H_2O_2 radicals in addition to ferric reducing antioxidant power. For superoxide radicals, only lysine and leucine showed strong positive contributions while SCAA had strong negative contributions to scavenging by the protein hydrolysates. In contrast, positively-charged amino acids strongly contributed negatively to ferric reducing antioxidant power and scavenging of DPPH and H_2O_2 radicals. Therefore, food protein hydrolysates containing appropriate amounts of amino acids with strong contribution properties could be potential candidates for use as potent antioxidant agents. We conclude that information presented in this work could support the development of low cost methods that will efficiently generate potent antioxidant peptide mixtures from food proteins without the need for costly peptide purification.

INTRODUCTION

Antioxidant enzymatic food protein hydrolysates and food-derived peptides have gained particular interest as potential ingredients for formulation of functional foods and natural health products. Such formulated products could be used for human health sustenance especially for prevention and management of chronic diseases induced or propagated by oxidative stress. There is abundant information in the literature on food protein hydrolysates, peptide fractions and purified peptides with antioxidant activities in various oxidative models *in vitro* and in cell cultures. The antioxidant activity of food-derived peptides is based on scavenging of free radicals or reactive species [1,2], which is predominantly based on proton-coupled single electron or hydrogen atom transfer mechanisms [3,4]. In addition, food protein-derived antioxidant peptides also exhibit their activity by chelating pro-oxidant transition metals, e.g., Fe^{2+} and Cu^{2+}, reducing ferric ion, inhibiting oxidation of biological macromolecules such as unsaturated fatty acids [1,5] and cellular regulation of gene expression of antioxidant proteins, e.g., heme oxygenase-1 and ferritin [6]. Antioxidant activities of free amino acids have been evaluated [7] and a number of amino acids have been proposed to contribute positively to the antioxidant activity of purified food-derived and synthetic natural peptides. These amino acids include *Trp*, *Tyr*, *Met*, *Cys*, *His*, *Phe* and *Pro* [1,7–10]. Till date, several studies have attributed the antioxidative properties of many food protein hydrolysates, peptide fractions and purified peptides to these amino acids, but the chemistry and mechanisms of action have not been studied in detail. It is generally agreed upon that peptides possess substantially better antioxidant activities than their parent proteins and constituent amino acids [1], possibly due to increased accessibility of the functional side chain (R-group) to the reactive species, and the electron-dense peptide bonds. However, previous studies have indicated that some amino acids may be more active than their parent peptides [11,12].

For practical cost-effective application of food-derived peptides in the formulation of health-promoting food products, important limiting factors to be considered include a combination of high yields and potency of the natural bioactive ingredients. The expensive procedures and low product yield often associated with food protein-derived peptide purification and synthesis underscores the need to develop natural enzymatic food protein hydrolysates and peptide fractions that will possess potent antioxidant activities without the need for further extensive processing. In order to achieve this goal, initial directions of approach should involve elucidation of amino acid requirements for potency of food protein hydrolysates, and development of enzymatic hydrolysis and simple processing methods for enrichment of the desired amino acid residues in the peptide mixtures based upon their unique physicochemical properties. Indeed, the presence of the proposed antioxidant amino acid residues (*Trp, Tyr, Met, Cys, His, Phe and Pro*) may promote the antioxidant activities of food protein hydrolysates. However, the contributions of other amino acid residues to the oxidative system and the relative contribution/interactions of the so-called antioxidant amino acids in the complex peptide mixtures remain unclear. Moreover, the specific role of food-derived peptide antioxidants can be influenced by the type of oxidative species and chemistry of the reaction medium (e.g., pH) [2,3], which influence the properties of functional amino acid side chain groups. In-depth understanding of the relationship between composition and antioxidant activity of food protein hydrolysates and peptides is desirable for directed efforts towards the discovery of safer functional food ingredients [1]. Therefore, the objective of this study was to evaluate the contributions of individual amino acid residues, groups of amino acids (based on similar physicochemical properties of the R-group) and amino acid structural properties (based on *3-z* scale amino acid descriptor) of food protein hydrolysates to antioxidant activities in four different oxidative assay systems using partial least squares (PLS) regression.

RESULTS AND DISCUSSIONS

PLS modeling is a widely used descriptive and predictive chemometrics approach for quantitative structure-activity relationship (QSAR) studies to elucidate how variation of molecular structures affect bioactivity of therapeutic agents, especially when working with high number of descriptor variables compared to the number of observations [13,14]. This method has been widely applied in food science research for developing models in QSAR studies of food-derived peptides such as bitter peptides [15,16], angiotensin converting enzyme inhibiting peptides [14,17–19], renin inhibiting peptides [20] and antimicrobial peptides [19]. Moreover, PLS modeling has been applied to the study of functional properties of food proteins and polypeptides; these are studies where the functional properties under investigation are impacted mostly by proportion of relevant amino acids rather than sequence [21]. Despite the wide application in food science, there is dearth of information in the literature on the use of chemometrics approaches, such as PLS method, in studying bioactivity of food protein hydrolysates that contain mixtures of peptides. Previous chemometric work on food-related bioactive compounds involved PLS analysis to elucidate the structural requirements for potency of synthesized antioxidant polyphenols in different chemical, cellular and enzymatic oxidative systems [22–24]; these methods often resulted in the discovery of compounds with more potent end-point antioxidative activities.

PLS modeling of amino acid parameters (X) and four different antioxidant activities (Y) of the food protein hydrolysates and fractions resulted in 12 models as shown in Table 1. The ferric reducing antioxidant property (FRAP) model (AA only) gave the best fit and predictive power whereas models for H_2O_2-scavenging displayed the lowest predictive powers (Figure 1 and Table 1). The R^2 values (Table 1) indicated that the PLS models explained 30% to 73% of the sum of squares in Y-variance for all the oxidative systems with up to 66% predictive ability (derived from Q^2_{cv}). The 12 models were theoretically validated initially by cross-validation during modeling and their predictive power also validated by permutation, where the bioactivity data were each randomly permuted a number of times but with unaltered X-variable followed by modeling of each permutation [13]. As shown in Table 1, repeated (20) rounds of permutation yielded cumulative R^2 (R^2_{cum}) intercept values of 0.006 to 0.314 and Q^2_{cv} intercept values of -0.223 to -0.008 for the 12 models, which are within suggested valid limits of R^2_{cum} intercept <0.4 and Q^2_{cv} intercept <0.05 for valid PLS models [25]. The t/u PLS score plots show relationships between X and Y variables in the AA + gAA + Σz_i (antioxidant activity + amino acid group + sum of z values) models for the four oxidative systems (Figure 1A–D).

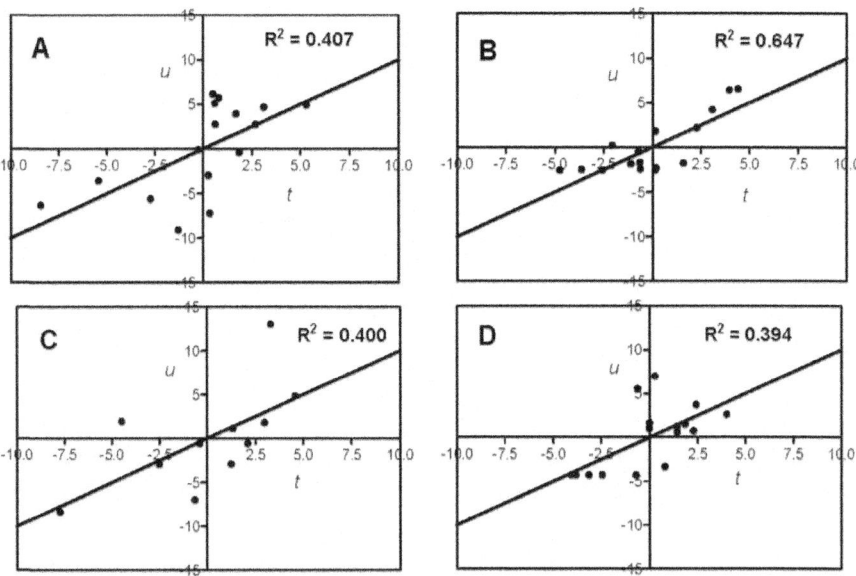

Figure 1: The t/u score plots of the partial least squares (PLS) models showing relationships between the antioxidant activity and amino acid descriptors (AA + gAA + Σz_i) of the food protein hydrolysates and peptide fractions; (**A**) DPPH radical scaveng-

ing; (**B**) ferric reducing antioxidant power (FRAP); (**C**) H_2O_2-scavenging, and (**D**) superoxide radical-scavenging.

With the exception of the weak fit for H_2O_2, the data showed good fit and must have contributed positively to obtaining valid models that we used to determine relationships between amino acids or groups of amino acids and contributions to antioxidant potential of the food protein hydrolysates.

Table 1: Summary of the partial least square regression models using amino acid compositions (AA), groups of amino acids (gAA) and sums of 3-z amino acid scales (Σz_i) of the food protein hydrolysates. The multiple correlation coefficient (R^2) estimates the model fit whereas the cross-validated correlation coefficient (Q^2_{cv}) indicates the models' predictive powers.

Antioxidant Property	N^a	X^b	A^c	$R^{2\,d}$	$Q^2_{cv}{}^e$	Permutation Test ^{f}	
						Int. R^2_{cum}	Int. Q^2_{cv}
DPPH-scavenging	16	AA + gAA+ Σz_i	1	0.407	0.327	0.241	−0.096
		AA	1	0.448	0.358	0.288	−0.108
		gAA	1	0.405	0.231	0.149	−0.043
		Σz_i	1	0.304	0.288	0.006	−0.127
Ferric reducing	16	AA + gAA + Σz_i	1	0.647	0.604	0.166	−0.208
		AA	1	0.728	0.668	0.223	−0.219
		gAA	1	0.536	0.531	0.073	−0.169
H_2O_2-scavenging	11	AA + gAA+ Σz_i	1	0.400	0.137	0.261	−0.122
		AA	1	0.467	0.136	0.314	−0.109
		Σz_i	1	0.340	0.232	0.088	−0.008
O_2^--scavenging	16	AA + gAA + Σz_i	1	0.394	0.179	0.212	−0.063
		gAA	2	0.602	0.142	0.166	−0.223

^{a} N, number of observations used for PLS analysis; ^{b} X, X-variables (descriptors) in the validated PLS models; ^{c} A, number of significant components used in PLS modeling; ^{d} R^2, multiple correlation coefficients; ^{e} Q^2_{cv}, cross-validation correlation coefficients; ^{f} Permutation test, R^2_{cum} and Q^2_{cv} intercepts were calculated by SIMCA-P software during model validation.

The importance or relative contribution of the amino acid descriptor variables (X) in PLS modeling was obtained from the VIP plots (Figure 2A–D). The X variables with VIP > 1.0 are regarded as important with above average contribution while those with VIP < 0.5 are unimportant; variables with 1.0 > VIP > 0.5 could be important or not depending on the size of the dataset [26]. For example, the sulphur-containing (SC) amino acids as a group as well the individual amino acids that include *Met,Cys*, *Leu*, and *Lys* are strong contributors (VIP values > 1.0) to the superoxide model (Figure 2D). In contrast, *Gly, Ala, His, Asx*, and *Ile* are poor contributors to the superoxide model due to VIP values of <0.5. For our present study, X-variables were regarded as important strong contributors only when their VIP > 1.0 and weak contributors with VIP of 0.5–1.0. The relative contribution of X-variables in the PLS models depends on the value of the coefficients relative to the origin in the loading space [13]; in other words, the higher the coefficients in both directions, the more the contribution of the X-variable in explaining or predicting Y [27]. The coefficients (Figure 3A–D) and VIP plots (Figure 2A–

D) indicated that the specific contribution of each or groups of amino acid and physicochemical properties (*3-z* scales) depends on the oxidative assay system. As shown in Figure 3A and Table 2, a high percentage composition of *Thr, Asx*, hydrophobic amino acids, as well as ↑Σz_3 (high electronic properties), ↓Σz_1 (low hydrophilicity or high hydrophobicity) and ↓Σz_2 (low bulk/molecular size) of the samples contributed positively to 2,2-diphenyl-1-picrylhydrazyl (DPPH)-scavenging. Interestingly, PCAA including *His* (highest VIP value) contributed negatively to the scavenging of DPPH. A similar pattern was also observed for H_2O_2-scavenging in addition to the strong positive contributions of *Cys,Phe, Leu* and *Pro* in this oxidative system (Figure 3C). Moreover, the SCAA (*Cys + Met*) were observed to be the strongest contributing amino acids for ferric reducing antioxidant power of the peptide mixtures whereas high amounts of *Lys* strongly reduced this activity (Figure 3B).

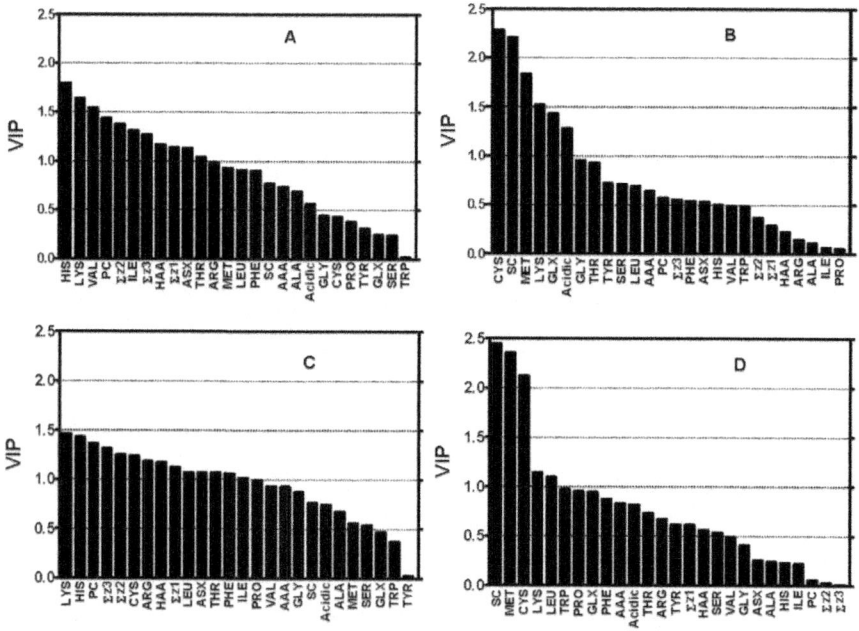

Figure 2: Variable Importance for the Projection (VIP) of the *3-z* scale models: (**A**) DPPH radical scavenging; (**B**) ferric reducing antioxidant power (FRAP); (**C**) H_2O_2-scavenging, and (**D**) superoxide radical-scavenging.

In contrast, *Lys* and *Leu* composition supported superoxide radical scavenging by the samples while SCAA *Met* and *Cys* (highest VIP values) strongly reduced this activity. Interestingly, low z_1 (high hydrophobicity) character seem to have strong positive contributions to scavenging of the free radicals like DPPH and superoxide anion radical (O_2^-) as well as H_2O_2 (a reactive oxygen species) but not FRAP. The results agree with previous reports that indicate hydrophobic amino acids are able to interact better (when compared to hydrophilic amino acids) with the lipophilic environments that contain these free radicals. Table 2 also shows other amino acid compositions and properties that weakly contributed positively or negatively to the antioxidant activities of the peptide samples in the four oxidative systems.

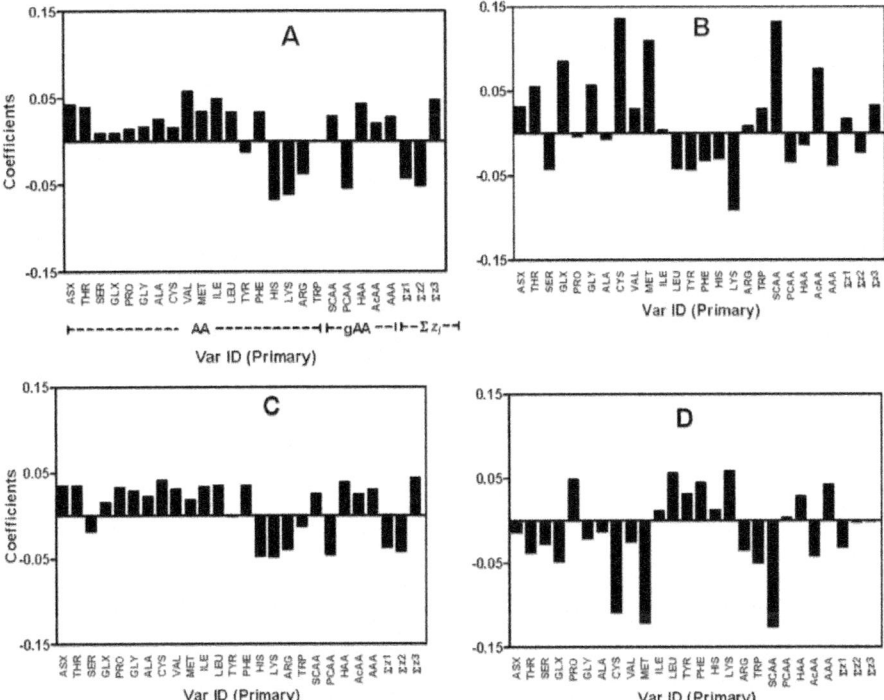

Figure 3: Coefficient plots of scaled and centered data of the partial least square regression models: (**A**) DPPH radical scavenging; (**B**) ferric reducing antioxidant power (FRAP); (**C**) H_2O_2-scavenging; and (**D**) superoxide radical-scavenging. The importance of a given X-variable is proportional to its distance (coefficient value) from the origin (zero). Above zero values indicate positive contributions while values less than zero indicate negative contributions.

Due to heterogeneity of reactive species implicated in human disease conditions, the structure and composition of amino acid residues required for potent antioxidant activity by food protein hydrolysates and peptides could depend on the type of reactive species and reaction conditions, e.g., pH and solvent.

Table 2: Summary of contribution of amino acid compositions and properties to antioxidant activities of food protein hydrolysates based on partial least square regression coefficients and VIP plots of the AA + gAA + Σz_i models; strong contributors with VIP > 1.0 are shown in bold; (weak contributors with 1.0 > VIP > 0.5 are shown in brackets).

Oxidative System	Positive Contributors	Negative Contributors
DPPH radical scavenging	*Asp + Asn (Asx)*, **Thr, Val, Ile,** **HAA** [a], ↑Σz_3 [b], ↓Σz_1 [c], ↓Σz_2 [d] (*Ala, Met, Leu, Phe, SCAA* [e], *AcAA* [f], *AAA* [g])	**His, Lys, Arg, PCAA** [h]
Ferric reducing	**Cys, Met, Glu + Gln (Glx),** **SCAA, AcAA** (*Asp + Asp, Thr, Gly,* ↑Σz_3)	**Lys** (*Ser, Leu, Tyr, Phe, His, PCAA, AAA*)
H$_2$O$_2$-scavenging	**Cys, Phe, Leu, Ile, Pro, Thr,** **Asx, HAA,** ↑Σz_3, ↓Σz_1, ↓Σz_2 (*Gly, Ala, Val, Met, SCAA, AcAA, AAA*)	**His, Lys, Arg, PCAA** (*Ser*)
Superoxide radical scavenging	**Lys, Leu** (*Pro, Phe, Tyr, HAA,* ↓Σz_1)	**SCAA, Met, Cys** (*Trp, Glx, AcAA, Thr, Arg, Ser*)

[a] HAA, Hydrophobic amino acids (*Pro + Ala + Cys + Val + Met + Ile + Leu + Tyr + Phe + Trp*); [b] Amino acids with ↑Σz_3 (electronic property) include *Cys > Pro > Asx > His > Trp*; [c] Amino acids with ↓Σz_1 (highly hydrophobic) include *Phe > Trp > Ile > Leu > Val > Met > Tyr*; [d] Amino acids with ↓Σz_2 (lower side chain bulk/molecular size) include *Gly > Val > Thr > Ala > Ile > Ser > Leu > Cys*; [e] SCAA, sulfur-containing amino acids (*Met + Cys*); [f] AcAA, Acidic amino acids (*Glx + Asx*); [g] AAA, Aromatic amino acids (*Phe + Tyr + Trp*); [h] PCAA, positively charged amino acids (*Lys + Arg + His*).

Consequently, high levels of particular amino acid residues can potentially increase or decrease the antioxidant activity of food protein hydrolysates depending on the reaction environment. Amino acid residues are major physiological targets of oxidants leading to the formation of stable and unstable oxidation products [28]; thus, the rate of transfer of the radicals to amino acids and stability of the reaction products can substantially influence potency of amino acids as antioxidants. Plausible mechanisms of radical scavenging or quenching activity of amino acid residues of proteins and peptides include proton or hydrogen atom donation by amino acids such as *Tyr*, *Trp* and *Cys* using their phenolic, indolic and sulfhydryl hydrogen, respectively [1,7,10]. The electron-dense side chain groups of *His*, *Trp* and *Met* also contribute to antioxidant properties of proteins and peptides [28]. In our study, the food protein hydrolysates in the dataset were composed of varying amounts of amino acid residues, which positively or negatively impacted their antioxidant activities based on the PLS models.

The positive contributions of SCAA (*Cys* and *Met*) to the antioxidant activity of the samples (Table 2) can be attributed to their S-groups, which are prone to oxidation by the reactive species leading to the formation of stable oxidation products, cystine and methione sulfoxide, respectively [1,7,28]. The sulfhydryl group of *Cys* also acts as a strong reducing agent, hence its strong positive contribution in reducing ferric ion. Aspartic acid (*Asx*) and glutamic acid (*Glx*) can also donate their acidic hydrogen atoms near neutral pH, and this may have resulted to their strong contributions to the antioxidant activities of the samples. Similarly, *Phe* played a positive role in H_2O_2-scavenging and this can be attributed to susceptibility of its aromatic ring to oxidation. It is not clear how the HAA residues and low Σz_1 (low hydrophilicity or high hydrophobicity) of the samples directly contributed to antioxidant activity of food protein hydrolysates and fractions, although HAA may have interacted with DPPH via hydrophobic forces thereby increasing the proximity of the radical to the active functional groups. Moreover, amino acids with high Σz_3 exhibited positive effects in the assays except superoxide radical-scavenging, and this can be attributed to their electron density. Amino acids with less bulky structures (low Σz_2) correlated positively with antioxidant activities of the samples (Table 2), suggesting that steric effects can influence DPPH-scavenging activity of peptides, considering the bulky structure of the synthetic radical. On the other hand, the negative effects of the PCAA could be due to the fact that they can accept hydrogen ion near physiological pH and exist in the protonated state. Surprisingly, *His*residue was observed to negatively contribute to antioxidant properties of the peptide mixtures, based on the PLS models (Table 2). *His* can act as both hydrogen acceptor and donor near physiological pH; thus, its particular role in the complex peptide mixture could not be elucidated using these models. Moreover, the current work did not take into consideration the amino acid sequence of peptides, which could be important for *His*-containing peptides. Previously, the presence of *His* in synthetic peptides was reported to promote antioxidative activity in a linoleic acid oxidation model but not DPPH and superoxide radical-scavenging [8].

EXPERIMENTAL SECTION

X-matrix

The *X*-matrix contained data for individual 18 amino acids AA (*Asx* for *Asn* + *Asp* and *Glx* for *Glu* + *Gln*) and groups of amino acids (gAA) such as sulphur-containing (SCAA-*Cys* and *Met*), positively charged (PCAA-*Arg*, *Lys* and *His*), acidic (AcAA-*Asx* and *Glx*), aromatic (AAA-*Phe*, *Tyr* and *Trp*) and hydrophobic amino acids (HAA-*Pro*, *Ala*, *Cys*, *Val*, *Met*, *Ile*, *Leu*, *Tyr*, *Phe* and *Trp*). The gAA were used because amino acids within each group have certain similarities (e.g., presence of sulphur or aromatic ring, *etc.*), which allow PLS modeling based on group characteristics. The results can be used to determine for example, whether aromatic rings contribute a greater to certain antioxidant property than acidic or sulphur groups. The X-matrix data for the amino acids of 16 samples were presented as percentage composition (Table 3) and the groups (Table 4) as sums of the respective data. The amino acid data were derived from previous studies that fractionated hempseed and pea protein hydrolysates based on molecular size [29], hydrophobic property [5] and net cationic property [30]. The amino acid3-z scale was also used as X-variable in the multivariate descriptor matrix. The 3-z scales was previously derived by Hellberg et al. [31] by principal component analysis from 29 physicochemical variables of amino acids and were interpreted to be related to hydrophilicity (z_1), steric properties or side chain bulk/molecular size (z_2) and electronic properties (z_3) (see supplementary material). Algebraic sums of each of the 3-z scores (Σz_i) were calculated for each sample (see supplementary material) as previously reported [21] using Equation 1.

Table 3: The % amino acid (AA) composition of the food protein hydrolysates used as variables in the X-matrix for partial least square regression analysis

Sample ID	ASX[a]	THR	SER	GLX[b]	PRO	GLY	ALA	CYS	VAL	MET	ILE	LEU	TYR	PHE	HIS	LYS	ARG	TRP
1	11.39	3.68	4.63	20.06	4.00	4.29	4.47	1.32	4.66	1.81	3.84	6.75	3.45	4.60	2.78	2.97	14.07	1.23
2	9.49	3.60	4.73	15.18	3.19	3.23	4.91	0.29	5.67	1.94	4.15	9.91	4.78	7.68	2.61	3.19	13.87	1.58
3	11.70	3.77	4.79	19.31	4.04	3.93	4.77	0.66	5.26	2.03	4.16	7.26	3.50	5.01	2.47	2.94	12.96	1.44
4	12.79	4.01	4.69	22.71	4.23	4.54	4.30	1.26	4.45	1.85	3.98	5.15	3.06	3.21	2.47	2.56	13.60	1.16
5	12.70	4.00	4.47	22.87	4.89	4.71	4.12	1.58	4.24	1.80	3.90	4.82	3.62	2.85	2.49	2.51	13.31	1.11
6	13.79	3.60	6.20	13.92	5.15	3.76	5.01	0.24	5.63	0.91	5.43	9.91	3.87	7.41	1.61	6.10	6.83	0.68
7	13.94	3.89	6.63	17.12	2.33	3.52	5.54	0.18	5.23	0.70	4.13	8.70	2.77	3.97	2.49	9.07	9.79	0.00
8	10.63	3.86	5.71	14.78	6.47	5.00	4.30	0.39	4.45	1.70	4.04	6.68	5.33	7.76	3.28	7.35	8.00	0.27
9	12.59	3.34	6.19	13.75	5.14	3.96	5.03	0.39	4.13	0.87	6.71	9.95	7.15	8.73	1.90	4.26	5.15	0.74
10	10.85	3.11	4.41	12.87	5.42	4.66	3.44	0.38	5.82	1.07	5.85	14.57	5.09	12.03	1.81	3.31	3.97	1.36
11	11.04	3.22	3.82	6.64	8.05	3.26	3.62	0.29	7.68	0.68	9.13	19.48	2.44	16.44	0.63	1.20	1.22	1.16
12	15.09	4.43	5.16	20.30	5.16	4.12	5.47	0.41	4.77	1.07	4.99	10.16	3.41	6.81	1.05	3.40	3.12	1.06
13	11.23	3.42	3.96	5.78	3.83	4.31	0.26	4.28	0.88	3.23	6.55	4.12	4.50	3.55	8.03	9.32	0.63
14	9.86	3.59	5.73	10.81	4.56	3.47	4.62	0.13	3.64	0.96	3.21	8.22	3.21	6.20	3.76	16.38	10.19	1.46
15	8.12	2.68	3.72	9.47	2.37	3.87	4.18	0.12	2.66	0.51	2.25	8.92	5.61	4.86	4.22	11.79	24.50	0.14
16	6.61	1.74	4.24	9.04	2.93	2.00	2.48	0.11	2.02	0.21	2.14	6.64	3.48	2.24	4.26	16.76	30.18	2.93

[a] GLX = glutamine + glutamate; [b] ASX = asparagines + aspartate.

Table 4: Amino acid groups (gAA) present in food protein hydrolysates used as variables in the X-matrix for partial least square regression analysis

$$\sum z_i = \sum_{X=1}^{18} z_{iX} c_X \tag{1}$$

X represents each of the 18 amino acids in the protein hydrolysates and fractions, and c represents their percentage composition; the z values for Asx and Glx were calculated as averages of the z values of their respective constituent amino acids.

Sample ID	X-Variables (gAA)				
	SCAA [a]	PCAA [b]	HAA [c]	AcAA [d]	AAA [e]
1	3.13	19.82	36.13	31.45	9.28
2	2.22	19.67	44.10	24.67	14.05
3	2.70	18.37	38.14	31.00	9.95
4	3.11	18.62	32.64	35.50	7.42
5	3.39	18.31	32.95	35.56	7.58
6	1.15	14.54	44.24	27.71	11.96
7	0.88	21.35	33.55	31.06	6.74
8	2.09	18.63	41.39	25.41	13.36
9	1.26	11.31	48.84	26.34	16.62
10	1.45	9.09	55.03	23.72	18.48
11	0.97	3.05	68.97	17.68	20.04
12	1.48	7.58	43.32	35.39	11.28
13	1.14	20.90	34.53	33.37	9.24
14	1.09	30.33	36.21	20.67	10.88
15	0.63	40.51	31.63	17.59	10.62
16	0.33	51.19	25.18	15.65	8.65

[a] SCAA, sulphur-containing amino acids (*Met* + *Cys*); [b] PCAA, positively charged amino acids (*Lys* + *Arg* + *His*); [c] HAA, Hydrophobic amino acids; [d] AcAA, Acidic amino acids (*Glx* + *Asx*); [e] AAA, Aromatic amino acids (*Phe* + *Tyr* + *Trp*).

Y-Matrix

The Y-matrix contained the antioxidant data for the various observations, specifically the ability of the food protein hydrolysates to scavenge nitrogen-centered 2,2-diphenyl-1-picrylhydrazyl (DPPH) radical, superoxide anion radical (O_2^-) and H_2O_2, and their ferric reducing antioxidant power (FRAP). These antioxidant data were chosen because of the physiological relevance of the reactive species (O_2^- and H_2O_2) and reducing power in chronic human disease conditions, apart from DPPH-scavenging, which has been widely used in the literature as primary evaluation of antioxidant (reducing) capacity of food protein hydrolysates. The bioactivity data were presented as percentage scavenging of the free radicals and H_2O_2, and as absorbance at 700 nm for ferric reducing antioxidant power (see supplementary material). In order to ensure consistency in data interpretation, the antioxidant and amino acid composition data used in this study were reported by the same research group.

PLS Modeling

Modeling of the antioxidant activities (Y) as a function of the amino acid descriptors (X) of the protein hydrolysates was computed by the PLS method using SIMCA-P version 11.0 (Umetrics AB, Umeå, Sweden); the PLS models were generated for the four oxidative assay systems using individual (AA, gAA and Σz_i) and a combination of all the descriptors. All variables were centered and scaled to unit variance to ensure equal contribution in the models. The PLS models were validated theoretically using a combination of cross-validation and permutation tests [13]. The multiple correlation coefficient (R^2) and cross-validation correlation coefficient (Q^2_{cv}) were computed by SIMCA-P software and used to represent model fit and predictive ability, respectively. The relative contribution of the amino acid (X) descriptors to the antioxidant activities of the samples was computed by the software and presented as the Variable Importance for the Projection (VIP) and coefficient plots.

CONCLUSIONS

The present work has shown that chemometrics approach using PLS models successfully elucidated specific contributions (positive or negative) of individual and groups of amino acid residues to the antioxidative properties of food protein hydrolysates; however, the effects depend on the oxidative assay system. Based on the PLS models, it was observed that previously reported antioxidant amino acid residues (especially *His*) had negative influence on the antioxidant activities (DPPH, H_2O_2, superoxide, ferric reducing) studied in this work. Overall, low hydrophilic property (high hydrophobicity) was a strong positive contributor to scavenging of free radicals (but not ferric reducing ability) by food protein hydrolysates. In contrast to previous assumptions in the literature, aromatic amino acids did not show strong contributions to antioxidant systems studied in this work (except for H_2O_2). However, the data cannot be used to preclude strong contributions of aromatic amino acids to other antioxidant systems that were not included in our report. This is because antioxidant activities of food protein hydrolysates could also be influenced by the amino acid sequence and interactions between neighboring residues; these factors were not part of the PLS analysis in this work. Data from this work can serve as background for further chemometrics study on bioactive food protein hydrolysates using larger uniformly generated datasets and determining the effect of amino acid sequence. Results from this study could contribute to proper understanding of the structure-function relationships of antioxidant food protein hydrolysates and peptides. The results could also enhance development of enzymatic tools and processing conditions that will concentrate antioxidant amino acids into highly potent fractions. Future studies to apply the results of this study may include optimization of enzymatic hydrolysis and processing conditions to enrich the final peptide product with the desirable amino acid residues. The choice of amino acids or proportion of amino acids that need to be dominant in potent antioxidant food protein hydrolysates will depend on the target antioxidant system.

ACKNOWLEDGMENTS

This research work was supported by a Discovery grant to REA from the Natural Sciences and Engineering Research Council of Canada (NSERC).

REFERENCES

1. Elias, RJ; Kellerby, SS; Decker, EA. Antioxidant activity of proteins and peptides. Crit. Rev. Food Sci. Nutr 2008, 48, 430–441.

2. Udenigwe, CC; Lu, YL; Han, CH; Hou, WC; Aluko, RE. Flaxseed protein-derived peptide fractions: Antioxidant properties and inhibition of lipopolysaccharide-induced nitric oxide production in murine macrophages. Food Chem 2009, 116, 277–284.

3. Prior, RL; Wu, X; Schaich, K. Standardized methods for the determination of antioxidant capacity and phenolics in foods and dietary supplements. J. Agric. Food Chem 2005, 53, 4290–4302.

4. Huang, D; Ou, B; Prior, RL. The chemistry behind antioxidant capacity assays. J. Agric. Food Chem 2005, 53, 1841–1856.

5. Pownall, TL; Udenigwe, CC; Aluko, RE. Amino acid composition and antioxidant properties of pea seed (Pisum sativum L.) enzymatic protein hydrolysate fractions. J. Agric. Food Chem 2010, 58, 4712–4718.

6. Erdmann, K; Grosser, N; Schipporeit, K; Schroder, H. The ACE inhibitory dipeptide Met-Tyrdiminishes free radical formation in human endothelial cells via induction of heme oxygenase-1 and ferritin. J. Nutr 2006, 136, 2148–2152.

7. Hernández-Ledesma, B; Dávalos, A; Bartolomé, B; Amigo, L. Preparation of antioxidant enzymatic hydrolysates from -lactalbumin and -lactoglobulin. Identification of active peptides by HPLC-MS/MS. J. Agric. Food Chem 2005, 53, 588–593.

8. Chen, HM; Muramoto, K; Yamauchi, F; Fujimoto, K; Nokihara, K. Antioxidant properties of histidine-containing peptides designed from peptide fragments found in the digests of a soybean protein. J. Agric. Food Chem 1998, 46, 49–53.

9. Chen, HM; Muramoto, K; Yamauchi, F; Nokihara, K. Antioxidant activity of designed peptides based on the antioxidative peptide isolated from digests of a soybean protein. J. Agric. Food Chem 1996, 44, 2619–2623.

10. Saito, K; Jin, DH; Ogawa, T; Muramoto, K; Hatakeyama, E; Yasuhara, T; Nokihara, K. Antioxidative properties of tripeptide libraries prepared by the combinatorial chemistry. J. Agric. Food Chem 2003, 51, 3668–3674.

11. Erdmann, K; Cheung, BWY; Schroder, H. The possible roles of food-

derived bioactive peptides in reducing the risk of cardiovascular disease. J. Nutr. Biochem 2008, 19, 643–654.

12. Kitts, DD; Weiler, K. Bioactive proteins and peptides from food sources. Applications of bioprocesses used in isolation and recovery. Curr. Pharm. Des 2003, 9, 1309–1323.

13. Wold, S; Sjöström, M; Eriksson, L. PLS-regression: a basic tool of chemometrics. Chemom. Intell. Lab. Syst 2001, 58, 109–130.

14. Pripp, AH; Isaksson, T; Stepaniak, L; Sørhaug, T; Ardö, Y. Quantitative structure-activity relationship modelling of peptides and proteins as a tool in food science. Trends Food Sci. Technol 2005, 16, 484–494.

15. Kim, HO; Li-Chan, ECY. Quantitative structure-activity relationship study of bitter peptides. J. Agric. Food Chem 2006, 54, 10102–10111.

16. Wu, J; Aluko, RE. Quantitative structure-activity relationship study of bitter di- and tri-peptides with angiotensin I-converting enzyme inhibitory activity. J. Pept. Sci 2007, 13, 63–69.

17. Wu, J; Aluko, RE; Nakai, S. Structural requirements of angiotensin I-converting enzyme inhibitory peptides: quantitative structure-activity relationship study of di- and tripeptides. J. Agric. Food Chem 2006, 54, 732–738.

18. Wu, J; Aluko, RE; Nakai, S. Structural requirements of angiotensin I-converting enzyme inhibitory peptides: quantitative structure-activity relationship modelling of peptides containing 4–10 amino acid residues. QSAR Comb. Sci 2006, 25, 873–880.

19. Yang, L; Shu, M; Ma, K; Mei, H; Jiang, Y; Li, Z. ST-scale as a novel amino acid descriptor and its application in QSAM of peptides and analogues. Amino Acids 2010, 38, 805–816.

20. Udenigwe, CC; Aluko, RE. Quantitative structure–activity relationship modeling of renin-inhibiting dipeptides. Amino Acids 2011. [CrossRef]

21. Siebert, KJ. Modeling protein functional properties from amino acid composition. J. Agric. Food Chem 2003, 51, 7792–7797.

22. Khalebnikov, AI; Schepetkin, IA; Domina, NG; Kirpotina, LN; Quinn, MT. Improved quantitative structure-activity relationship models to predict antioxidant activity of flavonoids in chemical, enzymatic, and cellular systems. Bioorg. Med. Chem 2007, 15, 1749–1770.

23. Om, A; Kim, JH. A quantitative structure-activity relationship model for radical scavenging activity of flavonoids. J. Med. Food 2008, 11, 29–37.

24. Roy, K; Mitra, I. Advances in quantitative structure-activity relationship models of antioxidants.Expert Opin. Drug Discov 2009, 4, 1157–1175.

25. Van der Voet, H. Comparing the predictive accuracy of models using a simple randomization test. Chemom. Intell. Lab. Syst 1994, 25, 313–323.

26. SIMCA-P 11 Software Analysis Advisor; version or edition; Umetrics AB: Umeå, Sweden, 2005.

27. Sandberg, M; Eriksson, L; Jonsson, J; Sjöström, M; Wold, S. New chemical descriptors relevant for the design of biologically active peptides. A multivariate characterization of 87 amino acids. J. Med. Chem 1998, 41, 2181–2491.

28. Davies, MJ. The oxidative environment and protein damage. Biochim. Biophys. Acta 2005, 1703, 93–109.

29. Girgih, AT; Udenigwe, CC; Aluko, RE. In vitro antioxidant properties of hempseed (Cannabis sativa L.) protein hydrolysate fractions. J. Am. Oil Chem. Soc 2011, 88, 381–389.

30. Pownall, TL; Udenigwe, CC; Aluko, RE. Effects of cationic property on the in vitro antioxidant properties of pea protein hydrolysate fractions. Food Res. Int 2011, 44, 1069–1074.

31. Hellberg, S; Sjöström, M; Skagerberg, B; Wold, SJ. Peptide quantitative structure-activity relationships, a multivariate approach. J. Med. Chem 1987, 30, 1126–1135.

Chapter 3

PROFILING OF FATTY ACIDS COMPOSITION IN SUET OIL BASED ON GC–EI-QMS AND CHEMOMETRICS ANALYSIS

Jun Jiang [1,2] and Xiaobin Jia [1,2],

[1]Affiliated Hospital on Integration of Chinese and Western Medicine, Nanjing University of Chinese Medicine, Xianlin Avenue 138#, Xianlin University City, Nanjing 210023, China

[2]Key Laboratory of New Drug Delivery System of Chinese Meteria Medica, Jiangsu Provincial Academy of Chinese Medicine, 100# Shizi Road, Nanjing 210028, China

ABSTRACT

Fatty acid (FA) composition of suet oil (SO) was measured by precolumn methylesterification (PME) optimized using a Box–Behnken design (BBD) and gas chromatography/electron ionization-quadrupole mass spectrometry (GC–EI-qMS). A spectral library (NIST 08) and standard compounds were used to identify FAs in SO representing 90.89% of the total peak area. The ten most abundant FAs were derivatized into FA methyl esters (FAMEs) and quantified by GC–EI-qMS; the correlation coefficient of each FAME was 0.999 and the lowest concentration quantified was 0.01 µg/mL. The range of recovery of the FAMEs was 82.1%–98.7% (relative standard deviation 2.2%–6.8%). The limits of quantification (LOQ) were 1.25–5.95 µg/L. The number of carbon atoms in the FAs identified ranged from 12 to 20; hexadecanoic and octadecanoic acids were the most abundant. Eighteen samples of SO purchased from Qinghai, Anhui and Jiangsu provinces of China were categorized into three groups by principal component analysis (PCA) according to the contents of the most abundant FAs. The results showed SOs samples were rich in FAs with significantly different profiles from different origins. The method described here can be used for quality control and SO differentiation on the basis of the FA profile.

INTRODUCTION

Suet oil (SO), a fatty oil obtained from the domestic goat (*Capra hircus* Linnaeus) or sheep (*Ovis aries* Linnaeus), has been used in the food industry [1] and the medicine industry [2]. SO is rich in unsaturated and saturated fatty acids (FAs) [3], which are involved in a number of important physiological processes. They provide energy to the cell and act as substrates in the synthesis of fats, lipoproteins, liposaccharides and eicosanoids [4]. Furthermore, SO can be used as an excipient for enhancing the efficacy of traditional Chinese medicines such as Epimedium (Berberidaceae). It was hypothesized that the beneficial effects of Epimedium could be attributed to promotion of the intestinal absorption of drugs by the formation of micelles owing to the action of its FA ingredients [5]. The quality of SO can affect safety and efficacy for clinical patients. There has been little research on the FA composition of SO, however, and there are quality control difficulties in the production of SO. It is important to establish qualitative and quantitative analytical methodology for determining the FA composition of SO.

To date, the methods used for separation and measurement of FAs are mainly chromatographic, including thin-layer chromatography [6], high-performance liquid chromatography [7,8], gas chromatography [9,10,11], supercritical fluid chromatography [12] and liquid chromatography with tandem mass spectrometry [13,14,15,16]. These methods cannot identify major chemical components rapidly and accurately. However, the gas chromatography/electron ionization-quadropole mass spectrometry (GC–EI–qMS) [17,18,19,20] technique coupled with the use of a professional database (NIST 08) can identify many compounds directly and accurately according to their fragment ions and abundance ratio [21,22,23]. In addition, GC–EI–qMS used in the selective ion monitoring (SIM) can identify target compounds rapidly and accurately despite interference from impurities [24], which is especially useful for the analysis of a lipid-based matrix, including SO.

FAs need to be derivatized before they can be analyzed by GC–MS because they have boiling points, which make gasification difficult. In many precolumn derivative methods [25], FAs are normally precolumn methylesterified (PME) into FA methyl esters (FAMEs) [26]. To ensure optimum conditions for methylesterification, the influence of important experimental parameters affecting the efficiency of methylesterification, including methyl reagent volume, temperature and time, were investigated using a Box–Behnken design (BBD) [27,28]. During optimization, the total peak area of the identified FAs was used to select the best conditions. This study developed and validated a method for the qualitative and quantitative profiling of the FA content in SOs for the first time.

SOs have been used widely in medicinal and culinary areas, but their authentication and standardization have encountered some problems owing to deliberate contamination with other animal or vegetable oils. It is difficult to identify the origins and species of SO accurately on the basis of appearance and morphology. Furthermore, SOs from different species or from different regions are not of uniform composition. In this study, a total of 18 batches of SO collected from three provinces in China were analyzed by GC–EI–qMS to determine their FA compositions and principal component analysis (PCA) was used to evaluate and classify these samples.

RESULTS AND DISCUSSION

Optimal Results and Statistical Analysis of Precolumn Methyl-esterified (PME)

By retaining only the factors statistically significantly different at $p \leq 0.05$, the following final equation in terms of uncoded factors was obtained:

Total peak area = $+3.954 \times 10^9 + 9.987 \times 10^8 A + 1.196 \times 10^9 B + 9.163 \times 10^8 C + 1.099 \times 10^9 AB + 8.977 \times 10^8 AC + 9.306 \times 10^8 B - 1.596 \times 10^9 A^2 - 1.232 \times 10^9 B^2 - 1.452 \times 10^9 C^2$.

(A)

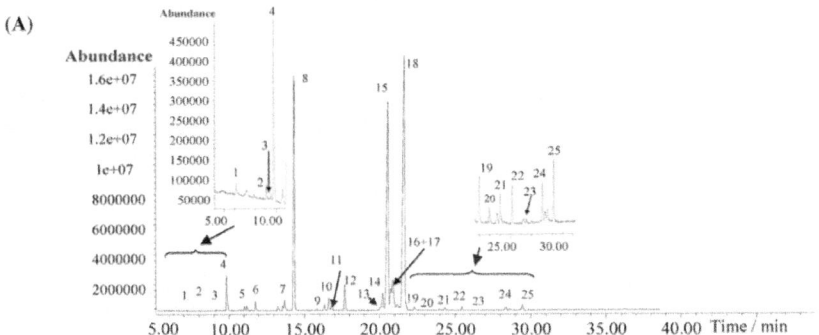

(B)

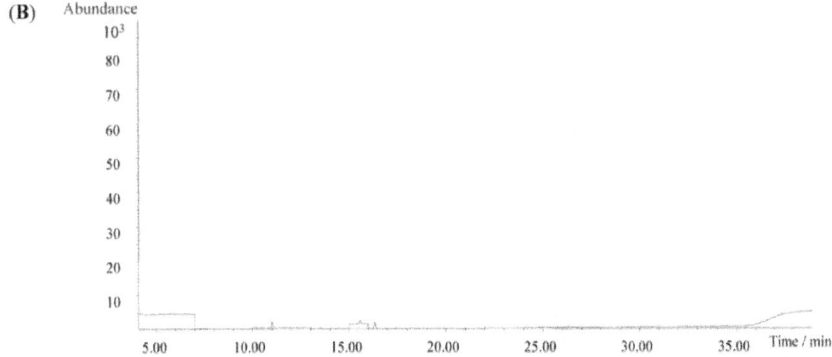

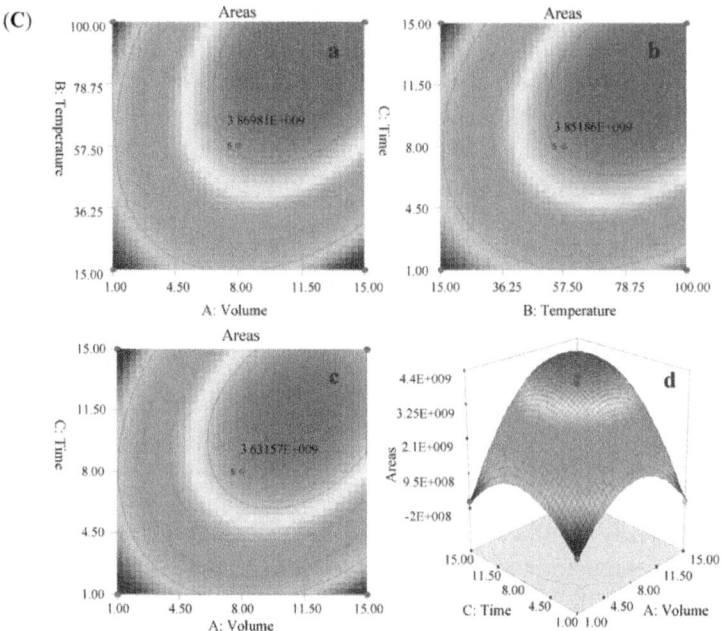

Figure 1: Optimization of precolumn methylesterified (PME) by Box–Behnken design (BBD)/GC–EI-qMS. (**A**) GC–EI-qMS chromatogram of the 25 fatty acid methyl esters (FAMEs) in suet oil (SO) sample under total ion chromatogram (TIC) mode. (**1**) Dodecanoic acid, methyl ester (DODME), (**2**) Methyl myristoleate, methyl ester, (**3**) Methyl 12-methyl-tridecanoate, methyl ester, (**4**) Tridecanoic acid, 12-methyl-, methyl ester, (**5**) Methyl tetradecanoate, methyl ester (MTEME), (**6**) Pentadecanoic acid, methyl ester (PENME), (**7**) (*Z*)-9-Hexadecenoic acid, methyl ester (9-HEME), (**8**) Hexadecanoic acid, methyl ester (HEXME), (**9**) Methyl 15-methylhexadecanoate, methyl ester, (**10**) *cis*-10-Heptadecenoic acid, methyl ester, (**11**) Heptadecanoic acid, methyl ester (HEPME), (**12**) (*Z,Z*)-9,12-Octadecadienoic acid, methyl ester (9,12-OCME), (**13**) Methyl 9-*cis*,11-*trans*-octadecadienoate methyl ester, (**14**) Methyl 10-*trans*,12-*cis*-octadecadienoate, (**15**) 9-Octadecenoic acid (*E*)-, methyl ester (9-OCME), (**16**) 9-Octadecenoic acid (*Z*)-, methyl ester, (**17**) 11-Octadecenoic acid, methyl ester, (**18**) Octadecanoic acid, methyl ester (OCTME), (**19**) *cis*-10-Nonadecenoic acid, methyl ester, (**20**) 10-Nonadecenoic acid, methyl ester, (**21**) Cyclopropaneoctanoic acid, 2-octyl-, methyl ester, (**22**) Nonadecanoic acid, methyl ester, (**23**) Methyl 8,11,14-eicosatrienoate, methyl ester, (**24**) *cis*-11-Eicosenoic acid, methyl ester, (**25**) Eicosanoic acid, methyl ester (EICME); (**B**) GC–EI-qMS chromatogram of representative blank samples under TIC mode; (**C**) Response surface plots (3-D) and contour (2-D) showing the total peaks areas with different methyl esterified condition. (**a**) 2-D panel of temperature-volum, (**b**) 2-D panel of time-temperature, (**c**) 2-D panel of time-volum, (**d**) 3-D response surface plots.

In all, 25 FA species can be identified from the chromatogram shown in Figure 1A. Comprehensive test results for response surface plots (3D) and contour plots (2D) show the total peak area was a maximum when the methylesterification conditions were: reagent volume 10 mL; temperature 60 °C; and time 10 min (Figure 1C).

Fatty Acids (FAs) Composition in Suet Oil (SO)

Identification of FAs was achieved by comparing molecular mass, ion fragments and abundance ratios in the NIST 08 spectral library. A typical total ion chromatogram obtained for SO samples is shown in Figure 1A.

The FAs in SO were investigated using optimized PME: 10 mL of BF_3–MeOH (14%, v/v), 60 °C and 10 min. The PME/GC–EI-qMS analysis of SO led to the identification of 25 different FAs (Table 1): including saturated FAs (dodecanoic acid, 12-methyl-tridecanoate, tridecanoic acid, 12-methyl-tetradecanoate, pentadecanoic acid, hexadecanoic acid, heptadecanoic acid, octadecanoic acid, cyclopropaneoctanoic acid, nonadecanoic acid, eicosanoic acid, cis-11-eicosenoic acid and 8,11,14-eicosatrienoate) and unsaturated FAs (myristoleate, 9-hexadecanoic acid, cis-10-heptadecanoic acid, (Z,Z)-9,12-octadecadienoic acid, 10-nonadecanoic acid, 10-trans,12-cis-octadecadienoate, (E)-9-octadecanoic acid, (Z)-9-octadecanoic acid, 11-octadecanoic acid, cis-10-nonadecanoic acid and 9-cis,11-trans-octadecadienoate).

In all, 25 FAs were identified (match > 90%). Hexadecanoic acid, octadecanoic acid and (E)-9-octadecanoic acid were the three most abundant and occupied 16.46%, 37.96% and 19.47% of the total peak area, respectively (Table 1).

Validation of Quantitative Analysis

Methylester derivatives of the ten most abundant FAs were purchased for use as standards. These FAs in the SO samples were quantified by derivatization into FAMEs, which were analyzed by GC–EI-qMS without significant matrix interference. Four fragment ions were monitored in the SIM mode for each compound. The best characteristic ion in the spectrum was selected for quantification of each FAME and the other three were used for confirmation.

The validity of the method was investigated by examination of the linearity, recovery, and limit of quantification (LOQ) for all FAMEs in this study. The ranges of concentration, regression equations, r^2(coefficient of determination), recovery, relative standard deviation (RSD) and LOQ for the target FAMEs are given in Table 2. Most of the FAMEs had good linearity ($r^2 > 0.999$); more importantly, the results showed a stabilized recovery of ten FAMEs in the range 82.1%–98.7% with optimized PME parameters (Table 2, Figure S1) and the LOQ values of these FAMEs ranged from 1.25 to 5.95 µg/L. These results indicated LOQ was sufficiently low to meet the requirements of determination of the FA composition of SOs. A chromatogram of these ten FAMEs in SOs obtained in the SIM mode is shown inFigure 2A.

Table 1: Fatty acids identified in SO sample using precolumn esterified/GC–EI-Qms

No.	RT (min)	Compounds Name	CAS No.	Mw [a]	Formula	Match (%)	RC [b] (%)
1	6.142	Dodecanoic acid, methyl ester	000111-82-0	214	$C_{13}H_{26}O_2$	98	0.11
2	9.016	Methyl myristoleate	056219-06-8	240	$C_{15}H_{28}O_2$	96	0.19
3	9.504	Methyl 12-methyl-tridecanoate	1000336-46-9	242	$C_{15}H_{30}O_2$	98	0.087
4	9.739	Tridecanoic acid, 12-methyl-, methyl ester	005129-58-8	242	$C_{15}H_{30}O_2$	94	0.25
5	11.071	Methyl tetradecanoate	000124-10-7	242	$C_{15}H_{30}O_2$	98	2.52
6	11.663	Pentadecanoic acid, methyl ester	007132-64-1	256	$C_{16}H_{32}O_2$	99	0.29
7	13.614	9-Hexadecenoic acid, methyl ester, (Z)-	001120-25-8	268	$C_{17}H_{32}O_2$	99	2.51
8	14.311	Hexadecanoic acid, methyl ester	000112-39-0	270	$C_{17}H_{34}O_2$	98	16.46
9	16.253	Methyl 15-methylhexadecanoate	1000336-34-2	284	$C_{18}H_{36}O_2$	99	0.64
10	16.531	cis-10-Heptadecenoic acid, methyl ester	1000333-62-1	282	$C_{18}H_{34}O_2$	99	0.98
11	16.723	Heptadecanoic acid, methyl ester	001731-92-6	284	$C_{18}H_{36}O_2$	99	1.06
12	17.507	9,12-Octadecadienoic acid (Z,Z)-, methyl ester	000112-63-0	294	$C_{19}H_{34}O_2$	99	1.61
13	19.492	Methyl 9-cis,11-trans-octadecadienoate	1000336-44-0	294	$C_{19}H_{34}O_2$	95	1.29
14	19.919	Methyl 10-trans,12-cis-octadecadienoate	1000336-44-2	294	$C_{19}H_{34}O_2$	96	0.10
15	20.119	9-Octadecenoic acid (E)-, methyl ester	001937-62-8	296	$C_{19}H_{36}O_2$	99	37.96
16	20.546	9-Octadecenoic acid (Z)-, methyl ester	000112-62-9	296	$C_{19}H_{36}O_2$	99	1.97
17	20.955	11-Octadecenoic acid, methyl ester	052380-33-3	296	$C_{19}H_{36}O_2$	99	2.05
18	21.992	Octadecanoic acid, methyl ester	000112-61-8	298	$C_{19}H_{38}O_2$	99	19.47
19	22.166	cis-10-Nonadecenoic acid, methyl ester	1000333-64-4	310	$C_{20}H_{38}O_2$	98	0.065
20	23.316	10-Nonadecenoic acid, methyl ester	056599-83-8	310	$C_{20}H_{38}O_2$	93	0.24
21	24.308	Cyclopropaneoctanoic acid, 2-octyl-, methyl ester	3971-54-8	310	$C_{20}H_{38}O_2$	99	0.42
22	25.287	Nonadecanoic acid, methyl ester	001731-94-8	312	$C_{20}H_{40}O_2$	99	0.26
23	26.829	Methyl 8,11,14-eicosatrienoate	1000336-38-1	320	$C_{21}H_{36}O_2$	97	0.034
24	28.262	cis-11-Eicosenoic acid, methyl ester	1000333-63-8	324	$C_{21}H_{40}O_2$	99	0.12
25	29.653	Eicosanoic acid, methyl ester	001120-28-1	326	$C_{21}H_{42}O_2$	99	0.20

[a] Mw = Molecular Weight (nominal values); [b] RC (%) = The relative content of total peak areas, the sum of the RC was 90.89%, the other 9.11% may contain inorganic elements, glycerin and something else.

Table 2: The linear regression equations, the correlation coefficient (r), limit of quantification (LOQ), recoveries of 10 fatty acids under GC–EI-qMS selective ion monitoring (SIM) conditions

NO. of Identified Fatty Acids	Compounds	Linear Regression Equations	Coefficient of Determination/r^2	Linear Range µg/mL	Qualitative/ Quantitative Ions	Abundance Ratio (%)	0.5 Times Spiked ($n=3$) Recovery (%)	RSDs (%)	1.0 Times Spiked ($n=3$) Recovery (%)	RSDs (%)	2.0 Times Spiked ($n=3$) Recovery (%)	RSDs (%)	LOQ (µg/mL, ×10⁻³)
1	DODME	$Y = 1.56 \times 10^4 X - 1.79 \times 10^3$	0.999	0.010–10.0	74 *:28:87:214	100:8:64:6	85.3	4.3	92.4	4.8	93.2	2.2	1.25
5	MTEME	$Y = 1.54 \times 10^4 X - 1.48 \times 10^3$	0.999	0.013–12.8	74 *:87:143:199	100:68:24:16	95.2	4.1	97.2	3.3	98.7	4.5	1.60
6	PENME	$Y = 1.86 \times 10^5 X - 4.74 \times 10^3$	0.999	0.011–12.0	74 *:87:143:213	100:68:20:18	91.3	6.2	92.8	4.5	95.8	4.2	1.40
7	9-HEXME	$Y = 3.10 \times 10^5 X - 1.13 \times 10^5$	0.999	0.020–20.0	55 *:74:87:236	100:68:50:23	83.6	4.8	94.6	3.5	96.6	2.4	2.50
8	HEXME	$Y = 2.62 \times 10^4 X - 1.579 \times 10^4$	0.999	0.048–50.0	74 *:87:143:227	100:70:20:14	87.5	5.2	95.4	5.4	96.9	3.8	5.95
11	HEPME	$Y = 2.93 \times 10^4 X - 6.54 \times 10^5$	0.999	0.013–13.0	74 *:87:143:241	100:70:22:15	87.6	6.8	90.8	5.0	92.7	4.1	1.63
12	9,12-OCME	$Y = 7.71 \times 10^5 X - 1.15 \times 10^2$	0.999	0.020–20.0	67 *:81:95:294	100:92:66:16	91.7	4.4	91.7	2.8	93.9	3.6	2.50
15	9-OCME	$Y = 5.02 \times 10^5 X - 3.62 \times 10^5$	0.999	0.020–20.0	55 *:41:81:222	100:62:40:24	84.8	5.1	88.4	4.2	97.2	4.4	2.50
18	OCTME	$Y = 3.31 \times 10^4 X - 1.24 \times 10^4$	0.999	0.030–32.0	74 *:87:143:255	100:72:23:14	85.5	4.3	86.7	5.2	88.8	5.1	3.75
25	EICME	$Y = 3.18 \times 10^4 X - 4.78 \times 10^3$	0.999	0.010–10.8	74 *:143:255	100:76:26:18	82.1	3.9	90.1	4.7	89.4	3.7	1.25

* Quantitative ion, LOQ was calculated as 10 times of the signal to noise ratio (10 S/N). 0.5, 1.0 and 2.0 times "Spiked" means the added standards amount was 0.5, 1.0 and 2.0 times of the initial content of samples.

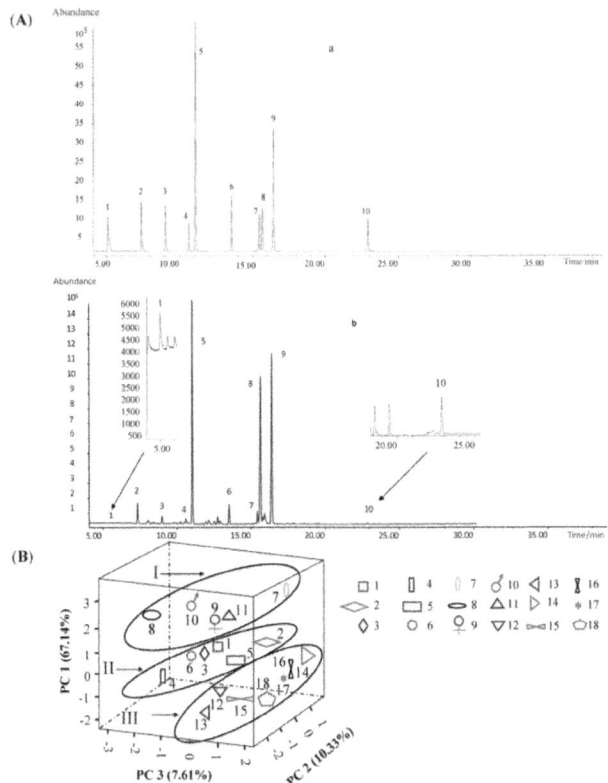

Figure 2: GC–EI-qMS chromatograms of the ten FAMEs standard mixture and sample under SIM mode and principal component analysis (PCA) of 18 SO samples. **(A)** (a) mixed standard solution (the concentration of (**1–10**) was 10.8, 47.6, 31.6, 13.0, 11.2, 12.8, 10.0, 20.0, 20.0 and 20.0 µg/mL, respectively), **(b)** SO sample solution. (**1**) DODME, (**2**) MTEME, (**3**) PENME, (**4**) 9-HEXME, (**5**) HEXME, (**6**) HEPME, (**7**) 9,12-OCME, (**8**) 9-OCME, (**9**) OCTME, (**10**) EICME; **(B)** The 3D scatter plot obtained by PCA of 18 SO samples.

Quantitative Results

The molecular species detected after methylesterification were FAMEs rather than FAs. Therefore, in order to quantify the FAs, the contents of FAMEs were converted into FAs by multiplication with the corresponding coefficient as follows:

(1)

where W is concentration and A is the molecular mass of the FAMEs/the molecular mass of the FAs.

The results for ten FAs in 18 batches of SO are given in Table 3. The contents of hexadecanoic acid and octadecanoic acid were 3.79%–13.22% and 3.41%–18.11%, respectively. The contents of (*E*)-9-octadecanoic acid in SO batches 3–6, 8, 12–15 and 18 were 4.48%–6.28%, second only to hexadecanoic acid and octadecanoic acid, whereas no other FA was detected. On the basis of the quantitative results for 18 batches of SO, the content of octadecanoic acid was greatest, followed by hexadecanoic acid, (*E*)-9-octadecanoic acid and tetradecanoate in that order.

Principal Component Analysis (PCA) of the SO Samples

In order to evaluate the variation between batches of SO, PCA was applied on the basis of the contents of the ten most abundant FAs. The first three principal components (PC1, PC2 and PC3) with >85% of the whole variance were extracted for analysis. PC1, PC2 and PC3 accounted for 67.14%, 10.33% and 7.61% of the total variance, respectively (Table S1). The remaining principal components had only a minor effect on the model and were discarded. The component loading matrix is given inTable S2 and Table S3. According to the loadings, PC1 had a good correlation with each of the ten FA compounds. The results mentioned above suggested that most of the compounds contributed to the classification of the samples. The scatter plots are shown in Figure 2B, where each sample is represented as one marker.

The dots of 18 samples were classified into group I, group II or group III in accord with their origin. Dots in groups II and III were relatively close to each other, indicating a close relationship among the six batches from Anhui and the seven batches from Jiangsu. The dots in group I were quite scattered, suggesting diversification of the five batches from Qinghai Province. These observations might be explained as follows. Firstly, the land area of Qinghai Province (722,300 km^2) is larger compared to Anhui Province (1,396,002 km^2) and Jiangsu Province (106,700 km^2), representing a greater area for diversity of the samples. Secondly, Qinghai, Anhui and Jiangsu provinces are considerably different environments with large differences in climate, which influences the differences in FA metabolism in domestic sheep and goats. Thirdly, Anhui and Jiangsu provinces are geographic neighbors, which is reflected in the similarities among samples from these two origins. Finally, the SO samples from Qinghai Province were from sheep, whereas those obtained from Anhui and Jiangsu provinces were from goats.

Table 3: Contents of ten FAs in 18 batches of Suet oil ($n = 3$)

FAs Compounds	DODME	MTEME	PENME	9-HEXME	HEXME	HEPME	9,12-OCME	9-OCME	OCTME	EICME
Coefficient "A" [a]	0.9348	0.9414	0.9454	0.9478	0.9482	0.9508	0.9524	0.9527	0.9531	0.9571
Batch					Content (g/100 g, %) [b]					
1	0.0240±0.0014	1.1476±0.0014	0.2124±0.0092	0.5716±0.0012	8.9067±0.1483	0.4996±0.0006	0.8667±0.0015	Nd	12.5609±0.1614	0.0933±0.0021
2	0.0284±0.0010	1.0827±0.0012	0.1787±0.0015	0.4596±0.0029	7.6098±0.0040	0.5022±0.0011	1.1867±0.0023	Nd	11.0827±0.0011	0.1040±0.0023
3	0.0284±0.0009	0.8684±0.0010	0.2009±0.0028	0.4044±0.0025	6.2578±0.0157	0.5289±0.0053	1.1218±0.0102	Nd	8.7600±0.3606	0.0942±0.0013
4	0.6196±0.0004	0.4862±0.0015	0.2098±0.0029	0.4204±0.0023	4.4382±0.0011	0.4851±0.0022	0.6587±0.0065	Nd	4.0124±0.0017	0.0471±0.0008
5	0.0231±0.0023	0.7707±0.0083	0.1991±0.0017	0.3191±0.0017	5.8071±0.0045	0.5707±0.0015	0.6604±0.0024	4.5991±0.0093	10.2213±0.0163	0.1467±0.0032
6	0.0418±0.0012	1.1912±0.0034	0.2844±0.0012	1.3378±0.0051	7.5067±0.0074	0.6187±0.0052	0.9227±0.0008	Nd	8.3093±0.0297	0.0978±0.0017
7	0.0400±0.0031	1.9422±0.0068	0.3040±0.0108	1.1564±0.0020	13.2151±0.0241	0.7956±0.0023	1.4640±0.0028	Nd	18.1129±0.0528	0.1662±0.0035
8	0.0328±0.0017	1.1182±0.0037	0.2729±0.0017	0.7662±0.0026	7.9458±0.0026	0.7218±0.0017	0.9048±0.0633	Nd	12.1111±0.0015	0.1582±0.0001
9	0.0400±0.0031	1.4516±0.0012	0.3902±0.0035	0.8649±0.0053	10.8133±0.0034	1.1004±0.0001	1.4924±0.0035	Nd	12.9360±0.0275	0.1111±0.0016
10	0.0280±0.0046	1.0978±0.0045	0.2516±0.0924	0.4222±0.0029	8.264±0.0394	0.6960±0.0027	1.0480±0.0042	Nd	13.0320±0.0337	0.1467±0.0040
11	0.0373±0.0035	1.1991±0.0031	0.4791±0.0043	0.4764±0.0040	10.0676±0.0045	0.9218±0.0017	0.8107±0.0059	Nd	14.2569±0.0029	0.1040±0.0016
12	0.0350±0.0039	1.0196±0.0041	0.2658±0.0039	0.7564±0.0031	6.7449±0.0017	0.4942±0.0633	0.7787±0.0041	6.2764±0.0023	6.0773±0.0036	0.5560±0.0034
13	0.0178±0.0040	0.4080±0.0046	0.1769±0.0034	0.3396±0.0044	3.7902±0.0047	0.3804±0.0018	0.5449±0.0039	5.4489±0.0048	3.4080±0.0051	0.0382±0.0012
14	0.0322±0.0035	1.1831±0.0040	0.2391±0.0045	0.4276±0.0046	8.3760±0.0164	0.5822±0.0029	1.2246±0.0167	5.2880±0.0282	11.2276±0.0060	0.087±0.0045
15	0.0240±0.0023	0.6462±0.0034	0.1458±0.0028	0.4647±0.0013	4.6827±0.0034	0.3671±0.0035	0.6240±0.0293	3.0649±0.0384	6.6560±0.0220	0.0773±0.0038
16	0.0267±0.0029	0.8071±0.0039	0.1636±0.0032	0.3363±0.0032	5.8738±0.0015	0.4347±0.0038	1.0738±0.0028	4.8996±0.0044	9.0898±0.0042	0.0989±0.0023
17	0.0267±0.0034	0.7911±0.0061	0.1564±0.0028	0.3209±0.0078	5.4720±0.0046	0.4329±0.0038	1.0169±0.0033	4.6578±0.0046	8.1529±0.0040	0.0862±0.0039
18	0.0166±0.0051	0.6969±0.0034	0.0907±0.0064	0.3413±0.0031	5.5209±0.0071	0.2480±0.0046	0.5013±0.0034	4.4756±0.0034	7.9644±0.0033	0.0622±0.0042

[a] A = Molecular Weight (FAMEs)/Molecular Weight (FAs). [b] W_{FAs} (%) = A × W_{FAMEs} (%), Nd = not detected.

EXPERIMENTAL SECTION

Materials

Methyl dodecanoic acid (DODME ≥ 98.0 (purity)), methyl tetradecanoate (MTEME ≥ 99.0), methyl pentadecanoic acid (PENME ≥ 98.0), methyl 9-hexadecenoic acid (Z) (9-HEME ≥ 99.0), methyl hexadecanoic acid (HEXME ≥ 99.0), methyl heptadecanoic acid (HEPME ≥ 99.0), 9,12-methyl octadecadienoic acid (Z,Z) (9,12-OCME ≥ 99.0), methyl 9-octadecenoic acid (E) (9-OCME ≥ 99.0), methyl octadecanoic acid (OCTME ≥ 98.0), methyl eicosanoic acid (EICME ≥ 99.0), boron trifluoride-methanol (14%, v/v), sodium hydroxide (NaOH) and sodium (NaCl) were purchased from Anpel Scientific Instrument Co., Ltd. (Shanghai, China). HPLC grade methanol, and n-hexane were obtained from Merck (Darmstadt, Germany).

Sample Material

Eighteen batches of SO samples were purchased from Qinghai (batches 7–11), Jiangsu (batches 12–18) and Anhui (batches 1–6) provinces, China between January and July 2013 (Figure S2). All samples were stored in darkness at temperatures <4 °C. For the blank sample, the n-hexane was used instead of SOs.

Box–Behnken Design for Optimization of PME Parameters

The application of an effective PME methodology requires optimization of the main parameters that influence the methylesterification process, including the volume of methyl reagent, temperature and time.

A Box–Behnken Design, a response surface methodology, was used in this study. Design Expert 7.0.0 software was used for analyzing the experimental data. The study type was Response Surface, the initial design was Box–Behnken, the design model was Quadratic and Blocks was No Blocks. A Box–Behnken statistical screening design with three independent variables (*A*, PME volume; *B*, PME temperature; *C*, PME time) was used to optimize the PME process for the qualitative and quantitative analysis. Statistically significant difference was set at $p \leq 0.05$. The r^2 value of the "Final Equation" > 0.995 indicated derived results were accurate. Data were expressed as mean ± standard deviation (SD) of triplicate determinations. Statistical calculations used Statistical Product and Service Solutions (SPSS) version 16.0 software (SPSS Inc., Chicago, IL, USA). One-way ANOVA was used for evaluating the statistical differences among samples.

PME Procedure

A 0.4 g sample of SO from batch 12 was weighed and placed into a 50-mL conical flask followed by 15 mL NaOH–MeOH (0.5 mol/L) then heated at 60 °C in a waterbath for 20 min until the yellow beads of SO disappeared completely after cooling. The flask contents were subjected to the PME procedure, in which 10 mL of boron trifluoride methanol (BF_3–MEOH, 14% *v/v*) was added to the flask then heated at 60 °C in a waterbath for 10 min. The mixture was cooled and then 10 mL of *n*-hexane and 10 mL of saturated NaCl were added. Samples 1.5 mL of supernatants were injected through a 0.45-μm pore size membrane before GC–EI-qMS qualitative analysis.

Sample Pretreatment for Quantitative Analysis

Eighteen batches of SO were treated as described in section 3.4 above. Sequentially, 25-μL was transferred into a 10-mL volumetric flask followed by addition of *n*-hexane to a final volume of 10 mL and then shaken. After passage through an organic 0.45-μm pore size filter, the treated samples were injected into the GC–EI-qMS for quantitative analysis.

Preparation of Standard Solutions

Stock solutions of the ten FAMEs (DODME, MTEME, PENME, 9-HEX-ME, HEXME, HEPME, 9,12-OCME, EICME, OCTME and 9-OCME) were prepared in *n*-hexane at concentrations of 20.0, 11.2, 20.0, 12.8, 31.6, 47.6, 10.8, 20.0, 13.0 and 10.0 μg/mL, respectively. Appropriate amounts of the above stock solutions were mixed and diluted into a series of concentrations with *n*-hexane to obtain the working solutions. All solutions were stored at <4 °C.

GC–EI-qMS Analysis Conditions

For separation, detection and identification of FAs, the qualitative and quantitative analyses were made with a GC–EI-qMS instrument (Agilent 7890/5975) coupled to an automatic sampler (Agilent 7693) and an electron impact ionization source (Agilent, Santa Clara, CA, USA). Water was purified by a Milli-Q Plus apparatus (Millipore, Bedford, MA, USA). The H2050R centrifugal apparatus was provided by the Hunan Saite xiangyi centrifuge instrument Co., Ltd. (Xiangya, China). Analytes were separated using a 30 m × 0.25 mm capillary column (HP-5 ms 0.25 μm film thickness; Agilent Technology, Santa Clara, CA, USA). The primary oven temperature protocol was: 150 °C for 1 min; increased to 200 °C at 5 °C/min; maintained at this temperature for 5 min; increased to 250 °C at a rate of 5 °C /min; maintained at this temperature for 5 min; increased to 300 °C at a rate of 5 °C/min; and maintained at this temperature for 10 min. The injection port temperature was 250 °C. The carrier gas was helium at a constant flow of 1 mL/min. The MS operating conditions in the splitless injection mode were as follows: ion source temperature 280 °C; electron energy 70 eV; emission current 250 μA; injection volume 0.2 μL; and solvent delay 4 min. The SIM mode was used for quantitative determination of FAs.

Method for PCA of Samples

PCA was done with SPSS 16.0 software (SPSS, Chicago, IL, USA) [29]. In this study, the contents of the ten FAs in the 18 SO samples were used as a data matrix with 18 rows and ten columns for PCA analysis after normalization. The first three PCs were extracted, and the scatter plot was obtained by plotting the scores of PC1 vs. PC2 and PC3.

CONCLUSIONS

The optimal conditions for methylesterification of FAs were obtained by a Box–Behnken Design, which identified 25 kinds of FAs in SO by GC–EI-qMS. In addition, ten FAs in 18 batches of SO were analyzed with good performance with regard to selectivity, recovery, precision and accuracy. Significant differences among origins in FA composition profiles and their contents were revealed. The method described here could be used in quality control and standardization of SOs and their products as well as providing supportive chemical information.

ACKNOWLEDGMENTS

This work was supported by the Natural Science Foundation of China (No. 81274088).

AUTHOR CONTRIBUTIONS

Jun Jiang designed research; Jun Jiang and Xiaobin Jia performed research and analyzed the data; Jun Jiang wrote the paper. Both authors read and approved the final manuscript.

REFERENCES

1.   Thurnhofer, S.; Hottinger, G.; Vetter, W. Enantioselective determination of anteiso fatty acids in food samples. Anal. Chem. 2007, 79, 4696–4701.

2.   Mattacks, C.A.; Sadler, D.; Pond, C.M. Site-specific differences in the action of NRTI drugs on adipose tissue incubated *in vitro* with lymphoid cells, and their interaction with dietary lipids.Comp. Biochem. Phys. 2003, 135, 11–29.

3.   Thurnhofer, S.; Vetter, W. A gas chromatography/electron ionization-mass spectrometry-selected ion monitoring method for determining the fatty acid pattern in food after formation of fatty acid methyl esters. J. Agric. Food Chem. 2005, 53, 8896–8903.

4.   Barzanti, V.; Maranesi, M.; Cornia, G.L.; Malavolti, M.; Mordenti, T.; Pregnolato, P. Effect of dietary oils containing different amounts of precursor and derivative fatty acids on prostaglandin E2 synthesis in liver, kidney and lung of rats. Prostag. Leukotr. Essent. 1999, 60, 49–54.

5.   Cui, L.; Sun, E.; Zhang, Z.H.; Tan, X.B.; Wei, Y.J.; Jin, X. Enhancement of epimedium fried with suet oil based on in vivo formation of self-assembled flavonoid compound nanomicelles.Molecules 2012, 17, 12984–12996.

6.   Ansorena, D.; Raes, K.; de Smet, S.; Demeyer, D. Analysis of fatty acid isomers in ruminant tissues by silver thin layer chromatography followed by gas chromatography. Meded. Rijksuniv. Gent Fak. Landbouwkd. Toegep. Biol. Wet. 2001, 66, 365–372.

7.   Zhang, S.; Sun, Y.; Sun, Z.; Wang, X.; You, J.; Suo, Y. Determination of triterpenic acids in fruits by a novel high performance liquid chromatography method with high sensitivity and specificity. Food Chem. 2014, 146, 264–269.

8.   Wang, A.; Li, G.; You, J.; Ji, Z. A new fluorescent derivatization reagent and its application to free fatty acid analysis in pomegranate samples using HPLC with fluorescence detection. J. Sep. Sci. 2013, 36, 3853–3859.

9.   Bielawska, K.; Dziakowska, I.; Roszkowska-Jakimiec, W. Chromatographic determination of fatty acids in biological

material. Toxicol. Mech. Methods 2010, 20, 526–537.

10. Li, A.; Ha, Y.; Wang, F.; Li, W.; Li, Q. Determination of thermally induced trans-fatty acids in soybean oil by attenuated total reflectance fourier transform infrared spectroscopy and gas chromatography analysis. J. Agric. Food Chem. 2012, 60, 10709–10713.

11. Bogusz, S.J.; Hantao, L.W.; Braga, S.C.; de Matos França, V.C.; da Costa, M.F. Solid-phase microextraction combined with comprehensive two dimensional gas chromatography for fatty acid profiling of cell wall phospholipids. J. Sep. Sci. 2012, 35, 2438–2444.

12. Hori, K.; Matsubara, A.; Uchikata, T.; Tsumura, K.; Fukusaki, E.; Bamba, T. High-throughput and sensitive analysis of 3-monochloropropane-1,2-diol fatty acid esters in edible oils by supercritical fluid chromatography/tandem mass spectrometry. J. Chromatogr. A 2012, 1250, 99–104.

13. Aslan, M.; Ozcan, F.; Aslan, B.; Yücel, G. LC–MS/MS analysis of plasma polyunsaturated fatty acids in type 2 diabetic patients after insulin analog initiation therapy. Lipidis Health Dis. 2013,12, 169.

14. Derogis, P.B.; Freitas, F.P.; Marques, A.S.; Cunha, D.; Appolinário, P.P.; de Paula, F. Detection and quantification of Hydroperoxy and Hydroxydocosahexaenoic acids as a tool for lipidomic analysis. PLoS One 2013, 8, e77561.

15. Le Faouder, P.; Baillif, V.; Spreadbury, I.; Motta, J.P.; Rousset, P.; Chêne, G. LC–MS/MS method for rapid and concomitant quantification of pro-inflammatory and pro-resolving polyunsaturated fatty acid metabolites. J. Chromatogr. B 2013, 932, 123–133.

16. Takahashi, H.; Suzuki, H.; Suda, K.; Yamazaki, Y.; Takino, A.; Kim, Y.I. Long-chain free fatty acid profiling analysis by liquid chromatography–mass spectrometry in mouse treated with peroxisome proliferator-activated receptor α agonist. Biosci. Biotechnol. Biochem. 2013, 77, 2288–2293.

17. Zeng, A.X.; Chin, S.T.; Nolvachai, Y.; Kulsing, C.; Sidisky, L.M.; Marriott, P.J. Characterisation of capillary ionic liquid columns for gas chromatography mass spectrometry analysis of fatty acid methyl esters. Anal. Chim. Acta 2013, 803, 166–173.

18. Valianpour, F.; Selhorst, J.J.; van Lint, L.E.; van Gennip, A.H.; Wanders, R.J.; Kemp, S. Analysis of very long-chain fatty acids using electrospray ionization mass spectrometry. Mol. Genet. Metab. 2003, 79, 189–196.

19. Byss, M.; Tríska, J.; Elhottová, D. GC–MS–MS analysis of bacterial fatty acids in heavily creosote-contaminated soil samples. Anal. Bioanal. Chem. 2007, 387, 1573–1577.

20. Oursel, D.; Loutelier-Bourhis, C.; Orange, N.; Chevalier, S.; Norris, V.; Lange, C.M. Identification and relative quantification of fatty acids in Escherichia coli membranes by gas chromatography/mass spectrometry. Rapid Commun. Mass Spectrom. 2007, 21, 3229–3233.

21. Catarina, L.S.; José, S.C. Profiling of volatiles in the leaves of Lamiaceae species based on headspace solid phase microextraction and mass spectrometry. Food Res. Int. 2013, 51, 378–387.

22. Mahinda, W.; Thava, V.; Feral, T.; Kevin, S. Volatile flavour composition of cooked by-product blends of chicken, beef and pork: A quantitative GC–MS investigation. Food Res. Int. 2001, 34, 149–158.

23. Diana, A.; Olga, G.; Iciar, A.; José, B. Analysis of volatile compounds by GC–MS of a dry fermented sausage: Chorizo de Pamplona. Food Res. Int. 2001, 34, 67–75.

24. Dodds, E.D.; McCoy, M.R.; Rea, L.D.; Kennish, J.M. Gas chromatographic quantification of fatty acid methyl esters: Flame ionization detection vs. electron impact mass spectrometry. Lipids2005, 40, 419–428.

25. Saliu, F.; Orlandi, M. In situ alcoholysis of triacylglycerols by application of switchable-polarity solvents. A new derivatization procedure for the gas chromatographic analysis of vegetable oils.Anal. Bioanal. Chem. 2013, 405, 8677–8684.

26. Igarashi, M.; Tsuzuki, T.; Kambe, T.; Miyazawa, T. Recommended methods of fatty acid methylester preparation for conjugated dienes and trienes in food and biological samples. J. Nutr. Sci. Vitaminol. 2004, 50, 121–128.

27. Box, G.E.P.; Wlson, K.B. On the experimental attainment of optimum conditions. J. R. Stat. Soc.1951, 13, 1–45.

28. Luo, C.; Chen, Y.S. Optimization of extraction technology of Se-enriched Hericium erinaceumpolysaccharides by Box–Behnken statistical design and its inhibition against metal elements loss in skull. Carbohydr. Polym. 2010, 82, 845–860.

29. Jiang, J.; Feng, L.; Li, J.; Sun, E.; Ding, S.M.; Jia, X.B. Multielemental composition of suet oil based on quantification by ultrawave/ICP-MS coupled with chemometric analysis. Molecules2014, 19, 4452–4465.

Chapter 4

COMPREHENSIVE AND COMPARATIVE METABOLOMIC PROFILING OF WHEAT, BARLEY, OAT AND RYE USING GAS CHROMATOGRAPHY-MASS SPECTROMETRY AND ADVANCED CHEMOMETRICS

Bekzod Khakimov , Birthe Møller Jespersen and Søren Balling Engelsen

Department of Food Science, Faculty of Science, University of Copenhagen, Rolighedsvej 30, Frederiksberg C, 1958 Copenhagen, Denmark

ABSTRACT

Beyond the main bulk components of cereals such as the polysaccharides and proteins, lower concentration secondary metabolites largely contribute to the nutritional value. This paper outlines a comprehensive protocol for GC-MS metabolomic profiling of phenolics and organic acids in grains, the performance of which is demonstrated through a comparison of the metabolite profiles of the main northern European cereal crops: wheat, barley, oat and rye. Phenolics and organic acids were extracted using acidic hydrolysis, trimethylsilylated using a new method based on trimethylsilyl cyanide and analyzed by GC-MS. In order to extract pure metabolite peaks, the raw chromatographic data were processed by a multi-way decomposition method, Parallel Factor Analysis 2. This approach lead to the semi-quantitative detection of a total of 247 analytes, out of which 89 were identified based on RI and EI-MS library match. The cereal metabolome included 32 phenolics, 30 organic acids, 10 fatty acids, 11 carbohydrates and 6 sterols. The metabolome of the four cereals were compared in detail, including low concentration phenolics and organic acids. Rye and oat displayed higher total concentration of phenolic acids, but ferulic, caffeic and sinapinic acids and their esters were found to be the main phenolics in all four cereals. Compared to the previously reported methods, the outlined protocol provided an efficient and high throughput analysis of the cereal metabolome and the acidic hydrolysis improved the detection of conjugated phenolics.

INTRODUCTION

Cereals such as wheat, barley, rye and oat are amongst the mostly grown agricultural food products worldwide and the most important cereal crops for human consumption in northern Europe. The detailed chemical and functional composition of these crops is defining their use for food and feed as well as their prices. Cereals are the most important study objects in foodomics studies seeking to optimize their health beneficial factors and/or reducing deleterious metabolites. While the gross chemical composition, such as carbohydrates, proteins, dietary fibers and micronutrient contents, are important characteristics of cereal products, recent studies showed that relatively low concentration secondary metabolites such as antioxidant phenolics, organic acids and phytosterols have a significant influence on the health and nutritional values of cereals [1,2]. The beneficial health effects associated with the consumption of cereals have been attributed to dietary fiber content [3] as well as phenolics that possess antioxidant, radical scavenging and cholesterol lowering properties [4,5,6,7]. Whole grain barley intake has proven to decrease the low-density lipoprotein (LDL) cholesterol in an intervention study involving hypercholesterolemic patients [8]. Moreover, phenolic acids were found to be important texturizing agents in cooking-extrusion of cereals [9] and recognized as the main antioxidant constituents of cereals [10].

Quantitative and qualitative analysis of both, secondary and primary metabolites (with molecular weight of up to 1500 Da) of grains are studied within cereal metabolomics. Cereal metabolomics offers an insight into the metabolic fluctuations of cereal cultivars that may reveal effects of genetic modifications as well as of biotic and abiotic stresses [11]. Recent studies have illustrated the power of cereal metabolomics to reveal effects of growth temperature [12], salt stress [13], drought stress [14], and biotic stress [15]. Cereal metabolomics is also a promising approach to reveal biochemical and genetic backgrounds of quality traits and may open new possibilities towards targeted breeding [16,17].

Comprehensive metabolomic profiling of cereals requires a reliable protocol that enables extraction of maximum metabolic information in a high-throughput and reproducible manner. Metabolomics studies performed for uncovering single and/or multiple internal and/or external effects on cereals aim to cover as broad range of metabolites as possible. However, due to the great physico-chemical diversity of cereal metabolites, it is in practice impossible to cover the whole cereal metabolome using a single protocol. The phytochemical composition, including phenolics of wheat [18,19,20], barley [21], oat [22] and rye [23] have been investigated in a number of studies within the HEALTHGRAIN diversity-screening program [24].

This study demonstrates the development of comprehensive GC-MS metabolomics protocol for profiling a broad range of phenolics and organic acids from whole grain flour samples, and applied on wheat, barley, rye and oat. Phenolics of cereals are primarily present in conjugated and bonded forms with carbohydrates, lipids and other cell membrane components that alter their solubility and thus bioavailability [21]. Analysis of phenolic content of cereals is mainly performed by basic hydrolysis of cereal extracts [18], which can only cleave ester bonds and stabilize de-esterification reactions. However, a substantial part of phenolics and other organic acids of cereals are conjugated through glycosidic and/or ether bonds to carbohydrates and other molecules. In contrast to basic hydrolysis, acidic hydrolysis allows the cleavage of not only ester bonds, but also glycosidic and ether bonds at an elevated temperature. The advantages of this approach have been demonstrated in polyphenol analysis of the wheat and rice grains [25,26].

In this study, a standardized, high-throughput and unbiased protocol was developed for GC-MS metabolomic profiling of free and conjugated phenolics and organic acids of whole-grain cereals using hydrochloric acid based hydrolysis followed by trimethylsilyl derivatization. The study demonstrates the first application of a novel trimethylsilylation method based on trimethylsilyl cyanide (TMSCN) for derivatization of cereal metabolites. When compared to other frequently used derivatization methods, the new protocol provides a more unbiased and broad-spectrum derivatization of metabolites and is able to provide reproducible metabolomics profiles of complex biological samples [27]. The obtained raw GC-MS data of cereals were processed by a semi-automated multi-way decomposition method, PARAFAC2 [28]. The PARAFAC2 processing of the raw GC-MS data lead to unambiguous deconvolution of elusive peaks such as, overlapped, retention time shifted and low s/n peaks and enable an automatic estimation of relative concentrations of detected peaks [29,30]. Metabolite extraction and GC-MS analysis of the cereal samples were performed within a bigger study, which involved a larger set of barley samples (manuscript in preparation). The main aim of this study was to demonstrate the performance of the protocol, using new technologies within metabolomics, and to show first results of a comparative application to the four major north European cereals: wheat, barley, rye and oat. To the best of our knowledge, this is the first study illustrating a comprehensive GC-MS profiling of phenolics and organic acids of cereals using exactly the same protocol across different cereals.

EXPERIMENTAL SECTION

Whole grain samples of wheat (*Tr. aestivum*, variety Bussard), barley (*H. vulgare*, variety Bomi), rye (*S. cereal*, variety Petkus) and oat (*A. sativa*, variety Sang) were purchased in Sepetember 2012 from the Danish bread cereal producing company Aurion (Hjørring, Denmark). All four cereals were grown under biodynamical conditions in Jutland during the season 2011/12.

Metabolite Extraction and Sample Derivatization

Cereal metabolites were extracted from 50 mg of milled grains that were soaked into 600 μL 85% methanol and vortexed for 20 s at 3000 rpm followed by 20 min incubation at 30 °C using a Thermomixer (Model 5436, Eppendorf, Hamburg, Germany) at 1400 rpm. After 3 min of centrifugation at $16,000 \times g$, the supernatant was transferred to a fresh 2 mL Eppendorf tube (Hamburg, Germany) and the remaining flour sample was extracted a second time using the same extraction procedure. Then, the combined extracts were completely dried under nitrogen gas flow at 40 °C and hydrolyzed by using 240 μL of 6 M hydrochloric acid at 96 °C for 1 h by stirring at 1400 rpm. The hydrolyzed extracts were transferred into a fresh 2 mL glass vials and phenolics and organic acids were extracted into diethyl ether. Ether-based extraction of phenolics and organic acids was performed twice, by addition of 800 μL diethyl ether and vortexing for 25 s. The obtained ether fractions were completely dried using nitrogen gas flow and re-solubilized in 200 μL 100% methanol. Aliquots, 90 microliter, of the final extracts were transferred into 200 μL glass inserts and completely dried under nitrogen gas flow, sealed and stored at −20 °C until GC-MS analysis. Each sample was spiked with an internal standard (IS) (5 μL of 0.2 mg mL^{-1} solution of ribitol). In order to avoid any moisture, the samples stored in the freezer were dried under reduced pressure before derivatization. Sample derivatization and injection were fully automated by using a Multi-Purpose Sampler (MPS, GERSTEL, Mülheim, Germany) with DualRait WorkStation integrated to a GC-MS system from Agilent (CA, USA). Each sample was individually derivatized by addition of 40 μL trimethylsilyl cyanide (TMSCN) and incubated for 40 min at 40 °C. Two replicate samples per cereal were analyzed in randomized order and the MPS autosampler allowed a sequential derivatization of all samples in the same manner by keeping the derivatization time constant, throughout the analysis.

GC-MS Data Acquisition

The GC-MS consisted of an Agilent 7890A GC and an Agilent 5975C series MSD. GC separation was performed on a Phenomenex ZB 5MSi column (30 m × 250 μm × 0.25 μm). A derivatized sample volume of 1 μL was injected into a cooled injection system (CIS port) using Solvent Vent mode at the vent pressure of 7 kPa until 0.3 min after injection at the vent flow of 100 mL min^{-1}. Detailed information on CIS and MPS parameters are described in Khakimov *et al.* 2013 [27]. Hydrogen was used as carrier gas, at a constant flow rate of 1.2 mL min^{-1}, and the initial temperature of CIS was set to 120 °C for 0.3 min followed by heating at 5 °C s^{-1} until reaching 320 °C and then held for 10 min. The GC oven program was as follows: initial temperature 40 °C, equilibration time 3.0 min, heating rate 12.0 °C min^{-1}, end temperature 300 °C, hold time 8.0 min and post run time 5 min at 40 °C. Mass spectra were recorded in the range of 50–500 *m/z* with a scanning frequency of 3.2 scans s^{-1}, and the MS detector was switched off during the 8.5 min of solvent delay time and after 25.5 min of the run time. The transfer line, ion source and quadrupole temperatures were set to 290, 230 and 150 °C, respectively. The mass spectrometer was tuned according to manufacturer's recommendation by using perfluorotributylamine (PFTBA).

Data Analysis

Initial analysis and visualization of the GC-MS data was performed using ChemStation software (Agilent, Germany). Retention indices of detected metabolites were calculated using the Van den Dool and Kratz equation and retention times of C10-C40 alkanes that were analyzed using the same GC-MS protocol [31]. The raw GC-MS data was imported from netCDF format to .mat files into Matlab® ver. R2012b (8.0.0.783) and data was manually divided into 121 smaller baseline separated intervals in retention time dimension. Each interval was modeled separately by PARAFAC2 as described previously [30]. PARAFAC2 modeled the three-way raw GC-MS data (elution time × mass spectra × samples) without any prior data pre-processing. The PARAFAC2 model outcomes: the elution profiles, which represent the TIC in the raw data, and spectral profiles, which represent the experimental EI-MS of deconvoluted peaks, were used for metabolite identification. The PARAFAC2 resolved mass spectrum of each peak was extracted and compared against NIST05 library (NIST, USA), Golm Metabolite Database [32]. Finally, PARAFAC2 concentration profiles, which represented relative concentrations of detected peaks were extracted and normalized according to the peak area of the internal standard (ribitol). The obtained metabolite table was used for exploring variations of phenolics in cereals and for principal component analysis (PCA) [33] after autoscaling of the data.

RESULTS AND DISCUSSION

GC-MS Metabolomic Profiling and PARAFAC2 Based Data Processing

The total ion current (TIC) chromatograms of the GC-MS data obtained from hydrolyzed extracts of the four cereals are illustrated in Figure 1. Just over 300 peaks with a s/n ratio >10 were detected from GC-MS profiles.

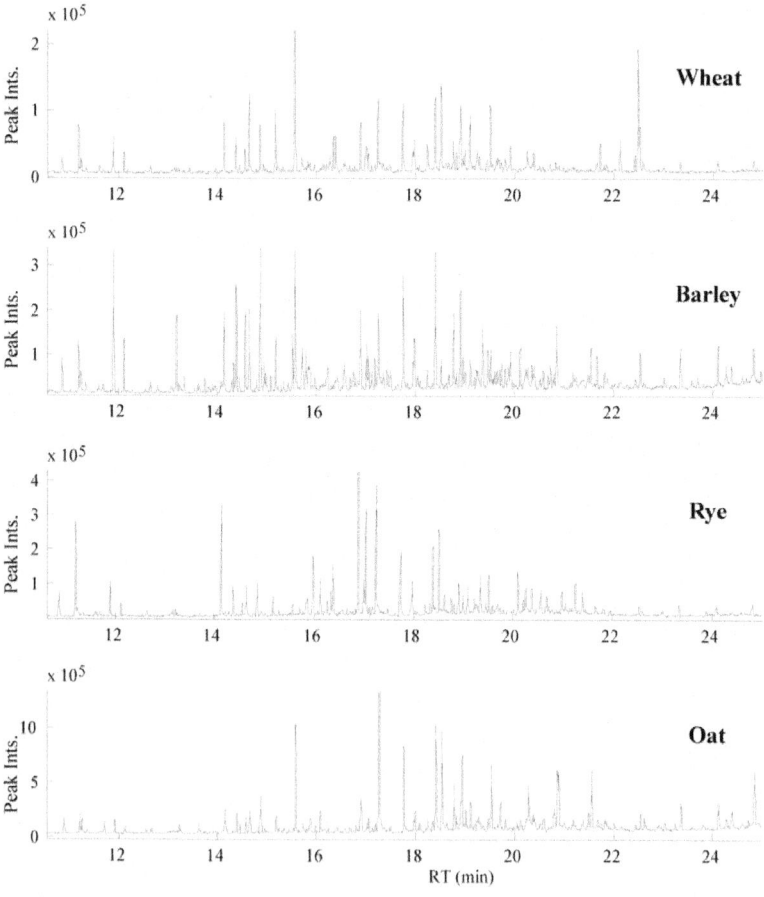

Figure 1: The total ion current (TIC) chromatograms of GC-MS data obtained on wheat, barley, rye and oat metabolite extracts.

Validated PARAFAC2 models of 121 intervals of the raw GC-MS data revealed 389 components including resolved peaks, shoulders of neighbor peaks and baseline. Then, each PARAFAC2 model was individually evaluated and components that represent baseline, artifact peaks such as column bleed and reagent derived peaks and shoulders of neighbor peaks were eliminated, resulting in 247 chromatographic peaks with unique retention indices and mass spectra. The PARAFAC2 modeling of GC-MS intervals representing vanillin, protocatechuic acid and β-resorcylic acid are demonstrated inFigure 2.

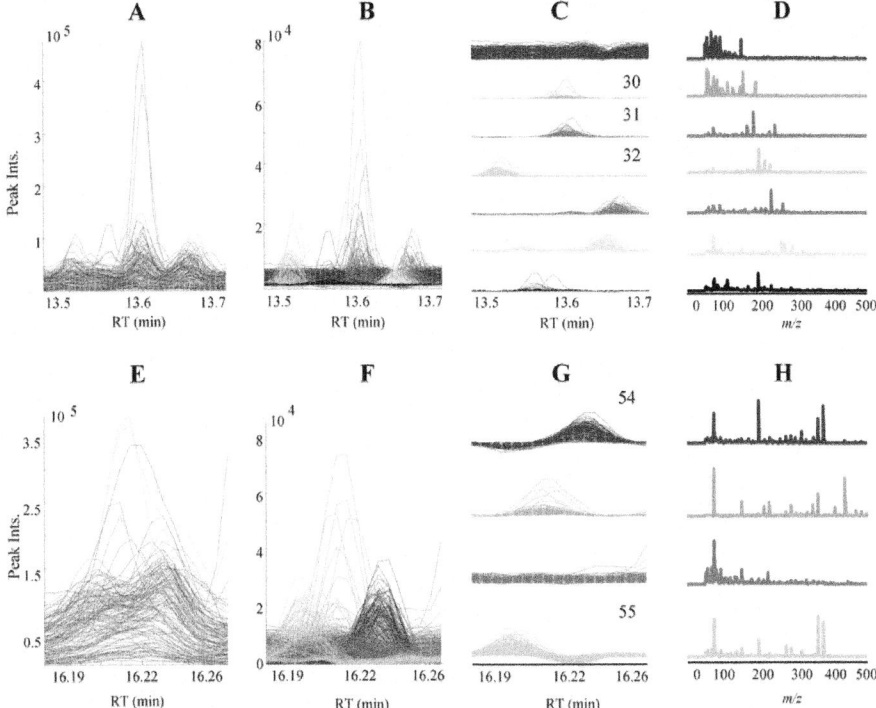

Figure 2: PARAFAC2 based processing of raw GC-MS data intervals. (**A**) and (**E**) are the TIC of raw GC-MS data intervals. (**B**) and (**F**) are the superimposed PARAFAC2 elution profiles of the raw GC-MS data intervals with seven and four components, respectively. (**C**) and (**G**) are subplots of (**B**) and (**F**), respectively. * Numbers of elution profiles correspond to the metabolites represented in Table 1. (**D**) and (**H**) are subplots of PARAFAC2 mass spectral profiles.

Comparison of RIs and PARAFAC2 resolved mass spectra of 247 resolved peaks against the NIST05 and Golm Metabolite Database resulted in the identification of 89 metabolites (Table 1) at level 2 as described in Metabolomics Standards Initiative report [34]. A total of 32 out of 89 identified metabolites were trimethylsilyl (TMS) derivatives of phenolic acids, their esters and aldehydes. In addition to the previously found phenolic acids from different barley genotypes [21], several other phenolics such as *p*-salicylic, gallic, gentisic, homovanillic and α-resorcylic acids and methyl esters of ferulic, caffeic, protocatechuic and sinapinic acids were identified. Small molecular organic acids, alcohols and their esters constituted 30 out of 89 identified metabolites. These included succinic, glyceric, maleic, fumaric, malic, pyroglutamic, azelaic acids and methyl esters of aconitic and citric acids that are part of the same or different metabolic pathways, and in addition, TMS-derivatives of 10 fatty acids and their esters, 6 sterols and a flavonoid, catechin-nTMS.

Table 1: A list of identified metabolites from wheat, barley, rye and oat flour samples by GC-MS. Metabolite identification was performed at level 2 as described in Metabolomics Standards Initiative report [34] and was based on RI and EI-MS library match (>80). [a]Metabolites with more than one isomers and/or TMS-derivatives; [b]tentatively identified

No	Metabolites	RT min	RI (r)	RI (c)
1.	Laevulic acid-1TMS	9.04	1030	1070
2.	Sorbic acid-1TMS	9.06	1009	1071
3.	Hepta-2,4-dienoic acid, methyl ester	9.28	1000	1080
4.	Octanol-1-1TMS	9.51	1101	1090
5.	Malonic acid-2TMS	9.99	1205	1207
6.	(3,3-Dimethyl-1-cyclohexen-1-yl)oxy]-1TMS	9.97	1110	1206
7.	Benzoic acid-1TMS	10.42	1228	1226
8.	3-Methyl-2-furoic acid-1TMS	10.38	1107	1224
9.	Glycerol-3TMS	10.88	1282	1246
10.	1,3-Dihydroxypropanone-2-2TMS	11.03		1249
11.	Succinic acid-2TMS	11.24	1292	1262
12.	Glyceric acid-3TMS	11.51	1199	1274
13.	Maleic acid-2TMS	11.55	1286	1275
14.	Fumaric acid-2TMS	11.60	1178	1278
15.	p-Hydroxybenzaldehyde-1TMS	11.85	1280	1289
16.	2-Hydroxyheptanoic acid-2TMS	11.83	1312	1288
17.	3-Hydroxybutanoic acid-2TMS	12.12	1403	1401
18.	Resorcinol-2TMS	12.2	1378	1404
19.	Trimethyl aconitate	12.50	1428	1419
20.	Citric acid, trimethyl ester	12.82	1442	1435
21.	3-Hydroxyanthranilic acid, methyl ester-1TMS	12.8		1434
22.	2,4-Dihydroxy-5-methylpyrimidine-2TMS	12.89	1403	1439
23.	5-Hydroxy-2-(hydroxymethyl)-4H-pyran-4-one-2TMS	13.08	1492	1448
24.	Maseptol-1TMS	13.12	1358	1450
25.	Malic acid-2TMS	13.19	1494	1453
26.	2-Hydroxycyclohexanecarboxylic acid-2TMS	13.23	1402	1456
27.	3-Hydroxyoctanoic acid-2TMS	13.35	1452	1462
28.	Pyroglutamic acid-2TMS	13.46	1466	1467
29.	Erythritol-4TMS	13.47		1467
30.	Dimethyl azelate	13.61	1485	1474
31.	4-Hydroxybenzeneacetic acid, methyl ester-1TMS	13.62	1458	1475
32.	Vanillin-1TMS	13.55	1469	1471
33.	Citric acid, trimethyl ester-1TMS	13.76		1482
34.	2-Furancarboxylic acid, 5-[(oxy)methyl]-1TMS	13.72	1540	1480
35.	4-Hydroxyphenylethanol-2TMS	13.92	1475	1490
36.	Anozol	14.15	1603	1601

Principal Component Analysis (PCA)

In order to explore the metabolomics data, PCA was performed on the metabolite table, including eight cereal samples in duplicates and 89 identified metabolites. PC1 *versus* PC2 scores plot of the PCA model (Figure 3A) show a clear separation of four different cereals explaining more than 60% variation of the data. The loadings plot of the corresponding model (Figure 3B) demonstrates a large spread of the 89 metabolites and revealed no clear groupings of metabolites classes. However, major part of the benzoic acid derived phenolics such as 3,5-dihydroxybenzoic, 3,4-dihydroxybenzoic and 3,4,5-trihydroxybenzoic acids are grouped on the upper left part of the loadings plot showing greater abundance in barley compared to the other cereals. In contrast to this, cinnamic acid derived phenolics such as ferulic, sinapinic and syringic acids are located on the bottom right corner showing greater concentrations in rye and wheat. Phenolics such as caffeic and 4-hydroxybenzoic acids have the highest concentrations in oat and significantly contribute to its separation from other cereals. However, detailed variations of phenolics and organic acids within and between cereal cultivars require a closer investigation of the data. In the following section, univariate comparisons of some metabolites are represented and the findings are compared to previous results reported in the literature.

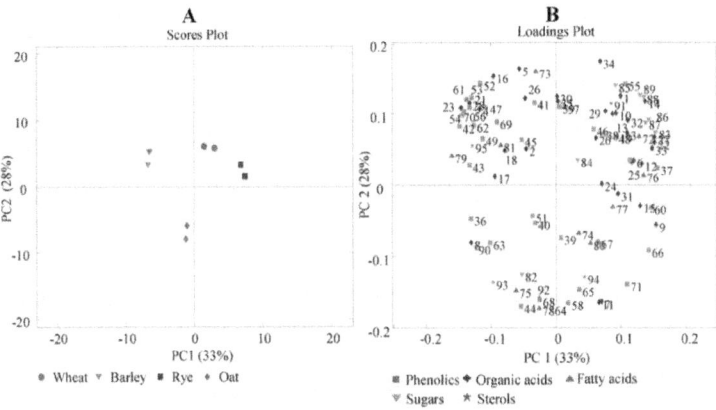

Figure 3: (**A**) scores and (**B**) loading plots of the three component PCA model developed using identified metabolite table. * Numbers in loadings plot correspond to the metabolites represented in Table 1.

Variation of Phenolics and Organic Acids in Cereals

Phenolic acid composition of wheat, barley, rye and oat were compared to previously reported data [18,21,22,23]. Figure 4 shows relative percentages of the nine most abundant, free and conjugated phenolic acids of cereals reported in previous studies and makes comparisons with the data obtained in the current study. In previous studies, the phenolic acids of cereals were extracted using 80% ethanol followed by hydrolysis of conjugated phenolics in 2 M sodium hydroxide solution and analyzed by LC-DAD. In the current study, free and conjugated phenolics were extracted using 85% methanol, hydrolyzed in 2 M solution of hydrochloric acid followed by GC-MS analysis and PARAFAC2 based data processing. These two methodologies in phenolic profiling of cereals result in several apparent compositional differences. However, it should be underlined that the compared cereal genotypes are different in the two studies and the goal of this study is not a comprehensive comparison of phenolics of cereal varieties, but to demonstrate the power of the standardized cereal metabolomics protocol developed.

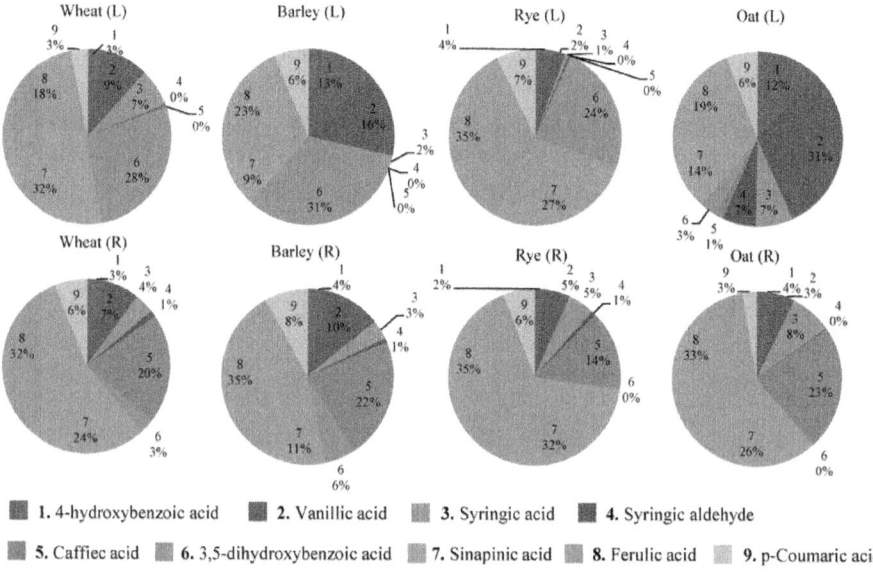

Figure 4: Comparison of relative percentages of the nine most abundant phenolic acids of cereals reported in the literature (L) with the results of the current study (R). In literature the following genotypes were studied: winter wheat (*Triticum aestivum* var. *aestivum*) [18], Dicktoo barley (USA) [21], Grandrieu rye (France) [23] and Bajka oat (Poland) [22].

Nine major phenolics of the cereals investigated in this study were compared with winter wheat (*Triticum aestivum* var. *aestivum*) [18], Dicktoo barley (USA) [21], Grandrieu rye (France) [23] and Bajka oat (Poland) [22] varieties (Figure 4). Figure 4 shows that the relative concentrations of caffeic acid consistently increased (14%–23%) in all cereal cultivars compared to the previous studies where its abundance was below 1%. Similarly, for wheat, barley and oat, concentrations of ferulic acid increased from approximately 20% to 33%, while the comparison is more consistent for the two rye varieties. These results suggest that in grains, a significant amount of caffeic and ferulic acids are present in conjugated forms that cannot be cleaved by alkaline hydrolysis. Thus, the most abundant phenolic acids in previous cereal metabolomics studies were ferulic, sinapinic and 3,5-dihydroxybenzoic acids, while in this study, ferulic, sinapinic and caffeic acids were the most abundant ones.

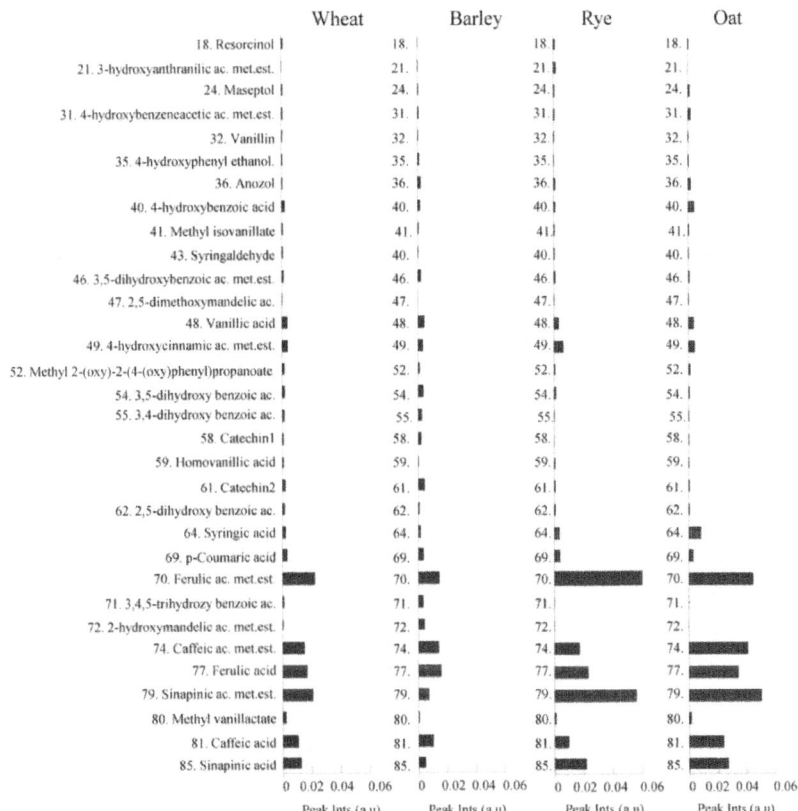

Figure 5: Relative concentrations of 32 phenolics detected from wheat, barley, rye and oat. Metabolites are numbered according to theTable 1.

Figure 5 and Figure 6 demonstrate relative concentrations of phenolics and organic acids/alcohols of wheat, barley, rye and oat genotypes investigated in this study. Figure 5 show that ferulic, caffeic and sinapic acids and their methyl esters are the most abundant metabolites among all other phenolics in the cereal samples. Moreover, the relative concentrations of the most abundant phenolics are found to be up to three times greater in rye and oat than in wheat and barley. Succinic and 3-hydroxybutanoic acids were the most abundant metabolites among all organic acids detected in the four different cereals (Figure 5). Relative concentrations of fumaric and 2-hydroxycyclohexanecarboxylic acids were significantly higher in rye, while concentrations of malic and ketoglutaric acids were highest in barley.

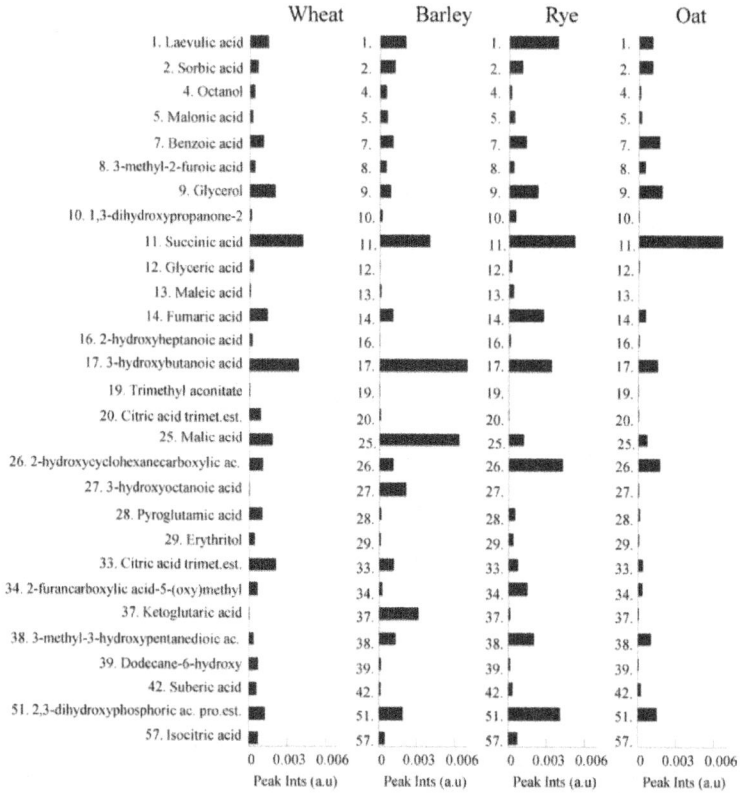

Figure 6: Relative concentrations of 29 organic acids/alcohols detected from wheat, barley, rye and oat. Metabolites are numbered according to the Table 1.

CONCLUSIONS

This paper outlines and demonstrates an optimized, relatively unbiased, comprehensive and high-throughput metabolomic profiling of whole-grain cereals based on new technologies developed within GC-MS metabolomics and chemometrics. A metabolite extraction protocol optimized towards phenolics and organic acids of whole-grains, and an unbiased and high-throughput protocol, was developed that allow processing of up to 60 samples per day.

The hydrochloric acid based hydrolysis allowed extraction of all major cereal phenolics, free and conjugated, and enabled the detection of 32 phenolic and 30 organic acids from 50 mg of flour. A novel trimethylsilylation method based on TMSCN allowed the detection of up to 300 metabolites from the GC-MS profiles. The multi-way decomposition method PARAFAC2 facilitated deconvolution of overlapping, retention time shifted and low s/n ratio peaks with high precision and in a semi-automated manner. The resolved mass spectra of deconvoluted peaks allowed the identification of 89 metabolites using NIST and Golm metabolite databases. Multivariate and univariate analysis of phenolic profiles of cereals revealed that ferulic, caffeic and sinapinic acids and their esters were the main phenolics of whole-grain samples across the four cereals studied. Rye and oat showed higher concentrations of the most abundant phenolics acids compared to wheat and barley. Comparison of the relative concentrations of the nine most abundant phenolics of cereals with previously reported data showed that the acidic hydrolysis significantly improved detection of caffeic acid. However, metabolite profiles of cereals highly depend on several factors such as genotype, growth conditions, harvest time and storage. Thus, essential secondary metabolite profile comparisons of different cereals as well as different varieties require a strictly controlled experimental design. This paper has demonstrated a new methodology that is ready to be applied in a larger metabolomic profiling studies that may reveal biological information related to phenolic and organic acids of whole-grain cereals. Moreover, the protocol developed can easily be modified for polar metabolite fractions, including mono- and di-saccharides and amino acids, of cereals by altering metabolite extraction method and the additional of a methoximation step in GC-MS derivatization.

ACKNOWLEDGEMENTS

Faculty of Science is acknowledged for support to the elite-research area "metabolomics and bioactive compounds" with a PhD stipendium to B. Khakimov and The Ministry of Science and Technology is acknowledged for a grant to University of Copenhagen (S.B. Engelsen) with the title "metabolomics infrastructure" under which the GC-MS was acquired.

AUTHOR CONTRIBUTIONS

B.K. B.M.J. and S.B.E. designed the study; B.K. conducted the GC-MS analysis. B.K. and S.B.E. performed the chemometric analysis and drafted the manuscript. All authors contributed to, read and approved the final manuscript.

REFERENCES

1. Zilic, S.; Sukalovic, V.H.T.; Dodig, D.; Maksimovic, V.; Maksimovic, M.; Basic, Z. Antioxidant activity of small grain cereals caused by phenolics and lipid soluble antioxidants. J. Cereal Sci2011, 54, 417–424.

2. Björck, I.; Östman, E.; Kristensen, M.; Anson, N.M.; Price, R.K.; Haenen, G.R.M.M.; Havenaar, R.; Knudsen, K.E.B.; Frid, A.; Mykkänen, H.; et al. Cereal grains for nutrition and health benefits: Overview of results from in vitro, animal and human studies in the HEALTHGRAIN project. Trends Food Sci. Technol. 2012, 25, 87–100.

3. Andersson, A.A.M.; Andersson, R.; Piironen, V.; Lampi, A.M.; Nystrom, L.; Boros, D.; Fras, A.; Gebruers, K.; Courtin, C.M.; Delcour, J.A.; et al. Contents of dietary fibre components and their relation to associated bioactive components in whole grain wheat samples from the HEALTHGRAIN diversity screen. Food Chem. 2013, 136, 1243–1248.

4. Amarowicz, R.; Zegarska, Z.; Pegg, R.B.; Karamac, M.; Kosinska, A. Antioxidant and radical scavenging activities of a barley crude extract and its fractions. Czech J. Food Sci. 2007, 25, 73–80. [Google Scholar]

5. Wood, P.J. Cereal beta-glucans in diet and health. J. Cereal Sci. 2007, 46, 230–238.

6. Mcintosh, G.H.; Whyte, J.; Mcarthur, R.; Nestel, P.J. Barley and wheat foods—Influence on plasma-cholesterol concentrations in hypercholesterolemic men. Am. J. Clin. Nutr. 1991, 53, 1205–1209. [Google Scholar]

7. Madhujith, T.; Shahidi, F. Antioxidative and antiproliferative properties of selected barley (Hordeum vulgarae L.) cultivars and their potential for inhibition of low-density lipoprotein (LDL) cholesterol oxidation. J. Agric. Food Chem. 2007, 55, 5018–5024.

8. Behall, K.M.; Scholfield, D.J.; Hallfrisch, J. Diets containing barley significantly reduce lipids in mildly hypercholesterolemic men and women. Am. J. Clin. Nutr. 2004, 80, 1185–1193. [Google Scholar]

9. Gibson, S.M.; Strauss, G. Implication of phenolic-acids as texturizing agents during cooking-extrusion cereals. Abstr. Pap. Am. Chem. Soc. 1991, 202, 150. [Google Scholar]

10. Vinson, J.A.; Erk, K.M.; Wang, S.Y.; Marchegiani, J.Z.; Rose, M.F. Total polyphenol antioxidants in whole grain cereals and snacks: Surprising sources of antioxidants in the US diet. Abstr. Pap. Am. Chem. Soc. 2009, 238, 246. [Google Scholar]

11. Khakimov, B.; Bak, S.; Engelsen, S.B. High-throughput cereal

metabolomics: Current analytical technologies, challenges and perspectives. J. Cereal Sci. 2014, 59, 393–418.

12. Soltesz, A.; Smedley, M.; Vashegyi, I.; Galiba, G.; Harwood, W.; Vagujfalvi, A. Transgenic barley lines prove the involvement of TaCBF14 and TaCBF15 in the cold acclimation process and in frost tolerance. J. Exp. Bot. 2013, 64, 1849–1862.

13. Widodo; Patterson, J.H.; Newbigin, E.; Tester, M.; Bacic, A.; Roessner, U. Metabolic responses to salt stress of barley (Hordeum vulgare L.) cultivars, Sahara and Clipper, which differ in salinity tolerance. J. Exp. Bot. 2009, 60, 4089–4103.

14. Manavalan, L.P.; Chen, X.; Clarke, J.; Salmeron, J.; Nguyen, H.T. RNAi-mediated disruption of squalene synthase improves drought tolerance and yield in rice. J. Exp. Bot. 2012, 63, 163–175.

15. Balmer, D.; Flors, V.; Glauser, G.; Mauch-Mani, B. Metabolomics of cereals under biotic stress: Current knowledge and techniques. Front. Plant Sci. 2013, 4, 82. [Google Scholar]

16. Fernie, A.R.; Schauer, N. Metabolomics-assisted breeding: A viable option for crop improvement? Trends Genet. 2009, 25, 39–48.

17. Bino, R.J.; Hall, R.D.; Fiehn, O.; Kopka, J.; Saito, K.; Draper, J.; Nikolau, B.J.; Mendes, P.; Roessner-Tunali, U.; Beale, M.H.; et al. Potential of metabolomics as a functional genomics tool.Trends Plant Sci. 2004, 9, 418–425.

18. Li, L.; Shewry, P.R.; Ward, J.L. Phenolic acids in wheat varieties in the HEALTHGRAIN diversity screen. J. Agric. Food Chem. 2008, 56, 9732–9739.

19. Fernandez-Orozco, R.; Li, L.; Harflett, C.; Shewry, P.R.; Ward, J.L. Effects of environment and genotype on phenolic acids in wheat in the HEALTHGRAIN diversity screen. J. Agric. Food Chem.2010, 58, 9341–9352.

20. Shewry, P.R.; Piironen, V.; Lampi, A.M.; Edelmann, M.; Kariluoto, S.; Nurmi, T.; Fernandez-Orozco, R.; Ravel, C.; Charmet, G.; Andersson, A.A.M.; et al. The HEALTHGRAIN wheat diversity screen: Effects of genotype and environment on phytochemicals and dietary fiber components. J. Agric. Food Chem. 2010, 58, 9291–9298.

21. Andersson, A.A.M.; Lampi, A.M.; Nystrom, L.; Piironen, V.; Li, L.; Ward, J.L.; Gebruers, K.; Courtin, C.M.; Delcour, J.A.; Boros, D.; et al. Phytochemical and dietary fiber components in barley varieties in the HEALTHGRAIN diversity screen. J. Agric. Food Chem. 2008, 56, 9767–9776.

22. Shewry, P.R.; Piironen, V.; Lampi, A.M.; Nystrom, L.; Li, L.; Rakszegi, M.; Fras, A.; Boros, D.; Gebruers, K.; Courtin, C.M.; et al. Phytochemical and fiber components in oat varieties in the HEALTHGRAIN diversity screen. J. Agric. Food Chem. 2008, 56, 9777–9784.

23. Nyström, L.; Lampi, A.M.; Andersson, A.A.M.; Kamal-Eldin, A.; Gebruers, K.; Courtin, C.M.; Delcour, J.A.; Li, L.; Ward, J.L.; Fras, A.; et al. Phytochemicals and dietary fiber components in rye varieties in the HEALTHGRAIN diversity screen. J. Agric. Food Chem. 2008, 56, 9758–9766.

24. Ward, J.L.; Poutanen, K.; Gebruers, K.; Piironen, V.; Lampi, A.M.; Nystrom, L.; Andersson, A.A.M.; Aman, P.; Boros, D.; Rakszegi, M.; et al. The HEALTHGRAIN cereal diversity screen: Concept, results, and prospects. J. Agric. Food Chem. 2008, 56, 9699–9709.

25. Arranz, S.; Calixto, F.S. Analysis of polyphenols in cereals may be improved performing acidic hydrolysis: A study in wheat flour and wheat bran and cereals of the diet. J. Cereal Sci. 2010,51, 313–318.

26. Sani, I.M.; Iqbal, S.; Chan, K.W.; Ismail, M. Effect of acid and base catalyzed hydrolysis on the yield of phenolics and antioxidant activity of extracts from germinated brown rice (GBR).Molecules 2012, 17, 7584–7594.

27. Khakimov, B.; Motawia, M.S.; Bak, S.; Engelsen, S.B. The use of trimethylsilyl cyanide derivatization for robust and broad-spectrum high-throughput gas chromatography-mass spectrometry based metabolomics. Anal. Bioanal. Chem. 2013, 405, 9193–9205.

28. Bro, R.; Andersson, C.A.; Kiers, H.A.L. PARAFAC2—Part II. Modeling chromatographic data with retention time shifts. J. Chemom. 1999, 13, 295–309.

29. Amigo, J.M.; Skov, T.; Coello, J.; Maspoch, S.; Bro, R. Solving GC-MS problems with PARAFAC2.Trac-Trends Anal. Chem. 2008, 27, 714–725.

30. Khakimov, B.; Amigo, J.M.; Bak, S.; Engelsen, S.B. Plant metabolomics: Resolution and quantification of elusive peaks in liquid chromatography-mass spectrometry profiles of complex plant extracts using multi-way decomposition methods. J. Chromatogr. A 2012, 1266, 84–94.

31. Vandendool, H.; Kratz, P.D. A generalization of retention index system including linear temperature programmed gas-liquid partition chromatography. J. Chromatogr. 1963, 11, 463.

32. Golm Metabolome Database. Available online: http://gmd.mpimp-golm.mpg.de/ (accessed on 5 November 2013).

33.   Hotelling, H. Analysis of a complex of statistical variables into principal components. J. Educ. Psychol. 1933, 24, 417–441.

34.   Sumner, L.; Amberg, A.; Barrett, D.; Beale, M.; Beger, R.; Daykin, C.; Fan, T.; Fiehn, O.; Goodacre, R.; Griffin, J.; et al. Proposed minimum reporting standards for chemical analysis.Metabolomics 2007, 3, 211–221.

Chapter 5

QUANTITATIVE ANALYSIS OF TOTAL AMINO ACID IN BARLEY LEAVES UNDER HERBICIDE STRESS USING SPECTROSCOPIC TECHNOLOGY AND CHEMOMETRICS

Yidan Bao[1], Wenwen Kong[1], Yong He[1,3], Fei Liu[1], Tian Tian[2] and Weijun Zhou[2]

[1]College of Biosystems Engineering and Food Science, Zhejiang University, Hangzhou 310058, China

[2]College of Agriculture and Biotechnology, Zhejiang University, Hangzhou 310058, China

[3]Cyrus Tang Center for Sensor Materials and Applications, Zhejiang University, Hangzhou 310058, China

ABSTRACT

Visible and near infrared (Vis/NIR) spectroscopy were employed for the fast and nondestructive estimation of the total amino acid (TAA) content in barley (*Hordeum vulgare* L.) leaves. The calibration set was composed of 50 samples; and the remaining 25 samples were used for the validation set. Seven different spectral preprocessing methods and six different calibration methods (linear and nonlinear) were applied for a comprehensive prediction performance comparison. Successive projections algorithm (SPA) and regression coefficients (RC) were applied to select effective wavelengths (EWs). The results indicated that the latent variables-least-squares-support vector machine (LV-LS-SVM) model achieved the optimal performance. The prediction results by LV-LS-SVM with raw spectra were achieved with a correlation coefficients (r) = 0.937 and root mean squares error of prediction (RMSEP) = 0.530. The overall results showed that the NIR spectroscopy could be used for determination of TAA content in barley leaves with an excellent prediction precision; and the results were also helpful for on-field monitoring of barley growing status under herbicide stress during different growth stages.

INTRODUCTION

Barley is one of the earliest cultivated cereal grains in the World, which is attracting renewed interest for its use in food and as a bioethanol feedstock [1]. It is a preferred grain for cultivation in many areas in the World due to its resistance to drought and ability to mature in climates with a short growing season [2]. Amino acid content is a very important physiological indicator which has a close relationship with the influence of environment stress during plant growing season. Recently, propyl 4-(2-(4,6-dimethoxypyrimidin-2-yloxy)benzylamino)benzoate (ZJ0273), a newly developed herbicide, has been applied to remove and control the weeds in barley fields. ZJ0273 is an ALS (acetolactate synthase)-inhibiting herbicide, which is considered to influence the formation of branch-chain amino acids (like aspartic acid, valine and proline) [3]. Hence, total amino acids (TAA) are basic physiological data and important parameters to understand the mechanism of herbicide effects on barley growth. The traditional amino acid detection method uses an automatic amino acid analyzer, which is laborious, time consuming, destructive and expensive. This method is not convenient for the fast and nondestructive detection of amino acids for field monitoring of plant growth information. Therefore, a rapid and practical method was necessary for the fast and accurate detection of amino acids.

Near infrared (NIR) spectroscopy is a common alternative analysis tool to traditional analytical methods. The NIR spectroscopy technique is rapid, and does not require labor-intensive sample processing, allowing for large-scale sampling [4]. It has developed rapidly in the past decades. In the agriculture field, NIR can be used to predict the neutral detergent fiber (NDF) and acid detergent fiber (ADF) of cereal residues from dryland cropping systems and is a useful tool to estimate residue decomposition potential [5]. Some researchers had shown the possibility of using NIR spectroscopy to analyze the β-glucan content in barley [6]. It is also possible to predict ergosterol content in whole barley samples using NIR [7]. The application of herbicides is an efficient and effective chemical weed control method to achieve optimal crop production [8], but herbicides also cause crop damage. Some physiological indicators are useful in evaluating the effect of herbicides [9]. This study was mainly focused on the feasibility of developing a rapid and effective method for the quantification of TAA in barley leaves using NIR spectroscopy to provide a new monitoring method for herbicide injury.

MATERIAL AND METHODS

Samples Preparation and Reflectance Measurements

Barley (*Hordeum vulgare* L.) used in our research was planted at the farm of Zhejiang University, Hangzhou (30°10'N, 120°12'E), China. The samples included 75 barley leaves, 50 for calibration and 25 for validation, and no single sample was used in both the calibration set and validation set at the same time. The calibration and validation set were randomly repeated several times in order to obtain a stable model. A new herbicide called ZJ0273 was applied during the seeding stage, the herbicide concentrations were 0, 50, 100, 500 and 1,000 mg/L, which are normally used for herbicide stress studies and practical field applications.

A Handheld FieldSpec spectrometer (Analytical Spectral Device, Boulder, CO, USA) was used within the 325–1,075 nm wavelength region for the reflectance spectral acquisition of all barley leaf samples. The resolution of this instrument is 1.5 nm. The reflectance mode was applied to obtain the spectra data of fresh barley leaves. The field-of-view (FOV) of the spectroradiometer is 25°. The distance between leaf sample and detector was 20 cm. Three replicate spectra were collected for each leaf sample, and the averaged spectrum obtained by averaging 30 scans per spectrum was used as the spectral data of each leaf sample. All spectra data were processed using the RS^3 software for Windows (Analytical Spectral Devices, Boulder, CO, USA) with a Graphical User Interface. The software used in this study included ASD View Spec Pro, Unscrambler V9.8 (CAMO AS, Oslo, Norway) and MATLAB V7.0 (The Math Works, Natick, MA, USA). The pretreatment of leaf samples and the protocol for amino acid extraction was based on the Lisiewska method [10]. The content of TAA in barley leaves was determined using a Hitachi automatic amino acid analyzer L-8900 (Hitachi High-Technologies Corporation, Tokyo, Japan) under common detection conditions.

Data Pre-Treatment

Previous studies showed that pre-treatment of measured spectral data was an important strategy to improve prediction performance [11]. In order to achieve the optimal spectral preprocessing method to predict TAA in barley, several different spectral preprocessing methods were compared. Seven different preprocessing methods were applied, including Savitzky-Golay smoothing (SG), standard normal variate (SNV), multiplicative scatter correction (MSC), first-derivative (1-Der), second-derivative (2-Der), de-trending and direct orthogonal signal correction (DOSC). SG smoothing, SNV, and MSC can be used for de-noising, light scatter correction, and light pathlength correction [12,13]. Derivatives were applied to correct the baseline shift [11]. De-trending seeks to remove nonlinear trends in spectral data [14]. DOSC corrected the major variance sources such as temperature effects, time influences and instrumental differences in spectral data [15]. The performance was determined by the prediction results in the later calibration stage.

Multivariate Analysis

Partial least squares (PLS) is a chemometrics method which is widely applied in NIR spectroscopic techniques. It is a bilinear modeling method. Latent variables (LVs) were used as the direct inputs of the PLS models to develop a relationship between the spectral data and TAA in barley leaves. A full cross-validation procedure was performed to test the model development.

In order to compare different modeling methods, a least squares-support vector machine (LS-SVM) model was built in this study. It is a powerful calibration method to handle linear and nonlinear problems with a good statistical basis [16]. The details of LS-SVM can be found in the literature [17,18]. Herein, the PLS and LS-SVM methods were compared to obtain the optimal prediction model of TAA in barley. PLS model can develop a linear relationship between the spectra data and TAA in barley. However, there is some useful nonlinear information in the spectra data which could be helpful to improve prediction performance. Therefore, LS-SVM was investigated to develop a model using both linear and nonlinear information in spectra data. LS-SVM applies linear equations using support vectors instead of quadratic programming problems to reduce the complexity of the optimization processes, which has advantages for multivariate analysis.

There are several indicators relating to the quality of developed models. Correlation coefficients (r) and root mean squares error of prediction (RMSEP) were considered as the main evaluation standards in this study. An ideal model should have a high r value closing to 1 and a low RMSEP value.

Selection of the Effective Wavelengths (EWs)

Normally, the full spectra might contain hundreds of variables, therefore, removing uninformative variables was an effective strategy to get better prediction and simpler models. The research by Wold has shown that using optimum wavelengths might be equally or more efficient than using full wavelengths in multivariate analysis [19]. Regression coefficients (RC) analysis and successive projections algorithm (SPA) were employed to select the effective wavelengths in this study. Regression coefficient (RC) by performing

PLS could be used as a way to select the effective wavelengths (EWs) [20]. The RC in the PLS model was used to calculate the response Y-variable from the X-variables. The coefficients gave an indication of which variables having the important impact on the response variables (Y). Large absolute values indicated the importance and the significance of the effect on the prediction of Y-variable. Successive projections algorithm (SPA) was a forward selection method which comprises three phases [21]. It starts with one wavelength, then incorporates a new one at each iteration, until a specified number of wavelengths is reached. With SPA, the informative variables with the least collinearity and redundancies could be selected. The selected EWs could be used as the direct input of the PLS and LS-SVM models.

Different Calibration Models

Different calibration methods were used for a better prediction of TAA in barley leaves under herbicide stress. Latent variables (LVs) were eigenvectors which were extracted during the building of the PLS model. Using LVs as the direct inputs of the PLS and LS-SVM models, the LV-PLS and LV-LS-SVM models were built. Based on the variables selected by SPA and RC, additional four different calibration models were developed, including SPA-PLS, RC-PLS, SPA-LS-SVM and RC-LS-SVM. The best model was achieved according to the prediction performance of the above mentioned calibration methods.

RESULTS AND DISCUSSION

Results of Full-Spectral Models

Figure 1 shows the original visible/near infrared reflectance spectra of 75 barley leaves. The trends of all samples with different herbicide concentrations were quite similar by visual inspection. There was a significant absorbance at around 680 nm caused by chlorophyll. The statistics of TAA in calibration and validation sets are shown in Table 1. Different PLS models were developed to find the optimal preprocessing methods. As the above-mentioned performance indicators, the correlation coefficients (r) and root mean squares error of prediction (RMSEP) were used to decide the quality of the calibration model.

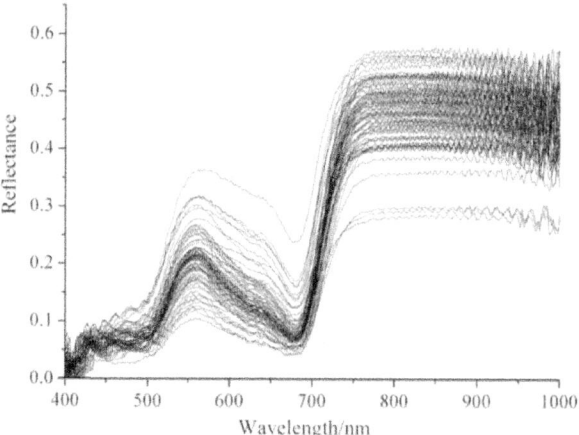

Figure 1: The original Vis/NIR reflectance spectra of barley leaves.

Table 1: Statistics of TAA in calibration and validation sets

Sample Set	Sample No.	Range (mg/g DW)	Mean (mg/g DW)	Standard deviation (mg/g DW)
Calibration	50	4.720–10.382	6.727	1.521
Validation	25	4.728–10.250	6.723	1.525
All	75	4.720–10.382	6.726	1.512

Table 2 includes the prediction results of TAA in validation set by the PLS models with eight preprocessing methods. A full cross-validation was applied during PLS calibration. Different latent variables (LVs) were used in PLS models related with different spectral preprocessing methods. The optimal PLS model was achieved by Raw spectra with $r = 0.879$ and RMSEP = 0.751. The next best PLS model was the de-trending spectra based model. Raw and de-trending were considered the optimal preprocessing methods in this study and were used in the further analysis.

On the other hand, the prediction results by the PLS models with the full-spectrum data were not so good, with none of the correlation coefficients of these prediction results exceeding 0.9. A possible reason was that the full-spectrum models contained too many variables (601), and some uninformative ones inevitably weakened the prediction performance of the models. Hence, further improvement should be done to give a smaller number of variables which carry the useful information to build more sensitive models.

Table 2: The prediction results of TAA in validation set by the PLS models with full-spectrum

Pretreatment	LV	r	RMSEP	Bias	Slope	Offset
Raw	6	0.879	0.751	−0.098	0.902	0.559
SG	6	0.868	0.790	−0.144	0.886	0.622
SNV	4	0.821	0.876	−0.026	0.783	1.432
MSC	4	0.814	0.893	−0.030	0.776	1.475
1-Der	6	0.823	0.866	0.031	0.769	1.582
2-Der	1	0.497	1.306	0.067	0.294	4.815
De-trending	6	0.875	0.759	−0.141	0.867	0.751
DOSC	1	0.835	0.909	−0.097	0.906	0.537

Selected EWs by SPA and RC

As mentioned above, SPA and RC were used for the selection of EWs, and the optimal preprocessing methods were also taken into consideration. In SPA, the maximum number of selected variables was set as 30 according to experience and previous literature [22]. Based on experience and preliminary studies, there were two basic principles using RC: (1) the absolute RC value of selected EWs should be larger than certain threshold value, and (2) these selected EWs should at certain peaks and valleys of the regression coefficient curve plot [20].

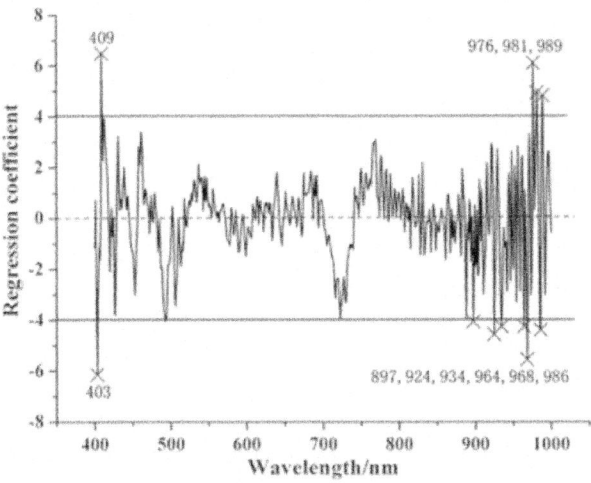

Figure 2: Selected effective wavelengths by regression coefficients.

Therefore, the threshold value was settled as ±4 in the RC analysis. The values of the regression coefficient which indicated the contribution of spectral (400–1,000 nm) to the calibration model were shown in Figure 2. Some obvious peaks and valleys could be found at certain wavelengths which were selected as the effective wavelengths. Table 3 shows the effective wavelengths which were selected by SPA and RC with two preprocessing methods, and the wavelengths selected by SPA were ranked in the order of importance.

Table 3: The selected EWs by SPA and RC

Pretreatment	Methods	No.	Selected EWs/nm
Raw	SPA	6	716, 976, 684, 982, 409, 407
	RC	8	409, 959, 968, 976, 982, 985, 988, 992
De-trending	SPA	7	747, 724, 888, 995, 415, 897, 922
	RC	11	403, 409, 897, 924, 934, 964, 968, 976, 981, 986, 989

Comparison of Six Calibration Models

Four different models were developed using the selected EWs by SPA and RC. Taking the selected LVs as direct inputs, two kinds of calibration models were built. In this study, these six linear and non-linear calibration models were developed to determine the TAA in barley leaves. Table 4 shows the calibration and validation results of the six models.

Table 4: The prediction results of total amino acid (TAA) content in barley leaves by different models

Models	Pretreatment	LV/EW/(γ, σ^2)	Calibration		Validation	
			r_c	RMSEC	r_v	RMSEP
LV-PLS	Raw	5/-/-	0.928	0.562	0.935	0.551
	De-trending	4/-/-	0.935	0.535	0.929	0.558
SPA-PLS	Raw	5/6/-	0.866	0.754	0.879	0.717
	De-trending	5/7/-	0.905	0.642	0.880	0.757
RC-PLS	Raw	3/8/-	0.693	1.085	0.625	1.205
	De-trending	4/11/-	0.880	0.716	0.862	0.779
LV-LS-SVM	Raw	6/-/(68.12, 271.15)	0.935	0.540	0.937	0.530
	De-trending	6/-/(8.91 × 10^6, 1.21 × 10^7)	0.936	0.533	0.930	0.309
SPA-LS-SVM	Raw	-/6/(1.16 × 10^6, 4.61 × 10^5)	0.869	0.744	0.872	0.737
	De-trending	-/7/(1.11 × 10^6, 4.74 × 10^5)	0.906	0.638	0.877	0.776
RC-LS-SVM	Raw	-/8/(2.06 × 10^6, 1.19 × 10^4)	0.837	0.827	0.360	1.553
	De-trending	-/11/(6.66, 45.84)	0.940	0.528	0.886	0.701

Compared with the above models, the PLS models achieved acceptable results in general. The performance of the LV-PLS and LV-LS-SVM models was better than that of other models in this study, which demonstrated that latent variables included more useful information for the determination of TAA content in barley leaves. The best prediction performance was achieved by the LV-LS-SVM (Raw) model, and the correlation coefficient and RMSEP in validation were 0.937 and 0.530. Comparing with the SPA-PLS model and the PLS model with full-spectrum, for raw spectral the correlation coefficient decreased by 0.02%, but the variables decreased by 99%; for de-trending spectral the correlation coefficient increased by 0.65%, while the variables decreased by 98.84% at the same time. The results indicated that the selected wavelengths carried most useful information of full-spectral, which was important for simplifying the model and developing portable instruments. On the other hand, the effective wavelengths selected by SPA performed better than those chosen by RC in this study, probably because the effective wavelengths selected by SPA were minimally redundant.

CONCLUSIONS

PLS and LS-SVM models were successfully developed from the Vis/NIR spectra for the fast determination of total amino acid (TAA) in barley leaves. This was important as a physiological indicator in crops during plant growth and herbicide stress. Raw and de-trending methods were the optimal preprocessing methods by the PLS models. The LV-LS-SVM models with Raw spectra achieved the best prediction performance for the validation set with $r = 0.937$ and RMSEP $= 0.530$. The results of this study indicated that NIR spectroscopy could be used for the determination of TAA content in barley leaves. The RC and SPA methods provided helpful approaches to determine the effective wavelengths, which was useful for the development of portable instrument or sensors for plant growth monitoring. Considering the limitation of samples used in this specific study, the results indicated the feasibility of using NIR spectroscopy to detect TAA in barley leaves under herbicide stress. More leaf samples with different growth stages and barley varieties would be taken into consideration to expand and develop more stable and robust models. This study supplied a new approach for the fast and accurate detection method of physiological parameters of barley growth.

ACKNOWLEDGMENTS

This work was supported by 863 National High-Tech Research and Development Plan (2011AA100705), Natural Science Foundation of China (31071332, 31201137), Science and Technology Department of

Zhejiang Province (2011C22070), China Postdoctoral Science Foundation (2012T50550, 2011M501009) and the Fundamental Research Funds for the Central Universities (2012FZA6005).

REFERENCES

1. Moreau, R.A.; Bregitzer, P.; Liu, K.; Hicks, K.B. Compositional equivalence of barleys differing only in low-and Normal-phytate levels. Agric. Food Chem. 2012, 60, 6493–6498.

2. HoltekØlen, A.K.; Uhlen, A.K.; Bråthen, E.; SahlstrØm, S.; Knutsen, S.H. Contents of starch and non-starch polysaccharides in barley varieties of different origin. Food Chem. 2006, 94, 348–358.

3. Chen, J.; Yuan, J.; Liu, J.D.; Fu, Q.M.; Wu, J. Mechanism of action of the novel herbicide ZJ0273. Acta Phys. Sin. 2005, 32, 48–52.

4. Stuth, J.; Jama, A.; Tolleson, D. Direct and indirect means of predicting forage quality through near infrared reflectance spectroscopy. Field Crops Res. 2003, 84, 45–56.

5. Stubbs, T.L.; Kennedy, N.; Fortuna, A.M. Using NIRS to predict fiber and nutrient content of dryland cereal cultivars. J. Agric. Food Chem. 2010, 58, 398–404.

6. Sohn, M.; Himmelsbach, D.S.; Barton, F.E.; Griffey, C.A.; Brooks, W.; Hicks, K.B. Near-infrared analysis of whole kernel barley. Comparison of three spectrometers. Appl. Spectrosc. 2008, 62, 427–432.

7. Borjesson, T.; Stenberg, B.; Schnurer, J. Near infrared spectroscopy for estimation of ergosterol content in barley: A comparison between reflectance and transmittance techniques. Cereal Chem. 2007, 84, 231–236.

8. Qasem, J.R. Weed control in cauliflower (Brassica oleracea var. Botrytis L.) with herbicides. Crop Prot. 2007, 26, 1013–1020.

9. Zobiole, L.H.S.; Bonini, E.A.; de Oliveira, R.S.; Kremer, R.J.; Ferrarese, O. Glyphosate affects lignin content and amino acid production in glyphosate-resistant soybean. Acta Physiol. Plant2010, 32, 831–837.

10. Lisiewska, Z.; Kmiecik, W.; Korus, A. The amino acid composition of kale (Brassica oleracea L. var. acephala), fresh and after culinary and technological processing. Food Chem. 2008, 108, 642–648.

11. Chu, X.L.; Yuan, H.F.; Lu, W.Z. Progress and application of spectral data pretreatment and wavelength selection methods in NIR analytical technique. Progr. Chem. 2004, 16, 528–542.

12. Dhanoa, M.S.; Lister, S.J.; Sanderson, R.; Barnes, R.J. The link between

multiplicative scatter correction (MSC) and standard normal variate (SNV) transformations of NIR spectra. J. Near Infrared Spectros. 1994, 2, 43–47.

13. Savitzky, A.; Golay, M.J.E. Smoothing and differentiation of data by simplified least squares procedures. Anal. Chem. 1964, 36, 1627–1639.

14. Barnes, R.; Dhanoa, M.; Lister, J. Standard normal variate transformation and detrending of near-infrared diffuse reflectance spectra. Appl. Spectrosc. 1989, 43, 772–777.

15. Westerhuis, J.A.; de Jong, S.; Smilde, A.K. Direct orthogonal signal correction. Chemometr. Intell. Lab. Syst. 2001, 56, 13–25.

16. Suykens, J.A.K.; Vandewalle, J. Least squares support vector machine classifiers. Neural Process. Lett. 1999, 9, 293–300.

17. Yang, F.; Tian, J.; Xiang, Y.H.; Zhang, Z.Y.; Harrington, P.B. Near infrared spectroscopy combined with least squares support vector machines and fuzzy rule-building expert system applied to diagnosis of endometrial carcinoma. Cancer Epidemiol. 2012, 36, 317–323.

18. Liu, F.; He, Y.; Wang, L. Comparison of calibrations for the determination of soluble solids content and pH of rice vinegars using visible and short-wave near infrared spectroscopy. Anal. Chim. Acta 2008, 610, 196–204.

19. Wold, J.P.; Jakobsen, T.; Krane, J. Hierarchical multiblock PLS and PC models for easier model interpretation and as an alternative to variable selection. J. Chemometr. 1996, 10, 463–482.

20. Liu, F.; Jiang, Y.H.; He, Y. Variable selection invisible/near infrared spectra for linear and nonlinear calibrations: A case study to determine soluble solids content of beer. Anal. Chim. Acta2009, 635, 45–52.

21. Araujo, M.C.U.; Saldanha, T.C.B.; Galvao, R.K.H.; Yoneyama, T.; Chame, H.C.; Visani, V. The successive projections algorithm for variable selection inspectroscopic multicomponent analysis.Chemometr. Intell. Lab. Syst. 2001, 57, 65–73.

22. Liu, F.; He, Y. Application of successive projections algorithm for variable selection to determine organic acids of plum vinegar. Food Chem. 2009, 115, 1430–1436.

Chapter 6

CONCOMITANT USE OF FOURIER TRANSFORM INFRARED ATTENUATED TOTAL REFLECTANCE SPECTROSCOPY AND CHEMOMETRICS FOR QUANTIFICATION OF MULTIPLE ADULTERANTS IN ROASTED AND GROUND COFFEE

Nádia Reis,[1] Adriana S. Franca,[1,2] and Leandro S. Oliveira[1,2]

[1]PPGCA, Universidade Federal de Minas Gerais, Avenida Antônio Carlos 6627, 31270-901 Belo Horizonte, MG, Brazil

[2]DEMEC, Universidade Federal de Minas Gerais, Avenida Antônio Carlos 6627, 31270-901 Belo Horizonte, MG, Brazil

ABSTRACT

This paper proposed the joint use of Fourier Transform Infrared Attenuated Total Reflectance Spectroscopy (FTIR-ATR) and Partial Least Square (PLS) regression for the simultaneous quantification of four adulterants (coffee husks, spent coffee grounds, barley, and corn) in roasted and ground coffee. Roasted coffee samples were intentionally blended with the adulterants, at adulteration levels ranging from 0.5 to 66% w/w. A robust methodology was implemented in which the identification of outliers was carried out. High correlation coefficients (0.99 for both calibration and validation) coupled with low degrees of error (0.69% for calibration; 2.00% for validation) confirmed that FTIR-ATR can be a valuable analytical tool for quantification of adulteration in roasted and ground coffee. This method is simple, fast, and reliable for the proposed purpose.

INTRODUCTION

New and challenging risks, such as adulteration, have emerged as food supply chains become increasingly global and complex, although fraud in the food sector has been an issue since ancient times. Food adulteration tends to be economically motivated and is achieved through addition, substitution, or removal of food ingredients. It is an issue that concerns not only consumers, but producers and distributors as well [1].

Coffee is one of the most valuable and most commonly consumed beverages in the world. Due to its high price, this commodity is usually targeted for adulteration. Impurities and adulterants are the most common concern. Any low-cost material of biological origin could be used as a potential adulterant in coffee [2]. Roasted and ground coffee presents physical characteristics (particle size, texture, and color) that are easily reproduced by roasting and grinding a variety of biological materials (cereals, seeds, parchments, etc.). As reported in previous works, coffee husks, sticks, spent coffee grounds, corn, barley, rice, and soybeans have been worldwide admixed with coffee for the sole purpose of adulteration [3, 4].

In order to develop analytical tools suitable to detect and identify adulteration in roasted and ground coffee, different techniques and procedures have been proposed, including UPLC [2], GC-MS [3], Direct Infusion Electrospray Ionization [5], HPAEC-PA [6], HPLC-DAD [7], UV-Vis, and Infrared Spectroscopy [4, 8–11]. Among these techniques, spectroscopic methods have gained attention in recent studies because they are fast, reliable, and simple to perform and usually do not require sample pretreatment, being thus appropriate for establishment of routine laboratory analysis.

In previous studies, we have shown that Diffuse Reflectance Fourier Transform Infrared Spectroscopy (DRIFTS) is suitable for identification, discrimination, and quantification of adulterants in roasted and ground coffee [4, 9, 10]. However, application of this method requires that the sample be mixed with KBr prior to analysis, and the amount employed for analysis is quite small, which could affect representativity, considering that adulterated coffee samples are inherently heterogeneous. Such problems could be minimized by employing Attenuated Total Reflectance (ATR) instead. ATR does not require any sample pretreatment and also allows the employment of larger samples [12]. Therefore, in the present study, we evaluate whether or not Fourier Transform Infrared Attenuated Total Reflectance Spectroscopy (FTIR-ATR) is a more effective technique for quantification of adulteration in roasted coffee than DRIFTS. Aside from employing a new measurement technique, we have also further improved our previous studies by increasing the range and number of adulterated samples.

MATERIAL AND METHODS

Samples

Arabica coffee, barley, and corn samples were acquired from local markets. Coffee husks were provided by the Minas Gerais State Coffee Industry Union (Sindicato da Indústria de Café do Estado de Minas Gerais, Brazil). Spent coffee grounds were provided by a local soluble coffee manufacturer (Café Brasília, Minas Gerais, Brazil).

Table 1: Mass composition of adulterated coffee samples

Samples	Adulteration level (%)	Mass fraction (%)				
		Coffee	Spent coffee grounds	Coffee husks	Barley	Corn
1	66	33.3		33.3		33.3
2	50	50		50		
3	50	50				50
4	40	60	10	10	10	10
5	40	60		20		20
6	40	60	20		20	
7	20	80	5	5	5	5
8	20	80		10		10
9	20	80	10		10	
10	10	90		5		5
11	10	90	5		5	
12	10	90	3.33	3.33		3.33
13	10	90	10			
14	10	90		10		
15	10	90			10	
16	10	90				10
17	1	99	1			
18	1	99		1		
19	1	99			1	
20	1	99				1
21	2	98		1		1
22	2	98	1		1	
23	2	98	1	1		
24	2	98			1	1
25	4	96	1	1	1	1
26	4	96			2	2
27	4	96	2	2		
28	8	92	2	2	2	2
29	8	92	4			4
30	8	92		4	4	
31	0.5	99.5				0.5
32	0.5	99.5		0.5		
33	0.5	99.5			0.5	
34	0.5	99.5	0.5			

Coffee beans (50 g), coffee husks (30 g), barley (50 g), and corn (30 g) samples were roasted in a convection oven (Model 4201D Nova Etica, S´ao Paulo, Brazil) at temperatures ˜ ranging from 200 to 260∘ C, under different time intervals. Roasting degrees (light, medium, and dark) were established by comparing luminosity ($L*$) values of the samples to measurements performed in commercially available coffee. A tristimulus colorimeter (Hunter Lab

Colorflex 45/0 Spectrophotometer, Hunter Laboratories, VA, USA) with standard D65 illumination and normal colorimetric observer angle of 10∘ was used in the color measurements. The established roasting degrees were defined as light ($23.5 < L^* < 25.0$), medium ($21.0 < L^* < 23.5$), and dark ($19.0 < L^* < 21.0$). Spent coffee grounds (three lots of 2 kg each) were washed with distilled water to remove impurities. Three 200 g samples were randomly selected from each lot and dried at 100∘ C for 5 h in order to reach moisture content levels similar to that of ground roasted coffee (~5 g/100 g). Further details on color measurements and roasting conditions are available in our previous study [9]. Pure coffee and adulterants (coffee husks, spent coffee ground, barley, and corn) were intentionally mixed, at adulteration levels ranging from 0.5 to 66 g/100 g, as described in Table 1.

FTIR Analysis

All measurements were performed in a dry controlled atmosphere (20 ± 0.5∘ C) employing a Shimadzu IRAffinity-1 FTIR Spectrophotometer (Shimadzu, Japan) with a deuterated L-alanine-doped triglycine sulfate (DLATGS) detector. A Pike sampling accessory (MIRacle), with zinc selenide window, was employed for the ATR measurements. All spectra were recorded in the range of 4000– 700 cm−1 with 4 cm−1 resolution and 20 scans and submitted to background subtraction (atmosphere spectra). Preliminary tests were performed to evaluate the effect of particle size (0.39 mm $<D<$ 0.5 mm; 0.25 mm $<D<$ 0.39 mm; 0.15 mm $<D<$ 0.25 mm; and $D < 0.15$ mm) on the quality of the spectra, and the best quality spectra (higher intensity and lower noise interference) were obtained for samples with $D < 0.15$ mm.

Statistical Analysis

MATLAB software, version 7.13 (MathWorks, Natick, MA, USA), and PLS Toolbox version 6.5 (Eigenvector Technologies, Manson, WA, USA) were employed for data analysis. PLS was employed for quantification of adulterants mixed in roasted coffee samples using the ATR spectra as chemical descriptors, with adulteration levels ranging from 0.5% to 66% in mass (see Table 1). The models were built with 170 spectra. The data were divided in two sets, calibration and validation, employing the Kennard-Stone algorithm, which promotes a data scan, selecting the more representative samples for the calibration set. The resulting calibration and validation sets were comprised of 102 and 68 spectra, respectively.

The data were submitted to two sequential evaluations. The first was focused on the efficiency of different data preprocessing applications. The second was related to the importance of the variables in the quantification process. In this step, different spectra ranges were evaluated in order to check if the use of specific region could improve the quality of the model.

The purpose of preprocessing is to linearize the response of variables and remove extraneous sources of variation (variance), which are not of interest in the analysis. Interfering variance appears in almost all real data because of systematic errors present in the experiment, requiring the model to work harder [13]. The data preprocessing methods tested were mean centering (1), Multiple Scatter Correction (MSC) followed by mean centering (2), MSC followed by first derivative, smoothing, and mean centering (3), Standard Normal Variates (SNV) followed by mean centering (4), SNV followed by first derivative, smoothing, and mean centering (5), absorbance normalization followed by mean centering (6), and first derivative followed by smoothing and mean centering (7).

Mean centering corresponds to subtraction of the average absorbance value of a given spectrum from each data point. Multiple scatter correction (MSC), originally developed to compensate the effects of light scattering in reflectance spectroscopy, has become a widely employed technique for removing general spectra drift features such as day-to-day intensity variations. Spectra derivatives are commonly used for baseline correction, because they provide visualization of small peaks that are difficult to detect in the original spectra. However its application also leads to a decrease in signal/noise ratio and thus a smoothing filter (Savitzky-Golay) was employed to provide noise reduction. SNV is applied to every spectrum individually; once the average and standard deviation of all the data points of the spectra are calculated, every data point is subtracted from the mean and divided by the standard deviation. Absorbance normalization consisted in dividing (i) the difference between the absorbance value at each data point and the minimum absorbance value by (ii) the difference between the maximum and minimum absorbance values [13, 14].

The optimal number of latent variables (LV) for each model was estimated by a cross-validation method (venetian blinds), based on the smallest value of root mean square error of cross-validation (RMSECV). Model performance was measured by evaluation of the root mean square errors for both calibration (RMSEC) and validation (RMSEP) sets, calculated as follows:

$$RMSEC = \sqrt{\sum_{i=1}^{I_C} \frac{(y_i - \hat{y}_i)^2}{I_C}},$$

$$RMSEP = \sqrt{\sum_{i=1}^{I_p} \frac{(y_i - \hat{y}_i)^2}{I_P}},$$

(1)

where y_i and $\hat{y}_i$ correspond to the real and predicted adulteration levels of sample i and I_C and I_p are the total number of samples in the calibration and prediction (validation) sets, respectively. The models with better prediction ability should present lower values of RMSEC and RMSEP.

Model optimization was performed by detection and elimination of outliers. Outliers correspond to samples that are very different from the rest of the data set, and their detection is crucial when developing multivariate models. In this study, outlier detection in the calibration set was based on the methodology proposed by Valderrama et al. [15], which is appropriate for detection of samples with extreme leverages, for example, large residuals in the X block (data) or large residuals in the Y block (model response). If a sample presents leverage (measure of the influence of each sample on the PLS model) larger than a limit value, it is considered an outlier. Such limit can be evaluated as three times the ratio between the number of latent variables and the number of samples [15]. The outliers of validation set were detected by jackknife residue test, as described by De Souza and Junqueira [16].

RESULTS AND DISCUSSION

Table 2 shows the results obtained for the PLS models based on the full-spectrum (4000–700 cm^{-1}) approach and employing the different preprocessing techniques cited in Section 2.3. For the obtained models, the LV number ranged from 4 to 8, and the RMSEC and RMSEP values ranged from 1.44 to 3.80 and from 2.42 to 3.56, respectively. Among the tested pretreatments, the ones that provided a significant improvement in model performance with the lowest RMSEC and RMSEP values were SNV followed by mean centering. This model was built with 8 LV that together explained 93.5% and 99.3% of the cumulative variance in X (spectra data) and in Y (adulterants concentration), respectively. The obtained RMSEC and RMSEP values were 1.44 and 2.42, respectively, and the correlation coefficient values of calibration (R_c) and validation (R_V) were 0.99 for both parameters (Table 2). It is noteworthy to mention that such model is more robust in comparison to the one obtained in our previous study [10] employing DRIFTS (LV = 10, RMSEC = 2.01, R_c = 0.99, RMSEP = 3.70, and R_V = 0.96).

Table 2: Performance results of full-spectrum PLS models based on different data preprocessing techniques

Data before treatment	LV	RMSEC (%)	R_c	RMSEP (%)	R_v
Mean centering (MC)	5	3.80	0.97	3.56	0.98
Multiple Scatter Correction (MSC) + MC	7	1.70	0.99	2.53	0.99
MSC + first derivatives + smoothing + MC	4	2.54	0.99	3.37	0.98
Standard Normal Variates (SNV) + MC	8	1.44	**0.99**	**2.42**	**0.99**
SNV + first derivatives + CM	6	1.78	0.99	2.68	0.99
Absorbance normalization + MC	8	1.67	0.99	2.52	0.99
First derivatives + MC	6	2.39	0.99	3.00	0.98

LV: latent variables; R_c: calibration correlation coefficient; R_v: validation correlation coefficient.

The next step was to evaluate if the selection of a specific spectral range could improve prediction accuracy, given that the full spectra could present some systematic variables that do not necessarily represent samples variance. For this reason, the plot of correlation coefficient that provided the main regions responsible for the quantification process is shown in Figure 1. The spectra regions that present greater contribution in the prediction process are characterized by having high absolute values of correlation coefficient. Analyzing the plot in Figure 1 it is possible to see that the highest values of correlation coefficient are concentrated in the range of 1134–700 cm^{-1} and that extending the spectra range from 700 up to 1735 cm^{-1} would still provide significant values of correlation coefficients, so both regions were tested. Comparing the data of Figure 1 with the data of Figure 2, in which the mean spectra of pure coffee and of the adulterants are shown, it is possible to check that the latter wavenumber range is characterized by vibrations of several types of bonds such as C–H, C–O, and C–N [17]. Chlorogenic acids, a class of phenolic compounds comprised of quinic acid esterified to a variety of trans-cinnamic acids, present strong absorption in the region of 1450–1000 cm^{-1}. Bands in the range 1085–1050 cm^{-1} can be assigned to axial C–O deformation of the quinic acid, in the range 1420–1330 cm^{-1} attributed to O–H angular deformation and C–O–C ester bond absorption in the 1300–1000 cm^{-1}range [18]. These chlorogenic acids are present in significantly greater amounts in coffee and its by-products than in barley and corn. Carbohydrates also exhibit several absorption bands in the range of 1500–700 cm^{-1}[19, 20], so it is expected that this class of compounds will contribute to many of the observed bands that occur in the spectra. Particularly, the skeletal mode vibrations of the glycosidic linkages in starch (present in corn and barley but not in the other samples) are usually observed in the 950–700 cm^{-1} wavenumber range, the so-called anomeric region of the spectrum [21]. Notice in Figure 2 that the sharp bands in the region of 950–700 cm^{-1} are coincident with the spectra of corn and barley but shifted in relation to the bands for the spectra of coffee, spent coffee, and coffee husks. These differences can be attributed to the different types of polysaccharides present in coffee and its adulterants. β-Glycosidic links are expected to appear in coffee and by-products in association with arabinogalactans, galactomannans, and cellulose, whereas α-glycosidic links should primarily appear in corn and barley due to the presence of starch. Other substances that naturally occur in coffee are reported to present absorbance bands in the range of 1700–1400 cm^{-1} [9]. Examples include caffeine (1700–1600 cm^{-1}) and trigonelline (1650–1400 cm^{-1}), as pointed out in the literature [22,23].

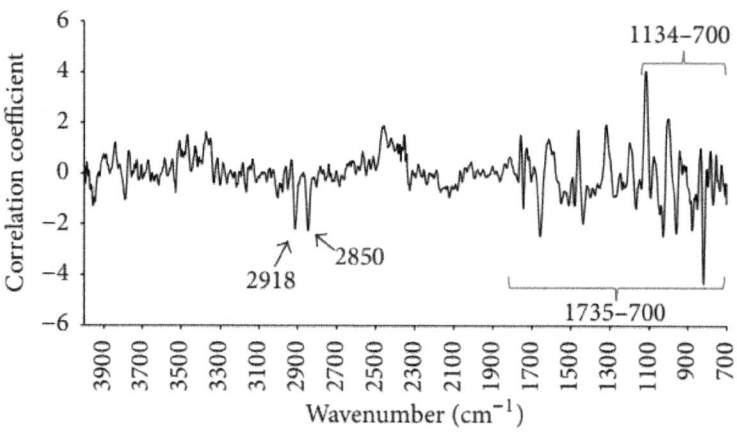

Figure 1: Full-spectrum regression coefficients (4000–700 cm^{-1}) of the PLS model based on the data submitted to SNV followed by mean centering.

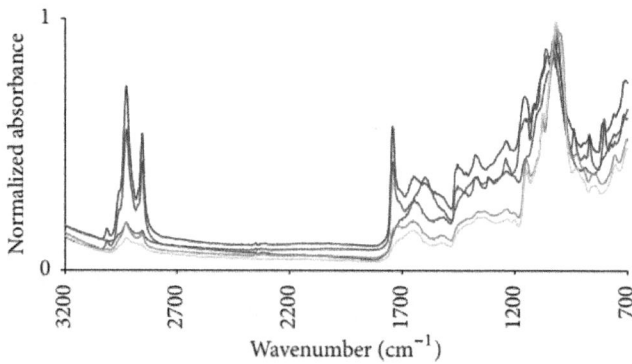

Figure 2: Average normalized ATR spectra obtained for roasted coffee (brown color), roasted coffee husks (pink color), spent coffee grounds (blue color), roasted barley (yellow color), and roasted corn (green color).

An evaluation of the coefficients shown in Figure 1 indicates that, besides the previously discussed regions, the only other peaks with significant values of correlation coefficient are 2918 and 2850 cm^{-1}. In the mean spectra shown in Figure 2, two significant absorption bands can be clearly seen between 2920 cm^{-1} and 2852 cm^{-1}, which are more intense in coffee and spent coffee grounds spectra. Such bands can be partly assigned to unsaturated and saturated lipids present in coffee, corn, and barley oils, which do not undergo changes during roasting, and, more specifically, the band at ~2852 cm^{-1} can be attributed to stretching of C–H bonds of methyl ($-CH_3$) group in the caffeine molecule [9]. This latter band is less evident in the spectra for barley and corn in comparison to the others, since corn and barley do not contain caffeine. The intensities of such bands are clearly affected by both levels of caffeine and lipids in coffee and primarily affected by caffeine in coffee husks (virtually devoid of oil) and by the lipids in roasted corn, roasted barley, and spent coffee grounds. The majority of the caffeine present in coffee is extracted during soluble coffee production whereas the lipid fraction is only partially extracted; thus spent coffee grounds may be considered to be devoid of caffeine but still containing significant amount of lipids.

In view of the aforementioned, the tested ranges were 4000–700 cm^{-1} (full spectra), 1735–700 cm^{-1}, and 1135–700 cm^{-1}. New models were built using these selected regions and the data were submitted to SNV and mean centering as preprocessing strategies. Table 3 shows the PLS results for this evaluation. As can be seen in Table 3, the selection of spectra ranges did not contribute to the prediction improvement; the model built with full spectra presented better prediction capacity than the other models.

Table 3: Performance results of PLS models based on different data ranges

Wavenumber range (cm^{-1})	LV	RMSEC (%)	R_c	RMSEP (%)	R_v
4000–700 (full spectra)	**8**	**1.44**	**0.99**	**2.42**	**0.99**
1735–700	7	1.74	0.99	2.77	0.98
1134–700	5	2.80	0.98	3.90	0.97

LV: latent variables; R_c: calibration correlation coefficient; R_v: validation correlation coefficient.

As the best PLS model obtained was built with full spectra and its data were submitted to SNV and mean centering, the next step was to optimize it by using the procedure for detection of outliers. The outliers were detected at 99% confidence level, and the results are summarized in Table 4. The optimization of the validation set was only performed after finishing the optimization of the calibration set. Besides, no more than three rounds of outlier detection (four models) should be performed, in order to avoid the "snowballing effect," when repetitive rounds continue to identify outliers [15]. As can be seen from Table 4, three rounds of outlier detection were performed. In the final model (4th), twenty-three outliers were detected in the calibration set (corresponding to 22% of the samples) and fourteen in the validation set (corresponding to 20.5% of the samples). Most of the outliers identified were associated with samples that presented the highest levels of adulteration (over 40%). The optimized model obtained after outliers removal consistedof 79 and 54 samples in the calibration and validation sets, respectively. It was built with 8 LV that together explained 88.7 and 99.5% of the accumulated variance in X (spectra data) and in Y (adulterants concentration), respectively. The RMSEC and RMSEP values were 0.69 and 2.00, respectively; the obtained correlation coefficients of calibration (R_c) and validation (R_v) were 0.99 for both. The curve of experimental values versus predicted values of the optimized model is shown in Figure 3(a). As can be seen by examination of the plot, this model is capable of predicting adulteration levels with accuracy. Residuals are randomly distributed about the mean value, which is satisfactorily close to zero, as is shown in Figure 3(b).

Table 4: Optimization of PLS model by detection and removal of outliers

Model	1st	2nd	3rd	4th
Number of calibration samples	102	92	86	**79**
Number of validation samples	68	68	68	**54**
LV	8	7	8	**8**
RMSEC (%)	1.44	1.19	0.82	**0.69**
RMSEP (%)	2.42	5.16	4.76	**2.00**
R_c	0.99	0.99	0.99	**0.99**
R_v	0.99	0.96	0.96	**0.99**

LV: latent variables; R_c: calibration correlation coefficient; R_v: validation correlation coefficient.

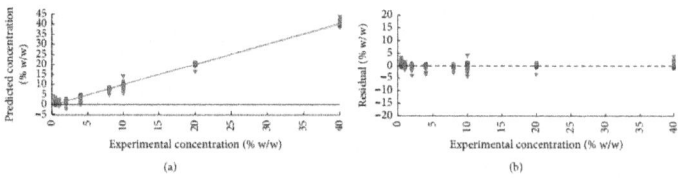

(a)   (b)

Figure 3: (a) Experimental versus predicted values of adulteration (% w/w) of coffee samples based on the optimized PLS model after outlier removal. (b) Residual versus adulteration levels (% w/w) of coffee samples based on the optimized PLS model after outlier removal.

A comparison of the model obtained in the present study with the one based on DRIFTS [10] is shown in Table 5. The types of adulterants and roasting conditions were the same in both studies, as well as the outlier removal procedure. The major difference, besides employing distinct measurements techniques (DR versus ATR), is that with FTIR-ATR we employed a larger number of samples at lower adulteration levels. A comparison of the models indicates that the one based on FTIR-ATR is more robust and presents better prediction abilities, with much lower RMSEC and RMSEP values. This in association with the fact that it employed a larger number of samples at low levels of adulteration indicates that FTIR-ATR is more appropriate for detection of adulteration in roasted and ground coffee.

Table 5: Comparison of the performances of models based on DRIFTS and FTIR-ATR

Optimized models	DRIFTS [10]	FTIR-ATR
Number of calibration samples	88	79
Number of validation samples	44	54
LV	10	8
RMSEC (%)	1.61	**0.69**
RMSEP (%)	2.34	**2.00**
R_c	0.99	**0.99**
R_v	0.98	**0.99**

LV: latent variables; R_c: calibration correlation coefficient; R_v: validation correlation coefficient.

CONCLUSION

PLS models of ATR spectra were successfully developed. The optimized model was built with full spectra (4000–700 cm^{-1}) that were submitted to SNV and mean centering as data preprocessing strategy. It was capable of predicting adulteration levels ranging from 0.5% to 40%. For this final model, the determination coefficients were 0.99 for both calibration and validation sets, and the errors observed during calibration and validation were quite low, 0.69% and 2.00%, respectively. It can be concluded that because the use of the full spectrum provided more robust models, the detection of adulteration and discrimination of adulterated and nonadulterated coffee samples cannot be attributed to a single class of components, rather being dependent on a variety of compounds, such as lipids, chlorogenic acids, caffeine, and polysaccharides. PLS and FTIR-ATR proved to be promising techniques, suitable for quantification of multiple adulterants in roasted and ground coffee.

ACKNOWLEDGMENTS

The authors acknowledge financial support from the following Brazilian government agencies: CAPES, CNPq, and FAPEMIG.

REFERENCES

1. J. C. Moore, J. Spink, and M. Lipp, "Development and application of a database of food ingredient fraud and economically motivated adulteration from 1980 to 2010," Journal of Food Science, vol. 77, no. 4, pp. R118–R126, 2012.

2. T. Cai, H. Ting, and Z. Jin-lan, "Novel identification strategy for ground coffee adulteration based on UPLC–HRMS oligosaccharide profiling," Food Chemistry, vol. 190, pp. 1046–1049, 2016.

3. R. C. S. Oliveira, L. S. Oliveira, A. S. Franca, and R. Augusti, "Evaluation of the potential of SPME-GC-MS and chemometrics to detect adulteration of ground roasted coffee with roasted barley," Journal of Food Composition and Analysis, vol. 22, no. 3, pp. 257–261, 2009.

4. N. Reis, A. S. Franca, and L. S. Oliveira, "Discrimination between roasted coffee, roasted corn and coffee husks by Diffuse Reflectance Infrared Fourier Transform Spectroscopy," LWT—Food Science and Technology, vol. 50, no. 2, pp. 715–722, 2013.

5. F. J. T. Aquino, R. Augusti, J. D. O. Alves et al., "Direct infusion electrospray ionization mass spectrometry applied to the detection of forgeries: roasted coffees adulterated with their husks,"Microchemical Journal, vol. 117, pp. 127–132, 2014.

6. L. M. Z. Garcia, E. D. Pauli, V. Cristiano, C. A. P. da Camara, I. S. Scarminio, and S. L. Nixdorf, "Chemometric evaluation of adulteration profile in coffee due to corn and husk by determining carbohydrates using HPAEC-PAD," Journal of Chromatographic Science, vol. 47, no. 9, pp. 825–832, 2009.

7. D. S. Domingues, E. D. Pauli, J. E. M. De Abreu et al., "Detection of roasted and ground coffee adulteration by HPLC by amperometric and by post-column derivatization UV-Vis detection," Food Chemistry, vol. 146, pp. 353–362, 2014.

8. D. F. Barbin, A. L. D. S. M. Felicio, D.-W. Sun, S. L. Nixdorf, and E. Y. Hirooka, "Application of infrared spectral techniques on quality and compositional attributes of coffee: an overview," Food Research International, vol. 61, pp. 23–32, 2014.

9. N. Reis, A. S. Franca, and L. S. Oliveira, "Performance of diffuse

reflectance infrared Fourier transform spectroscopy and chemometrics for detection of multiple adulterants in roasted and ground coffee,"LWT—Food Science and Technology, vol. 53, no. 2, pp. 395–401, 2013.

10. N. Reis, A. S. Franca, and L. S. Oliveira, "Quantitative evaluation of multiple adulterants in roasted coffee by Diffuse Reflectance Infrared Fourier Transform Spectroscopy (DRIFTS) and chemometrics,"Talanta, vol. 115, pp. 563–568, 2013.

11. U. T. C. P. Souto, M. F. Barbosa, H. V. Dantas et al., "Identification of adulteration in ground roasted coffees using UV–Vis spectroscopy and SPA-LDA," LWT—Food Science and Technology, vol. 63, no. 2, pp. 1037–1041, 2015.

12. G. Rytwo, R. Zakai, and B. Wicklein, "The use of ATR-FTIR spectroscopy for quantification of adsorbed compounds," Journal of Spectroscopy, vol. 2015, Article ID 727595, 8 pages, 2015.

13. B. M. Wise, N. B. Gallagher, and W. Windig, Chemometrics Tutorial for PLS_Toolbox and Solo, Eigenvector Research, Wenatchee, Wash, USA, 2006.

14. K. E. Kramer, R. E. Morris, and S. L. Rose-Pehrsson, "Comparison of two multiplicative signal correction strategies for calibration transfer without standards," Chemometrics and Intelligent Laboratory Systems, vol. 92, no. 1, pp. 33–43, 2008.

15. P. Valderrama, J. W. B. Braga, and R. J. Poppi, "Variable selection, outlier detection, and figures of merit estimation in a partial least-squares regression multivariate calibration model. A case study for the determination of quality parameters in the alcohol industry by near-infrared spectroscopy," Journal of Agricultural and Food Chemistry, vol. 55, no. 21, pp. 8331–8338, 2007.

16. S. V. C. De Souza and R. G. Junqueira, "A procedure to assess linearity by ordinary least squares method," Analytica Chimica Acta, vol. 552, no. 1-2, pp. 23–35, 2005.

17. R. M. Silverstein, F. X. Webster, and D. J. Kiemle, Spectrometric Identification of Organic Compounds, Wiley, New Jersey, NJ, USA, 2005.

18. D. Pujol, C. Liu, J. Gominho et al., "The chemical composition of exhausted coffee waste," Industrial Crops and Products, vol. 50, pp. 423–429, 2013.

19. R. Briandet, E. K. Kemsley, and R. H. Wilson, "Approaches to adulteration detection in instant coffees using infrared spectroscopy and chemometrics," Journal of the Science of Food and Agriculture, vol. 71,

no. 3, pp. 359–366, 1996.

20. E. K. Kemsley, S. Ruault, and R. H. Wilson, "Discrimination between Coffea arabica and Coffea canephora variant robusta beans using infrared spectroscopy," Food Chemistry, vol. 54, no. 3, pp. 321–326, 1995.

21. R. Kizil, J. Irudayaraj, and K. Seetharaman, "Characterization of irradiated starches by using FT-Raman and FTIR spectroscopy," Journal of Agricultural and Food Chemistry, vol. 50, no. 14, pp. 3912–3918, 2002.

22. J. S. Ribeiro, T. J. Salva, and M. M. C. Ferreira, "Chemometric studies for quality control of processed brazilian coffees using drifts," Journal of Food Quality, vol. 33, no. 2, pp. 212–227, 2010.

23. M. Szafran, J. Koput, Z. Dega-Szafran, and M. Pankowski, "Ab initio and DFT calculations of the structure and vibrational spectra of trigonelline," Journal of Molecular Structure, vol. 614, no. 1–3, pp. 97–108, 2002.

Chapter 7

IDENTIFICATION OF IMITATION CHEESE AND IMITATION ICE CREAM BASED ON VEGETABLE FAT USING NMR SPECTROSCOPY AND CHEMOMETRICS

Yulia B. Monakhova[1,2,3] Rolf Godelmann,[1] Claudia Andlauer,[1] Thomas Kuballa,[1] and Dirk W. Lachenmeier[1,4]

[1]Chemisches und Veterinäruntersuchungsamt (CVUA) Karlsruhe, Weissenburger Strasse 3, 76187 Karlsruhe, Germany

[2]Department of Chemistry, Saratov State University, Astrakhanskaya Street 83, 410012 Saratov, Russia

[3]Bruker Biospin GmbH, Silbersteifen, 76287 Rheinstetten, Germany

[4]Ministry of Rural Affairs and Consumer Protection, Kernerplatz 10, 70182 Stuttgart, Germany

ABSTRACT

Vegetable oils and fats may be used as cheap substitutes for milk fat to manufacture imitation cheese or imitation ice cream. In this study, 400 MHz nuclear magnetic resonance (NMR) spectroscopy of the fat fraction of the products was used in the context of food surveillance to validate the labeling of milk-based products. For sample preparation, the fat was extracted using an automated Weibull-Stoldt methodology. Using principal component analysis (PCA), imitation products can be easily detected. In both cheese and ice cream, a differentiation according to the type of raw material (milk fat and vegetable fat) was possible. The loadings plot shows that imitation products were distinguishable by differences in their fatty acid ratios. Furthermore, a differentiation of several types of cheese (Edamer, Gouda, Emmentaler, and Feta) was possible. Quantitative data regarding the composition of the investigated products can also be predicted from the same spectra using partial least squares (PLS) regression. The models obtained for 13 compounds in cheese (R^2 0.75–0.95) and 17 compounds in ice cream (R^2 0.83–0.99) (e.g., fatty acids and esters) were suitable for a screening analysis. NMR spectroscopy was judged as suitable for the routine analysis of dairy products based on milk or on vegetable fat substitutes.

INTRODUCTION

Due to industry efforts to provide low-cost foods or due to general ethical considerations against cow's milk consumption [1], imitation dairy products have recently appeared on the market [2–5]. Cheese analogues or imitation cheese are cheese-like products in which milk fat, milk protein, or both are partially or completely replaced with nonmilk-based components such as soy [2], starch [6], or vegetable replacers [3]. Other alternative products for consumers with cow milk intolerance [7] based on goat [8, 9] or sheep milk [9] can also be found on the market. Vegetable oils and fats are most commonly used as cheap substitutes for milk fat to manufacture imitation cheese or imitation ice cream. While not being harmful to health, the imitation products may be of lesser nutritional quality (e.g., by lower calcium content) and contain several artificial flavors and food colors [10]. Unfortunately, such imitation products may be offered without the necessary labeling, which is a deception of the consumer. Pizza topping is a good example of such a possibility [10]. It has therefore become necessary to develop a reliable technique able to detect such products in the market.

Chromatographic methods are the most popular choice for analysis of organic substances in cheese. For example, gas chromatography (GC) with mass-spectrometric detection [11–13] or flame ionization detection [11] and high-performance liquid chromatography [9, 13, 14] were previously applied. With these methods, precise and diverse information about volatile profiles of the particular type of cheese could be obtained. This has been done, for example, for Reggianito Argentino cheese [11], Italian mountain cheese (Bitto) [15], Majorcan cheese [16], Kuflu Turkish cheese [13], and different varieties of goat and sheep cheese [9, 12]. However, due to the matrix complexity of dairy products chromatographic analysis usually involves pretreatment steps such as solid-phase extraction (SPE) [17] or headspace sorptive extraction (HSSE) combined with thermal desorption (TD) [15]. Therefore, it can be concluded that chromatographic techniques are accurate but laborious, expensive, and time consuming.

Other methods based on spectroscopic techniques are also available. These include Fourier transform infrared (FTIR) spectroscopy [18–20], visible-near infrared reflectance spectroscopy [21], near infrared (NIR) spectroscopy [20, 22], atomic absorption spectroscopy [23], inductively coupled plasma optical emission spectrometry [24, 25], and fluorescence spectroscopy [20]. Fluorescence spectroscopy is the most sensitive method but only few compounds give rise to fluorescence signals. In FTIR and NIR spectra, strong and broad signals of water prevent the informative characterization of dairy products.

Among spectroscopic techniques in the area of food analysis, NMR is currently on the rise [26]. Previous application areas include beer [27], juice [28], grapes [29], infant formulas, [30] or pine nuts [31]. The application of NMR spectroscopy to cheese and ice cream analysis has been also presented in several studies.^{1}H NMR was used to investigate the influence of packaging on the degradation of soft cheese [32]. Full ^{1}H NMR assignments of signals of the water fractions of different types of cheese were recently provided [25, 33,34]. ^{1}H NMR is also able to provide reliable qualitative and quantitative analysis of amino acids [35] and biogenic amines [36] in cheese. ^{31}P and ^{23}Na NMR were used for the investigation of both phosphate and sodium ion distribution in semihard cheese [37]. A time-domain nuclear magnetic resonance (TD-NMR) was applied to the quick determination of moisture profiles during cheese drying [38]. Regarding NMR analysis of ice cream, TD-NMR was previously used for the investigation of the aggregation state (liquid or solid) of water and fat [39–41]. Another article utilized site-specific natural isotope fractionation NMR to detect adulteration of vanillin in ice cream [42].

Despite the mentioned diverse studies about composition of dairy products, there are only few articles dealing with the detection of their adulteration. For example, it was demonstrated that it is possible to identify the presence of cow milk in buffalo mozzarella by the use of electrophoretic mobility of cow and buffalo casein [43]. A method based on triacylglycerol composition obtained with GC-FID of cheese was also proposed to detect the levels of foreign fat [44]. Other techniques based on the determination of particular markers were also reviewed [4]. All of them are based on time-consuming chromatographic measurements and, what is more important, are able to detect only specific types of adulteration. In the view of these facts, NMR seems to be promising to provide accurate classification of dairy products according to the raw material origin. Therefore, the main objective of this research was to investigate the ability of NMR spectroscopy to differentiate milk fat products from vegetable fat substitutes. Cheese and ice cream were chosen as examples.

EXPERIMENTAL SECTION

Samples

A total of 109 cheese samples and 112 ice cream samples based on milk fat were analyzed. The products were either purchased at local stores in Karlsruhe, Germany, or submitted to the CVUA Karlsruhe for official food control purposes in Baden-Württemberg, Germany. Samples were selected to cover all possible imitation products available on the German market and a wide composition variability of milk fat products. Furthermore, imitation products based on vegetable fat (or vegetable fat/milk fat mixture) were analyzed (n=1cheese and n=1 ice cream). All samples were subjected to the standard GC/MS analysis that confirmed the labeling information in every case.

Sample Preparation and Validation

Sample preparation of cheese and ice cream was conducted by the German reference Weibull-Stoldt methodology for fat hydrolysis and extraction. The hydrolysis of the sample was conducted using the automated hydrolysis system HYDROTHERM (Gerhardt Analytical Systems, Königswinter, Germany) as shown in Figure 1. Briefly, a representative average sample (at least 200 g) is minced and homogenized. Then 10 g of the homogenized sample is weighed and put into the digestion beaker for automated hydrochloric acid hydrolysis. After the addition of hydrochloric acid (4 mol/L, 150 mL), the liquid is then quickly brought to boil and simmered for about 1 hour. At the end of hydrolysis the digestion mixture is diluted with hot water (100 mL) to the double amount and then is immediately filtered through pleated filter, which has been moistened automatically by the system with water (number of moisture cycles = 3 and water amount per cycle = 40 mL). After the program has finished, the filter is placed on a watch glass and dried for up to 1.5 h at$^\circ$ $103 \pm 2^\circ C$ in a drying oven. After cooling off, the fat is extracted using Soxhlet extraction with petroleum ether.

After finishing, the extraction flasks are dried in the drying oven for 60 minutes at $103 \pm 2^\circ C^\circ$. Then, they are placed in a desiccator, left to cool down to room temperature. After a constant weight was achieved, the fat phase was ready for NMR analysis, for which 200 mg of the fat fraction is mixed with 0.80 mL of $CDCl_3$ containing 0.1% tetramethylsilane (TMS). 0.6 mL of the mixture is poured into an NMR tube and directly measured.

To investigate the reproducibility of the sample preparation, two different imitation cheese samples were prepared twice and several resonances were integrated: 9.76–9.74 ppm (triplet), 4.33–4.30 ppm (doublet), and 2.80–2.72 ppm (triplet). The reproducibility was then calculated as relative standard deviation (RSD) between replicates.

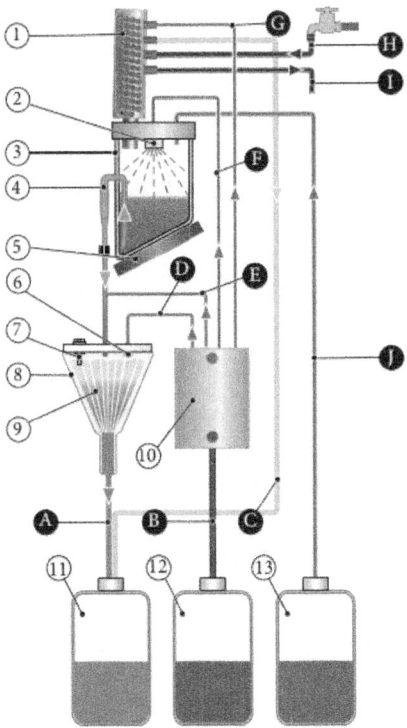

Figure 1: Schematic illustration of the automatic sample hydrolysis process necessary for Weibull-Stoldt fat extraction (reproduced with permission from Gerhardt Analytical Systems). 1 Condenser, 2 shower, 3 hydrolysis beaker, 4 sample transfer device, 5 hotplate, 6 shower for filter, 7 level sensor funnel, 8 funnel, 9 folded filter, 10 hot water generator, 11 tank for sample waste, 12 tank for H_2O, and 13 tank for HCl, A sample drainage, B distilled water addition, C air ventilation for condenser, D hot water addition-filter moisture, E hot water addition-rinsing sample transfer, F hot water addition-rinsing hydrolysis beaker, G hot water addition-rinsing condenser, H cooling water inlet, I cooling water outlet, J hydrochloric acid addition.

NMR Measurements at 400 MHz

All NMR measurements were performed on a Bruker Avance 400 Ultrashield spectrometer (Bruker Biospin, Rheinstetten, Germany) equipped with a 5 mm SEI probe with Z-gradient coils, using a Bruker Automatic Sample Changer (B-ACS 120). All spectra were acquired at 300.0 K. The data were acquired automatically under the control of ICON-NMR (Bruker Biospin, Rheinstetten, Germany), requiring about 12 min (^{1}H NMR) and 30 min (^{13}C NMR) per sample. All NMR spectra were phased and baseline corrected.

NMR Spectra Acquisition

^{1}H NMR spectra were acquired using the Bruker 1D zg pulse sequence with 128 scans (NS) and 2 prior dummy scans (DS). The sweep width (SW) was 20.5503 ppm and the time domain (TD) of the free induction decays (FIDs) was 131 k, acquisition time (AQ) was 7.97 s, and the repetition time (D1) was 1.0 s. Receiver gain (RG) value was set to 8.0. For acquisition of ^{13}C NMR spectra, the Bruker pulse sequence zgpg was used. After the application of 4 DS, 8 FIDs [(NS = 1024)] were collected into a TD of 131072 (131 k) complex data points using a 238.8728 ppm SW and a RG of 2050 [(AQ = 1.38 s and D1 = 2.00 s)].

Nontargeted Analysis and Chemometrics

The resulting spectra were analyzed using the software Unscrambler X version 10.0.1 (Camo Software AS, Oslo, Norway). We tested several spectral regions for calculation: aliphatic (0.25–3 ppm), midfield (3–6 ppm), aromatic (6–10 ppm) as well as the 0.25–6 ppm region with 0.01 ppm bucket width. Details on the bucketing process of NMR spectra for multivariate data analysis were previously described [27].

The technique of cross-validation was applied to determine the number of principal components (PCs) needed. For cheese spectra differentiation, we used 7 PCs (explained variance 97%) for ^{1}H NMR and 8 PCs (explained variance 98%) for ^{13}C NMR. The PCA model for ice cream spectra required 8 PCs (explained variance 97%). Using PLS regression, the NMR spectra were correlated with reference GC analysis data. PCA and PLS models were validated via full cross-validation. Furthermore, the PLS models were evaluated via test set validation (n=10), and results are compared with those obtained from a standard GC method.

RESULTS AND DISCUSSION

Sample Preparation and Spectra Analysis

Cheese and ice cream cannot be directly measured with liquid-state NMR, so that a sample preparation has to occur aiming to provide a measurable liquid solution. According to the literature, the most meaningful information about discrimination between types of cheese is contained in the fat fraction [45–47]. For this reason, we decided to apply the fat obtained with the Weibull-Stoldt methodology for our NMR analysis. According to Weibull-Stoldt, the sample is first hydrolyzed to free the fat, and then the fat fraction was extracted from the rest of the sample using a solvent. The use of the Weibull-Stoldt fat had also the advantage that this methodology is already conducted for nearly all samples of cheese and ice cream that reach our laboratory, as the labeled fat content on the package has to be controlled. The Weibull-Stoldt fat was also used for standard GC analysis. To simplify the Weibull-Stoldt protocol, we applied an automated device for the hydrolysis, which is the first system worldwide that was recently commercialized for this purpose. Prior to its application, we conducted this procedure using manual hydrolysis according to the Weibull-Stoldt method. The NMR spectra obtained with both methods showed the same fatty acid profile; however, the automated device was considerably more efficient as it is possible to prepare 6 samples at once without human intervention.

To demonstrate the reproducibility of this method, replicate measurements of different samples were performed. The relative standard deviations (RSD) between the two measurements were found to range between 0.1% and 2.1% (9.76–9.74 ppm), 1.0% and 1.2% (4.33–4.30 ppm), and 0.7 and 3.1% (2.80–2.72 ppm) for imitation cheese samples. The data indicated that the sample preparation procedure is adequately reproducible to facilitate a comparison between different cheese and ice cream samples.

Figure 2(a) showed the ^{1}H NMR spectrum of a representative sample of Gouda cheese. The signal of triglycerides and fatty acids dominated the spectrum [46]. Imitation cheese displayed a similar fatty acid profile (Figure 2(b)). By inspection of these spectra, we found the differences in the intensity of resonances relative to methyl (1.00–0.90 ppm) and bis-allylic protons (5.00–4.90 ppm) between the two groups of the products. In the ^{13}C NMR spectra, differences in the 173–170 ppm region (butyric acid) can also be observed. The same findings were valid for ice cream samples. Nevertheless, it can be concluded that NMR spectra of cheese and ice cream are very complex and a strong overlap of the resonances occurs. In the following, dairy products properties were uncovered from the NMR spectra using multivariate data analysis.

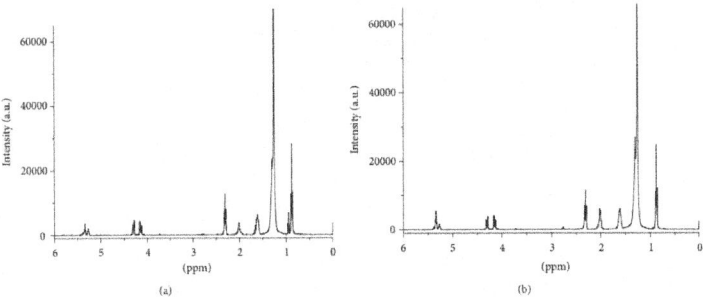

Figure 2: ^{1}H NMR spectra of Gouda cheese (a) compared to an imitation cheese based on vegetable fat (b).

Nontargeted Analysis

The spectroscopic data were visualized either through PCA scatter plots, in which each point represents an individual sample, or through loadings plots, which permit the identification of the most important spectral regions to separate the clusters and, therefore, reveal markers (compounds that are responsible for differentiation).

At first, PCA was performed on NMR spectra of cheese samples. The best grouping of similar samples was observed in the PCA scores plots of PC1-PC2 (^{1}H NMR spectra, Figure 3(a)) and PC3-PC6 (^{13}C NMR spectra, Figure 3(b)). On both plots, the imitation cheese samples were clearly separated from all of the remaining ones and were clustered in the range of negative values of PC1 (^{1}H NMR) or positive values of PC3 (^{13}C NMR). Furthermore, on both plots two especially conspicuous imitation samples were observed (marked with stars on Figures 3(a) and 3(b)). These two products represented tzatziki (a traditional Greek appetizer), which consists of both milk fat with vegetable fat and olive oil addition. Additionally, one outlier was located in the positive values of PC1 (Figure 3(a)). In addition to vegetable fat, this sample contained also about 3% of milk fat (as proven by GC analysis). Cheese made from milk (cow, goat, and sheep) was clustered around 0 in both PCA scores plots. Overall, we think that PCA in the aliphatic ^{1}H NMR region (Figure 3(a)) provided better differentiation of cheese samples. In this case, the samples in the imitation cluster were located closer to each other and the distance between milk fat/vegetable fat clusters was larger than that obtained with ^{13}C NMR data. Furthermore, vegetable fat/milk fat mixtures can also be recognized with ^{1}H NMR spectra.

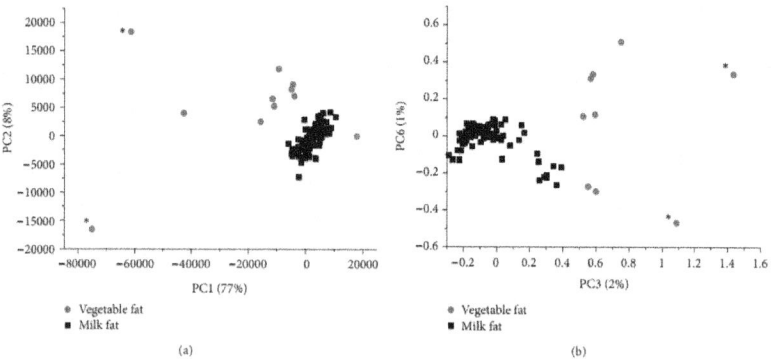

Figure 3: Scatter plot of the PCA scores for ^{1}H NMR ((a), 3.0–0.25 ppm) and ^{13}C NMR ((b), 200–0.25 ppm) for cheese samples (stars denote tzatziki samples).

The PCA scatter plot of ¹H NMR spectra (3–0.25 ppm) of ice cream samples was shown in Figure 4. In this case, seven PCs were found sufficient for differentiation. Unlike cheese samples, for which a good discrimination was observed between the first two PCs (Figure 3(a)), for ice cream the best model was constructed between PC4 and PC7. For ice cream samples, therefore, the variability in minor compound concentrations (such as alcohols and long-chain fatty acids) influenced the discrimination.

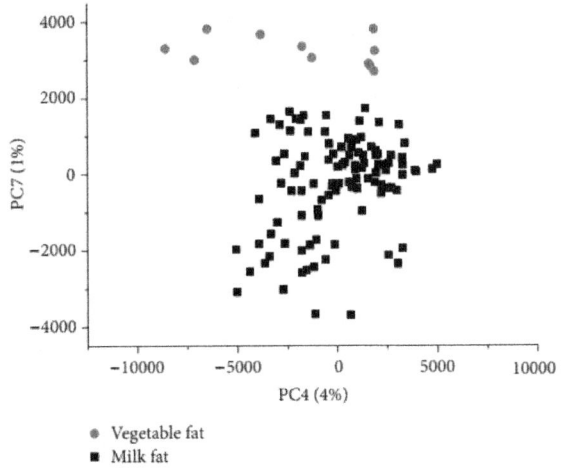

Figure 4: Scatter plot of the PCA ¹H NMR scores in the 3.0–0.25 ppm region for ice cream samples.

The chemical shifts and the associated functional groups that were responsible for the differentiation of the dairy products can be identified using the loadings plots. In the loadings plot, each chemical shift was plotted against its importance in discriminating the samples. In our case, the spectral regions 1.00–0.90 and 5.00–4.90 ppm were found to be important in the milk fat/vegetable fat product differentiation (for both cheese and ice cream). These regions consisted of the signals from methyl groups of different compounds and olefinic protons of all unsaturated chains [46]. The buckets at 2.32 and 2.30 ppm (most probably methylenic protons bonded to C2 of all fatty acid chains) [46] were also important for ice cream products.

Due to the low number of different imitation products currently available on the German market (probably because of a recent media campaign against these products [10]), we were not able to analyze our data with classification methods such soft independent modeling of class analogy (SIMCA) or linear discriminant analysis (LDA), which would require a larger dataset for training and validation. However, new samples can be distinguished by adding them to the developed PCA model.

While the differences of fat material used for cheese manufacture can be seen within the first several PCs (Figure 3(a)), higher PCs could uncover further clustering. To do this, we removed the imitation products and repeated the PCA. It can seen from Figure 5 that grouping in respect to the cheese types (Edamer, Gouda, Feta, and Emmentaler) is observed. It is not surprising because every cheese type was produced differently and, therefore, had a unique fatty acid profile. The two separate Gouda clusters were separated probably due to different ripening times similar to what was previously found for Italian Parmigiano Reggiano cheese [34]. It should be noted that ^{13}C NMR cannot provide such clear differentiation (only Emmentaler, Gouda, and Feta can be classified).

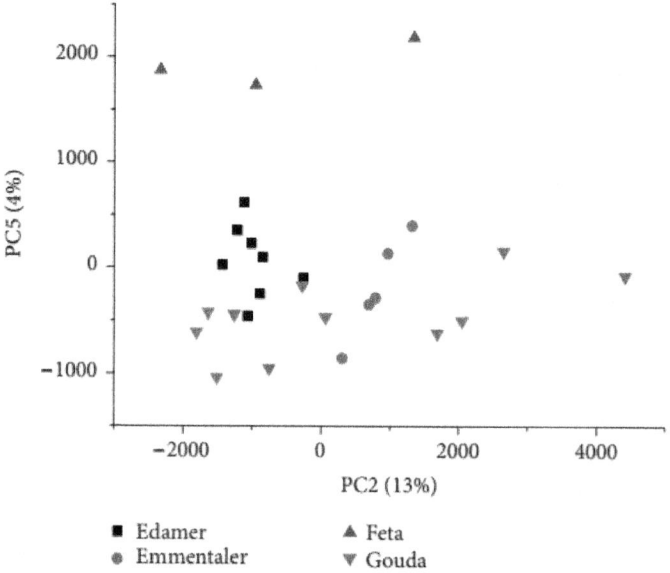

Figure 5: PCA of cheese types in the aliphatic region (3.0–0.25 ppm).

A recent review discussed the potential of different techniques coupled with chemometric analysis for the determination of the quality and the authenticity of dairy products, from which NMR plays an important role [47]. In the study of Rodrigues et al. [33] it was shown that metabolic profiling obtained by NMR combined with multivariate analysis allows to distinguish cheese samples in terms of maturation time, as well as added probiotic and prebiotic substances. PCA analysis was also performed on ^{1}H NMR spectra of Italian Parmigiano

Reggiano cheese to control time of ripening [34]. In the same research, the authors were able to provide geographical differentiation of cheese samples with Partial Least Squares-Discriminant Analysis (PLS-DA). ^{1}H and ^{13}C NMR coupled with PCA was used to differentiate PDO Asiago cheese produced in different areas [46]. High-resolution magic angle spinning (HR MAS) NMR together with PCA was able to distinguish Emmental cheese samples according to geographical region [48]. It should be mentioned, however, that data sets in all of these studies involved not more than 30 samples and focused on one specific group or origin. Therefore, our investigation is the first to apply ^{1}H NMR spectroscopy with multivariate methods to characterize a large number of commercial cheese and ice cream samples. Furthermore, to the best of our knowledge, no previous studies evaluated the performance of NMR spectroscopy to reveal vegetable fat adulteration of cheese and ice cream so far.

Quantitative Prediction of Dairy Product Composition

Besides the qualitative classification of our samples, in order to perform quality control of dairy products, it is also necessary to quantify certain compounds (e.g., saturated and unsaturated fatty acids and their esters). GC is among the most common methods for determining the fatty acid composition of cheese and ice cream [9, 11, 12, 15, 17]. ^{1}H NMR spectroscopic methods based on direct integration were also proposed for this purpose [45]. However, only a limited number of fatty acids can be quantified because they display similar and overlapping signals in the NMR spectra (Figure 2), which make simple quantification by integration of a distinct peak not possible. Moreover, in some cases the results can strongly deviate from the reference values [45].

To overcome these problems, we used Partial Least Squares regression (PLS) to correlate NMR spectra (in the 6– 0.25 ppm range) to the data of reference GC analysis. Results (i.e., root mean square error (RMSE), correlation coefficient (R^2), the number of PLS factors as well as NMR range used) of the best-fitting PLS models for ice creams were listed in Table 1. Fourteen of seventeen models exhibited correlation coefficients greater than 0.90. The correlation coefficients for butyric acid ($R = 0.89$), octadecanoic acid ($R = 0.83$), and nonadecanoic acid ($R = 0.89$) also appeared to be adequate for a screening procedure. The separate PLS models constructed for cheese showed slightly lower but comparable performance (R^2 values were in the range of 0.75–0.95). Inadequate PLS models ($R^2 < 0.50$) were only obtained for pentadecanoic acid, margaric acid, octadecanoic acid, and nonadecanoic acid for cheese samples (in comparison to ice cream, Table 1) due to the small concentrations of these compounds.

Table 1: PLS correlation between data of reference GC analysis and NMR spectra (6.0–0.25 ppm) for ice cream (n= 99)

Analytes	Reference range	PLS factors	Calibration		Validation	
			RMSE[a]	R^2	RMSE	R^2
Butyric acid (C4:0) (%)	0–25	7	0.21	0.89	0.24	0.85
Caproic acid (C6:0) (%)	0–1.9	7	0.15	0.87	0.17	0.84
Octanoic acid (C8:0) (%)	0–8.4	4	0.33	0.93	0.37	0.92
n-Capric acid (C10:0) (%)	0–6.7	4	0.28	0.91	0.33	0.87
Dodecanoic acid (C12:0) (%)	0–43.5	6	1.2	0.97	1.6	0.96
Tetradecanoic acid (C14:0) (%)	0–20.3	5	0.70	0.94	0.84	0.92
Myristoleic acid (C14:1) (%)	0–1.1	6	0.072	0.91	0.089	0.87
Hexadecanoic acid (C16:0) (%)	9.7–40.6	6	1.1	0.96	1.39	0.94
Palmitoleic acid (C16:1) (%)	0–1.7	6	0.09	0.94	0.11	0.91
Oleic acid (C18:1) (%)	5.2–60.7	4	1.0	0.99	1.5	0.97
Pentadecanoic acid (C15:0) (%)	0–1.6	7	0.099	0.93	0.12	0.90
Margaric acid (C17:0) (%)	0–0.8	7	0.04	0.92	0.05	0.89
Octadecanoic acid (C18:0) (%)	3.9–36	5	1.0	0.83	1.2	0.76
Nonadecanoic acid (C19:0) (%)	0–1.1	7	0.02	0.89	0.04	0.78
cis, cis-9.12-Octadecadienoic acid (C18:2) (%)	0.3–22.8	7	0.25	0.99	0.39	0.97
Methyl butanoate (g/100 g fat)	0–3.8	7	0.11	0.99	0.18	0.97
Hexanoic acid methyl ester (g/100 g fat)	0.04–2.24	7	0.07	0.98	0.09	0.96

[a] Root mean-squared error (RMSE) values are expressed in the same units as the analytes.

The PLS prediction models described previously were validated with an independent set of ten randomly chosen samples (only the acceptable models with $R^2 > 0.90$ were considered). The average relative deviations of the predicted values from the GC ones were in the range of 2–10% and 5–15% for ice cream and cheese samples. These data clearly demonstrated the reliability of the models and the potential of this technique for simultaneous quantitative analysis of dairy product characteristics along with the nontargeted control.

Therefore, we have extended quantitative NMR spectroscopy to 17 compounds that have to be analyzed during food control. In principle, NMR spectra contain the same information as GC but can be gathered much faster and more efficiently. The results emphasized the capability of NMR to rapidly and reliably predict cheese and ice cream characteristics.

CONCLUSIONS

Traditional analytical strategies to uncover adulteration of food rely on targeted analysis (in the course of which only certain marker compounds are analyzed). This approach has obstacles on many points, in particular, starting from time of analysis, use of sophisticated analytical equipment, or usage of expensive and/or toxic reagents. What is more important is that new forms of adulteration could not be uncovered as in the case of melamine adulteration of milk by applying the unspecific Kjeldahl assay [30]. A demand exists for rapid, accurate, and cheaper methods for direct quality measurement of food and food ingredients, which should include a nontargeted approach. Spectroscopic techniques combined with chemometric methods are a possible solution of this problem. From the wide range of spectroscopic methods, the high amount of spectral information of NMR is ideally suited for nontargeted analysis. This approach has only recently entered routine analysis in food control institutions. Examples are the investigation of infant formulas [30], pine nuts [31], or milk and milk substitutes [49]. This study has further shown that NMR is an efficient tool to detect fraud in the dairy product industry, protecting consumers against improper practices, and guaranteeing fair trade.

The method of automated sample preparation (Weibull Stoldt hydrolysis) used in this paper appears to be extremely efficient, especially when dealing with a high number of samples to be analyzed using multivariate methods. This procedure allowed the collection of high-quality spectra with distinct spectral features that were consistent within each sample. The developed technique could be further applied to solve other classification problems (differentiation between milk fat from different animals besides cows, or geographic discrimination) and shows great promise as a rapid tool for cheese analysis.

ACKNOWLEDGMENTS

The authors warmly thank H. Heger, M. Böhm, B. Siebler, and J. Geisser for excellent technical assistance. G. Raiber and M. Kranz (Gerhardt Analytical Systems) are thanked for their support in introducing the automatic Weibull-Stoldt hydrolysis to the authors' routine analysis. The views expressed in this paper do not necessarily reflect those of the Ministry of Rural Affairs and Consumer Protection.

REFERENCES

1. Beardsworth and T. Keil, "The vegetarian option: varieties, conversions, motives and careers,"Sociological Review, vol. 40, pp. 253–293, 1992.

2. N. Sutar, P. P. Sutar, and G. Singh, "Evaluation of different soybean varieties for manufacture of soy ice cream," International Journal of Dairy Technology, vol. 63, no. 1, pp. 136–142, 2010.

3. I. W. Rosnani, I. N. Aini, A. M. M. Yazid, and M. H. Dzulkifly, "Flow properties of ice cream mix prepared from palm oil: anhydrous milk fat blends," Pakistan Journal of Biological Sciences, vol. 10, no. 10, pp. 1691–1696, 2007.

4. H. Bachmann, "Cheese analogues: a review," International Dairy Journal, vol. 11, no. 4–7, pp. 505–515, 2001.

5. R. Cunha, A. I. Dias, and W. H. Viotto, "Microstructure, texture, colour and sensory evaluation of a spreadable processed cheese analogue made with vegetable fat," Food Research International, vol. 43, no. 3, pp. 723–

729, 2010.

6.   N. Noronha, E. D. O›Riordan, and M. O›Sullivan, "Replacement of fat with functional fibre in imitation cheese," International Dairy Journal, vol. 17, no. 9, pp. 1073–1082, 2007.

7.   M. Montalto, V. Curigliano, L. Santoro et al., "Management and treatment of lactose malabsorption,"World Journal of Gastroenterology, vol. 12, no. 2, pp. 187–191, 2006.

8.   Sanchez-Macias, M. Fresno, I. Moreno-Indias et al., "Physicochemical analysis of full-fat, reduced-fat, and low-fat artisan-style goat cheese," Journal of Dairy Science, vol. 93, no. 9, pp. 3950–3956, 2010. ·

9.   K. Raynal-Ljutovac, G. Lagriffoul, P. Paccard, I. Guillet, and Y. Chilliard, "Composition of goat and sheep milk products: an update," Small Ruminant Research, vol. 79, no. 1, pp. 57–72, 2008.

10. Rehm, "Käse ist nicht gleich Käse," Ernährung im Fokus, vol. 11, pp. 538–541, 2011.

11. V. Wolf, M. C. Perotti, S. M. Bernal, and C. A. Zalazar, "Study of the chemical composition, proteolysis, lipolysis and volatile compounds profile of commercial Reggianito Argentino cheese: characterization of Reggianito Argentino cheese," Food Research International, vol. 43, no. 4, pp. 1204–1211, 2010.

12. M. Poveda, E. Sánchez-Palomo, M. S. Pérez-Coello, and L. Cabezas, "Volatile composition, olfactometry profile and sensory evaluation of semi-hard Spanish goat cheeses," Dairy Science and Technology, vol. 88, no. 3, pp. 355–367, 2008.

13. A. Hayaloglu, E. Y. Brechany, K. C. Deegan, and P. L. H. McSweeney, "Characterization of the chemistry, biochemistry and volatile profile of Kuflu cheese, a mould-ripened variety," Lebensmittel-Wissenschaft & Technologie, vol. 41, no. 7, pp. 1323–1334, 2008.

14. S. Bonetta, J. D. Coïsson, D. Barile et al., "Microbiological and chemical characterization of a typical Italian cheese: Robiola di Roccaverano," Journal of Agricultural and Food Chemistry, vol. 56, no. 16, pp. 7223–7230, 2008.

15. S. Panseri, I. Giani, T. Mentasti, F. Bellagamba, F. Caprino, and V. M. Moretti, "Determination of flavour compounds in a mountain cheese by headspace sorptive extraction-thermal desorption-capillary gas chromatography-mass spectrometry," Lebensmittel-Wissenschaft & Technologie, vol. 41, no. 2, pp. 185–192, 2008.

16. Castell-Palou, C. Rosselló, A. Femenia, and S. Simal, "Application

of multivariate statistical analysis to chemical, physical and sensory characteristics of Majorcan cheese," International Journal of Food Engineering, vol. 6, no. 2, article 9, 2010.

17. S. Hauff and W. Vetter, "Quantification of branched chain fatty acids in polar and neutral lipids of cheese and fish samples," Journal of Agricultural and Food Chemistry, vol. 58, no. 2, pp. 707–712, 2010. ·

18. Subramanian, W. J. Harper, and L. E. Rodriguez-Saona, "Rapid prediction of composition and flavor quality of cheddar cheese using ATR-FTIR spectroscopy," Journal of Food Science, vol. 74, no. 3, pp. C292–C297, 2009.

19. N. A. Kocaoglu-Vurma, A. Eliardi, M. A. Drake, L. E. Rodriguez-Saona, and W. J. Harper, "Rapid profiling of swiss cheese by attenuated total reflectance (ATR) infrared spectroscopy and descriptive sensory analysis," Journal of Food Science, vol. 74, no. 6, pp. S232–S239, 2009.

20. M. Andersen, M. B. Frøst, and N. Viereck, "Spectroscopic characterization of low- and non-fat cream cheeses," International Dairy Journal, vol. 20, no. 1, pp. 32–39, 2010.

21. Lucas, D. Andueza, E. Rock, and B. Martin, "Prediction of dry matter, fat, pH, vitamins, minerals, carotenoids, total antioxidant capacity, and color in fresh and freeze-dried cheeses by visible-near-infrared reflectance spectroscopy," Journal of Agricultural and Food Chemistry, vol. 56, no. 16, pp. 6801–6808, 2008.

22. G. Cozzi, J. Ferlito, G. Pasini, B. Contiero, and F. Gottardo, "Application of near-infrared spectroscopy as an alternative to chemical and color analysis to discriminate the production chains of Asiago d›Allevo cheese," Journal of Agricultural and Food Chemistry, vol. 57, no. 24, pp. 11449–11454, 2009. ·

23. R. Moreno-Rojas, P. J. Sánchez-Segarra, F. Cámara-Martos, and M. A. Amaro-López, "Multivariate analysis techniques as tools for categorization of Southern Spanish cheeses: nutritional composition and mineral content," European Food Research and Technology, vol. 231, no. 6, pp. 841–851, 2010.

24. Lante, G. Lomolino, M. Cagnin, and P. Spettoli, "Content and characterisation of minerals in milk and in Crescenza and Squacquerone Italian fresh cheeses by ICP-OES," Food Control, vol. 17, no. 3, pp. 229–233, 2006.

25. M. A. Brescia, M. Monfreda, A. Buccolieri, and C. Carrino, "Characterisation of the geographical origin of buffalo milk and mozzarella cheese by means of analytical and spectroscopic determinations," Food

Chemistry, vol. 89, no. 1, pp. 139–147, 2005.

26.   G. Le Gall and I. J. Colquhoun, Food Authenticity and Traceability, M. Lees eds, Woodhead Publishing, Cambridge, UK, 2003.

27.   W. Lachenmeier, W. Frank, E. Humpfer et al., "Quality control of beer using high-resolution nuclear magnetic resonance spectroscopy and multivariate analysis," European Food Research and Technology, vol. 220, no. 2, pp. 215–221, 2005.

28.   P. S. Belton, I. J. Colquhoun, E. K. Kemsley et al., "Application of chemometrics to the ^{1}H NMR spectra of apple juices: discrimination between apple varieties," Food Chemistry, vol. 61, no. 1-2, pp. 207–213, 1998.

29.   Forveille, J. Vercauteren, and D. N. Rutledge, "Multivariate statistical analysis of two-dimensional NMR data to differentiate grapevine cultivars and clones," Food Chemistry, vol. 57, no. 3, pp. 441–450, 1996.

30.   W. Lachenmeier, H. Eberhard, F. Fang et al., "NMR-spectroscopy for nontargeted screening and simultaneous quantification of health-relevant compounds in foods: the example of melamine," Journal of Agricultural and Food Chemistry, vol. 57, no. 16, pp. 7194–7199, 2009.

31.   H. Köbler, Y. B. Monakhova, T. Kuballa et al., "Nuclear magnetic resonance spectroscopy and chemometrics to identify pine nuts that cause taste disturbance," Journal of Agricultural and Food Chemistry, vol. 59, no. 13, pp. 6877–6881, 2011.

32.   R. Lamanna, I. Piscioneri, V. Romanelli, and N. Sharma, "A preliminary study of soft cheese degradation in different packaging conditions by ^{1}H-NMR," Magnetic Resonance in Chemistry, vol. 46, no. 9, pp. 828–831, 2008.

33.   Rodrigues, C. H. Santos, T. A. P. Rocha-Santos, A. M. Gomes, B. J. Goodfellow, and A. C. Freitas, "Metabolic profiling of potential probiotic or synbiotic cheeses by nuclear magnetic resonance (NMR) spectroscopy," Journal of Agricultural and Food Chemistry, vol. 59, no. 9, pp. 4955–4961, 2011.

34.   R. Consonni and L. R. Cagliani, "Ripening and geographical characterization of Parmigiano Reggiano cheese by ^{1}H NMR spectroscopy," Talanta, vol. 76, no. 1, pp. 200–205, 2008.

35.   S. De Angelis Curtis, R. Curini, M. Delfini, E. Brosio, F. D›Ascenzo, and B. Bocca, "Amino acid profile in the ripening of Grana Padano cheese: a NMR study," Food Chemistry, vol. 71, no. 4, pp. 495–502, 2000.

36.   chievano, K. Guardini, and S. Mammi, "Fast determination of histamine

in cheese by nuclear magnetic resonance (NMR)," Journal of Agricultural and Food Chemistry, vol. 57, no. 7, pp. 2647–2652, 2009.

37. Gobet, C. Rondeau-Mouro, S. Buchin et al., "Distribution and mobility of phosphates and sodium ions in cheese by solid-state ^{31}P and double-quantum filtered ^{23}Na NMR spectroscopy," Magnetic Resonance in Chemistry, vol. 48, no. 4, pp. 297–303, 2010. ·

38. Castell-Palou, C. Rosselló, A. Femenia, J. Bon, and S. Simal, "Moisture profiles in cheese drying determined by TD-NMR: mathematical modeling of mass transfer," Journal of Food Engineering, vol. 104, no. 4, pp. 525–531, 2011.

39. T. Lucas, D. Le Ray, P. Barey, and F. Mariette, "NMR assessment of ice cream: effect of formulation on liquid and solid fat," International Dairy Journal, vol. 15, no. 12, pp. 1225–1233, 2005. ·

40. T. Lucas, M. Wagener, P. Barey, and F. Mariette, "NMR assessment of mix and ice cream. Effect of formulation on liquid water and ice," International Dairy Journal, vol. 15, no. 10, pp. 1064–1073, 2005. ·

41. Mariette and T. Lucas, "NMR signal analysis to attribute the components to the solid/liquid phases present in mixes and ice creams," Journal of Agricultural and Food Chemistry, vol. 53, no. 5, pp. 1317–1327, 2005.

42. S. Remaud, Y. Martin, G. G. Martin, and G. J. Martin, "Detection of sophisticated adulterations of natural vanilla flavors and extracts: application of the SNIF-NMR to method vanillin and p-hydroxybenzaldehyde," Journal of Agricultural and Food Chemistry, vol. 45, no. 3, pp. 859–866, 1997. ·

43. F. Locci, R. Ghiglietti, S. Francolino et al., "Detection of cow milk in cooked buffalo Mozzarella used as Pizza topping," Food Chemistry, vol. 107, no. 3, pp. 1337–1341, 2008.

44. Fontecha, I. Mayo, G. Toledano, and M. Juárez, "Triacylglycerol composition of protected designation of origin cheeses during ripening. Authenticity of milk fat," Journal of Dairy Science, vol. 89, no. 3, pp. 882–887, 2006.

45. G. Knothe and J. A. Kenar, "Determination of the fatty acid profile by ^{1}H-NMR spectroscopy,"European Journal of Lipid Science and Technology, vol. 106, no. 2, pp. 88–96, 2004.

46. Schievano, G. Pasini, G. Cozzi, and S. Mammi, "Identification of the production chain of Asiago d›Allevo cheese by nuclear magnetic resonance spectroscopy and principal component analysis,"Journal of Agricultural and Food Chemistry, vol. 56, no. 16, pp. 7208–7214, 2008.

47.   R. Karoui and J. De Baerdemaeker, "A review of the analytical methods coupled with chemometric tools for the determination of the quality and identity of dairy products," Food Chemistry, vol. 102, no. 3, pp. 621–640, 2007.

48.   Shintu and S. Caldarelli, "Toward the determination of the geographical origin of emmental(er) cheese via high resolution MAS NMR: a preliminary investigation," Journal of Agricultural and Food Chemistry, vol. 54, no. 12, pp. 4148–4154, 2006.

49.   Y. B. Monakhova, T. Kuballa, J. Leitz, et al., "NMR spectroscopy as a screening tool to validate nutrition labeling of milk, lactose-free milk, and milk substitutes based on soy and grains," Dairy Science and Technology, vol. 92, no. 2, pp. 109–120, 2012.

Chapter 8

THE ROLE OF VISIBLE AND INFRARED SPECTROSCOPY COMBINED WITH CHEMOMETRICS TO MEASURE PHENOLIC COMPOUNDS IN GRAPE AND WINE SAMPLES

Cozzolino Daniel

School of Agriculture, Food and Wine, Faculty of Sciences, The University of Adelaide, Waite Campus, PMB 1 Glen Osmond, Adelaide, SA 5064, Australia

ABSTRACT

The content of phenolic compounds determines the state of phenolic ripening of red grapes, which is a key criterion in setting the harvest date to produce quality red wines. Wine phenolics are also important quality components that contribute to the color, taste, and mouth feel of wines. Spectroscopic techniques (e.g., near and mid infrared) offer the potential to simplify and reduce the analytical time for a range of grape and wine analytes. It is this characteristic, together with the ability to simultaneously measure several analytes in the same sample at the same time, which makes these techniques very attractive for use in both industry and research. The objective of this mini review is to present examples and to discuss different applications of visible (VIS), near infrared (NIR) and mid infrared (MIR) to assess and measure phenolic compounds in grape and wines.

INTRODUCTION

Phenolic compounds, abundant in plants, are of considerable interest and have received more and more attention in recent years due to their bioactive functions [1,2,3,4]. Polyphenols are amongst the most desirable phytochemicals due to their antioxidant activity. These components are known as secondary plant metabolites, having antimicrobial, antiviral and anti-inflammatory properties along with their high antioxidant capacity [1,2,3,4]. Efforts have been made to develop highly sensitive and selective analytical methods for the determination and characterisation of polyphenol compounds where extensive research has been conducted on the different extraction and separation methods or techniques, as well as improving the chromatographic and spectral techniques used (e.g., NMR, infrared, UV, visible spectroscopy) [1,2,3,4].

Wine phenolics are important quality components that contribute to the color, taste, and mouth feel of wines [2,3,4,5]. Although phenolic compounds found in wine can also originate from microbial and oak sources, the majority of the phenolic constituents found in wine are grape-derived [1,2,3,4,5]. Grape growers and winemakers inherently understand the importance of grape and wine phenolics to overall wine quality, yet increasingly, few advances in the understanding of grape and wine phenolic chemistry have been made [2,3,4,5]. Wine is mainly composed of water, alcohol and other minor chemical components such as proteins, sugars, phenolic and volatile compounds that are present at low concentration (mg/100 g) [6,7,8,9,10,11,12]. The content of phenolic compounds determines the state of phenolic ripening of red grapes and is a key criterion in setting the harvest date to produce quality red wines. Grape and wine phenolics are structurally diverse, from simple molecules to oligomers and polymers usually designated as tannins [2,3,4,5]. They have an important impact on the organoleptic properties of wines; that is why their analysis and quantification are of primordial importance. The extraction of phenolics from grapes and from wines is the first step involved in analysis followed by several analytical methods that have been developed for the determination of total content of phenolic, while chromatographic and spectrophotometric analyses are continuously improved in order to achieve adequate separation of phenolic molecules, their subsequent identification and quantification [2,3,4,5]. The phenolic composition of grapes at the harvest time is a key factor determining their quality, and thus the quality of the finished wine [2,3,4,5]. The chemical methods used for the determination of seed and skin phenol content and extractability are generally slow because they require a preliminary extraction. Therefore, an evaluation of these parameters could be highly interesting for the oenological sector [6,7,8,9,10,11,12].

Today, public demand for high levels of quality and safety in grape and wine production requires high standards in quality assurance and process control methods, and this demand in turn requires appropriate tools for analysis during and after production. Desirable features of such tools should include speed, ease-of-use, minimal or no sample preparation, and the avoidance of sample destruction. These features are the main characteristics of a range of spectroscopic methods including mid infrared (MIR) and near-infrared (NIR) spectroscopy [6,7,8,9,10,11,12].

There is an increasing interest among the research and industry communities in the use of molecular spectroscopy methods (e.g., NIR, MIR, Raman) that has been expanded over the past two decades due to the many advantages these techniques offer [6,7,8,9,10,11,12]. Some of these advantages are related to the non-destructive nature of the method, the minimal or no sample preparation required and the speed of the analysis. Techniques using molecular spectroscopy are based upon the overtones and vibrations of the atoms of a molecule when passing infrared (IR) radiation through a tested sample [6,7,8,9,10,11,12]. In the IR region, various fundamental molecular vibrations, including those generated from C-H, O-H, N-H, C=O, and other functional groups can be detected [6,7,8,9,10,11,12]. When a sample is irradiated with IR light, it absorbs the light with frequencies matching characteristic vibrations of particular functional groups, whereas the light of other frequencies will be transmitted or reflected [6,7,8,9,10,11,12]. In this manner, the biochemical components of grapes and wines will determine the amount and frequency of absorbed light and the quantity of reflected or transmitted light can be used to infer the chemical composition of that sample [5,6,7,8,9,10]. Chemical bonds present in the organic matrix of most of the samples vibrate at specific frequencies, which are determined by the mass of the constituent atoms, the shape of the molecule, the stiffness of the bonds, and the periods of the associated vibrational coupling [6,7,8,9,10,11]. A specific vibrational bond is absorbed in the IR spectral region where diatomic molecules have only one bond that may stretch (e.g., the distance between two atoms may increase or decrease).

The use of NIR spectroscopy as an analytical technique is characterised by low molar absorptivity and scattering, which produce a nearly effortless evaluation of a sample [6,7,8,9,10,11]. Spectral "signatures" in the MIR result from the fundamental stretching, bending, and rotating vibrations of the sample molecules, whilst NIR spectra result from complex overtones and high frequency combinations at shorter wavelengths. Spectral peaks in the MIR frequencies are often sharper and better resolved than in the NIR domain, while the higher overtones of the O-H (oxygen-hydrogen), N-H (nitrogen-hydrogen), C-H (carbon-hydrogen) and S-H (sulphur-hydrogen) bands from the MIR wavelengths are still observed in the NIR region, although much weaker than the fundamental frequencies in the MIR. The existence of combination bands (e.g., CO stretch and NH band in protein), gives rise to a crowded NIR spectrum with strongly overlapping bands [6,7,8,9,10,11]. A major disadvantage of this characteristic overlap and complexity in the NIR spectra has been the difficulty of quantification and interpretation of data from NIR spectra. These same characteristics have the advantage that can reduce the need for using a large number of wavelengths during the analysis. In recent years, new instrumentation and computer algorithms have taken advantage of this complexity and have made the technique much more powerful and simple to use [5,6,7,8,9,10]. The MIR region of the electromagnetic spectrum lies between 4000 and 400 cm^{-1} and can be segmented into four broad regions: the X-H stretching region (4000–2500 cm^{-1}), the triple bond region (2500–2000 cm^{-1}), the double bond region (2000–1500 cm^{-1}), and the fingerprint region (1500–400 cm^{-1}) [6,7,8,9,10,11]. Such characteristic absorption bands are associated with major components of the sample matrix. Absorptions in the fingerprint region are mainly caused by bending and skeletal vibrations, which are particularly sensitive to large wavenumber shifts, thereby minimising against unambiguous identification of specific functional groups [6,7,8,9,10,11]. The application of FT-MIR in the routine analysis of grapes and wines is of special analytical interest due to the presence of sharp and specific absorption bands for constituents [6,7,8,9,10,11]. With the recent development of sampling accessories attached to a wide range of IR spectrophotometers, such as attenuated total reflectance (ATR) cells, improvements in routine IR analysis, by simplifying sample handling and avoiding measurement problems often found using transmission cells, have been achieved [6,7,8,9,10,11]. In conventional FT-MIR analysis, samples are analysed through a short-path length transmission cell [6,7,8,9,10,11]. The advantage of the transmission cell is that it provides very accurate and reproducible spectroscopic measurements while the main drawbacks are related with issues such as filling and cleaning the cell, variation of sample path length due to window wear and turbidity of the sample. The use of ATR cells minimise or avoid these issues allowing the analysis of a broad range of samples such as grapes and wine juice [6,7,8,9,10,11].

The use of NIR spectroscopy in the wine industry dates back to some early work by Kaffka and Norris while much of the NIR work in the wine industry has concentrated on ethanol analysis [6,7,8,9,10,11,12]. Information about constituents of grape juice and must, as well as wine can be used for management and decision support systems in order to improve, monitor and adapt grape and wine production systems to new challenges. Objective quality measures will allow vineyard managers to target required quality levels and will allow rewarding managers for quality, in terms of quality related grape payment systems where large areas of new plantings coming on stream will apply a correction to the fruit supply and demand situation, placing further urgency on the requirement to determine quality levels. The procedure reported here seems to have much potential for fast and reasonable cost analysis. The results of this work show that the models developed using NIR technology together with chemometric tools allow the quantification of total phenolic compounds and the main families of phenolic compounds in grape skins throughout maturation.

The objective of this mini review is to present examples and to discuss different applications of visible (VIS), near infrared (NIR) and mid infrared (MIR) to assess and measure phenolic compounds in grape and wines.

ANALYSIS OF PHENOLIC COMPOUNDS IN GRAPES, SKINS AND SEEDS

The use of both visible (VIS) and NIR spectroscopy in the wavelength range between 450 and 980 nm was explored to predict anthocyanin in Nebbiolo grapes grown in Italy [13]. Partial least squares (PLS) regression models were developed for ripening parameters and for phenolic ripening indexes in both fresh berries and homogenised samples. Using homogenised grape samples, the authors reported good correlations for ripening index based on phenolic content (R = 0.80) [13]. The use of NIR spectroscopy was also reported to measure total anthocyanins in Cabernet Sauvignon, Carmenere, Merlot, Pinot Noir, and Chardonnay wine grape varieties [14]. These authors also suggested that for the predictions of total anthocyanins, a better reference method should be used to develop robust PLS calibration models for grapes [14]. An optical portable VIS and NIR system (JAZ, Ocean Optics, Dunedin, FL, USA) was reported to predict different ripening parameters in fresh berries [15]. PLS regression was used as method to develop calibration models for extractable anthocyanins yielding a coefficient of determination (R^2) of 0.74, while less accurate models were obtained for total anthocyanins [15]. Similar results were reported by the same authors using grape samples sourced from different wine regions in Chile [16]. Red grape homogenates (n = 620) were analyzed using a combination of VIS and NIR (400–2500 nm) spectroscopy [17]. The spectra and the analytical data were used to develop PLS calibration models to predict dry matter (DM) content and condensed tannins (CT) [17]. The R^2 in cross-validation and the SECV were 0.92% and 0.83% w/w for DM and 0.86 and 0.46 mg/g epicatechin equivalents for CT, respectively [17]. The standard error in prediction (SEP) reported for CT was 0.89 mg/g epicatechin equivalent [17].

The potential of NIR spectroscopy to determine the content of phenolic compounds in intact red grapes has been reported and evaluated using a fibre-optic probe as well as a transport quartz cup [18]. Reference values for phenolic compounds were obtained using HPLC-DAD-MS, and modified (M) PLS regression was used as algorithm to develop the quantitative models for flavanols, phenolic acids, anthocyanins and total phenolic compounds [18]. According to these authors, NIR spectroscopy appeared to have an excellent potential for the quantification of total phenolic compounds in grape skins throughout maturation [18]. The validation of these models showed that the best results were obtained for the determination of flavonols (differences between HPLC and NIR of 7.8% using grapes and 10.7% using grape skins) [18]. Good statistics in the external validation were also obtained for the determination of total phenolic compounds (differences of 11.7% using grapes and 14.7% using grape skins) [18].

The feasibility of using FT-NIR spectroscopy to predict the extractable content of phenolic compounds directly in intact grape seeds was reported [19]. Calibration models were based on the correlation of spectral data with the phenolic composition determined by reference chemical methods on 40 grape samples [19]. The effect of season (vintage) was also evaluated and the results showed that the predictive accuracy improved only for spectrophotometric indices when samples from two years were simultaneously considered [19]. The calibration statistics showed that the models developed were sufficiently robust for quantitative purposes in terms of the SEP obtained for total flavonoids, pro-anthocyanidins, low molecular weight flavanols, catechin, epicatechin, procyanidin (SEP < 15%), as well as galloylation percentage [19]. The use of FT-NIR has been also evaluated and compared with instrumental texture parameters associated with the content of total phenols and extractability predictors in intact grape seeds [20]. This study was carried out using Cabernet-Sauvignon seeds from grapes harvested at six different advanced physiological stages throughout ripening and calibrated by flotation to reduce the in-field heterogeneity inside each sample [20]. The best prediction of phenol content in the seeds, performed directly on intact seeds, was found using FT-NIR spectroscopy in transmittance mode [20]. The SEP for total phenol content was less than 8%, while that for phenol extractability was worse [20]. High correlations were reported for the measurement of anthocyanins prediction in Canaiolo grape samples between NIR spectroscopy and total anthocyanis content using PLS regression. These authors reported an excellent performance in cross validation R^2 of 0.90 and a SECV of 45.15 mg/kg [21].

The feasibility of FT-MIR spectroscopy combined with PLS regression to quantify phenolic compounds in red grapes during ripening was reported [22]. The routine reference methods used to quantify these compounds such as total phenolic compounds, total anthocyanins, and condensed tannins were based on UV-VIS spectroscopy [22]. In order to take into account the high natural variability of grapes when building the calibration models, the authors collected fresh grapes from six varieties, at different phenolic ripening states, which were harvested during three vintages [22]. The statistics reported by the authors for the prediction of total phenolic were a root mean square error of prediction (RMSEP) of 4.3% and a ratio deviation in prediction (RPD) of 4.5, for total anthocyanins RMSEP of 5.9% and RPD of 3.5, and for condensed tannins RMSEP of 5.8% and RPD of 3.8 [22]. In another study, procyanidins were extracted with a mixture of methanol and acetone in water from seeds sourced from white and red wine grape varieties and analysed using FT-MIR spectroscopy [23]. A fractionation by graded methanol/chloroform precipitations produced a total of 26 samples that were characterised using thiolysis as pre-treatment followed by HPLC-UV and MS detection [23]. The average degree of polymerisation (DPn) of the procyanidins in the samples ranged from 2–11 flavan-3-ol residues [23]. PLS regression models for the determination of DPn, yield a RMSECV of 11.7%, with a R^2 of 0.91 and a RMSEP of 2.58 [23]. According to the authors, the application of orthogonal projection to latent structures (O-PLS) improves the interpretation of the regression models [23].

In recent years, new applications of VIS and NIR have been developed by means of hyperspectral imaging [24,25,26,27,28]. Hyperspectral images of intact grapes during ripening were recorded using a NIR hyperspectral imaging system (900–1700 nm) [24]. Spectral data were correlated with grape skin total phenolic concentration using MPLS regression, as well as using different spectral pre-treatments [24]. The calibration statistics obtained using MPLS and a combination of red and white wine grape samples were R^2 and SECV of 0.89 and 1.23 mg/g for total phenolic concentration [24]. Separate calibration models for red and white grape samples were also developed and compared by the authors [24]. The results obtained showed a good potential for a fast and inexpensive screening of these parameters in intact grapes [24]. The determination of the total anthocyanin content in skins of Cabernet Sauvignon grapes produced in Shaanxi province (China) using hyperspectral imaging was reported [25]. The hyperspectral images of intact grapes during ripening were collected using NIR hyperspectral imaging covering the spectral range between 900 and 1700 nm [25]. Calibrations were developed using MPLS as an algorithm, and an R of 0.86 and SEP values of 2.62 and 3.05 mg/g for the measurement of non-acylated and total anthocyanins in wine grape skin samples were reported by the authors [25]. NIR hyperspectral imaging has been used to determine flavanols in seeds of red (cv. Tempranillo) and white (cv. Zalema) grapes [26,27]. As reference measurements, the flavanol content was estimated using the p-dimethylaminocinnamaldehyde (DMACA) method [26,27]. Not only total flavanol content was evaluated but also the quantity of flavanols that would be extracted into the wine during winemaking [26,27]. Calibrations were developed using PLS regression yielding a R^2 of 0.73 for total flavanol content and R^2 of 0.85 for flavanols extracted using a model wine solution. Higher R^2values (0.88) were reported by the authors when grape cultivars were analysed separately [26,27]. The potential of NIR hyperspectral imaging to determine anthocyanins in intact grape has been evaluated [28]. The R^2 values reported for the concentration of total anthocyanins in red grapes were 0.65. According to the authors, the correlation value obtained was better than the value reported in another recent scientific work which estimated anthocyanin values grapes form the Cabernet Sauvignon variety [28].

ANALYSIS OF PHENOLIC COMPOUNDS IN WINE

The combination of FT-MIR and attenuated total reflectance (ATR) was explored for the determination of total phenolic, flavonoid content and antioxidant capacity (DPPH and FRAP assays) in Moscatel dessert wines (n = 56) [29]. Prediction models were developed for the referred parameters using PLS regression. The R^2 values in the calibration models ranged from 0.67–0.87 [29]. The root mean square errors of calibration (RMSEC) and cross validation (RMSECV) as well as the relative errors of prediction (REP) were calculated [29]. The minimum errors of prediction were obtained for total flavonoid content (0.2%) and maximum values (22%) for antioxidant capacity. The authors concluded that the proposed method may be used for rapid screening of total phenolic and flavonoid contents in Moscatel dessert wines [29].

The use of FT-MIR spectroscopy combined with chemometrics was also evaluated as method for correlating the spectral response of a sample to its compositional phenolic profile, as a valid alternative method to the standard UV-VIS technique [30]. In this study, the evaluation of FT-MIR combined with PLS regression was reported for the determination of 12 anthocyanins (3-O-glucosides of delphinidin, cyanidin, petunidin, peonidin and malvidin, as well as acetic acid esters and p-coumaric acid esters of petunidin, peonidin and malvidin and caffeic acid ester of malvidin) and three sums (sum of non-acylated anthocyanins, sum of acetylated anthocyanins and sum of coumaroylated anthocyanins) in red wines [31]. Reference values of anthocyanin concentrations by reverse-phase HPLC-DAD were used to calibrate the models [31]. A principal component analysis (PCA) method was applied to the reference values where a differentiation of wine samples by wine type (young wines of 2005, young wines of 2004 and crianza and reserva wines) was reported by the authors [31]. Most of the anthocyanins and their sums have been predicted with a SEP of 15%–30% for young wines. The results reported by the authors suggested that the model built using FT-MIR spectroscopy was adequate for the rapid determination of total anthocyanin content in young wines of the current vintage. However, a careful robust external validation of the technique is required in order to maintain the prediction errors within control limits for routine analysis [31]. Color components of commercial red wines such as total wine color, polymeric pigments, total anthocyanins, and copigmentation index were investigated using FT-MIR spectroscopy [32]. The composition of red wines showed great difference in terms of total color (5.07 +/− 1.95 AU at 520 nm) compared to the copigmentation index (0.66 +/− 0.58 AU at 520 nm). The prediction of total wine color, total anthocyanins, and polymeric pigments showed a good correlation of R^2 0.82; however, the copigmentation index yielded low correlation coefficients (R^2 0.57) [32].

A rapid method to quantify phenolic compounds all during the red winemaking process using FT-MIR spectroscopy and chemometrics was reported [33]. Reference values were obtained using UV-VIS spectroscopy, where total phenolic compounds (TPC), total anthocyanins (TA), and condensed tannins (CT) were measured [33]. The spectral regions selected by the authors for each model were between 979 and 2989 cm^{-1}, and the optimized calibration models yield good calibration statistics for the different parameters evaluated ($R^2 > 0.95$ and RPD > 4.0 for TPC; $R^2 > 0.90$ and RPD > 3.0 for TA; $R^2 < 0.8$ and RPD < 3.0 for CT). It was concluded by the authors that FT-MIR spectroscopy together with multivariate calibration could be a rapid and valuable tool for wineries to carry out the monitoring of phenolic compound extraction during winemaking [33].

The ability of an electronic tongue (ET) based on FT-MIR spectroscopy as a gustative sensor was assessed by emulating the responses of a tasting panel for the gustative mouthfeel "tannin amount" [34]. The FT-MIR spectra were modeled against the sensory responses evaluated in 37 red wines by means of PLS regression models. According to the authors, the iterative predictor weighting IPW-PLS technique showed the best results with the smallest RMSEC and RMSECV values (0.07 and 0.13, respectively) using 20 selected wavenumbers [34]. The use of spectroscopic analysis shows that Madeira wine age, produced from a known grape variety, can be predicted with good accuracy from its volatile and phenolic composition, as well as from the spectra collected in a UV-VIS instrument [35]. The PLS regression models estimated were able to predict wine age with a RMSECV of 0.9, 1.1, and 1.4 years, respectively. The sample-specific prediction intervals computed also allowed for the analysis of differences between observed and predicted values, and confirmed the interesting wine age prediction abilities of the proposed methodologies [35]. According to the authors, a compromise between model accuracy and cost of analysis can be reached in order to decide which methodology to use [35]. As a function of a particular application scenario, the more time-consuming and complex techniques such as GC-MS or HPLC-DAD delivered more accurate results [35]. However, satisfactory calibration and prediction statistics were obtained using UV-VIS spectroscopy as an analytical method

[35]. The use of FT-MIR spectroscopy allows fast measurement of different wine components, but quantification of tannins is difficult due to interferences from spectral responses of other wine components [36]. Four different variable selection tools were investigated for the identification of the most important spectral regions which would allow quantification of tannins from the spectra using PLS regression. The study included the development of a new variable selection tool, iterative backward elimination of changeable size intervals PLS regression. The spectral regions identified by the different variable selection methods were not identical, but all included two regions (1485–1425 and 1060–995 cm^{-1}). The spectral regions identified from the variable selection methods were used to develop calibration models. All four variable selection methods identified regions that allowed an improved quantitative prediction of tannins (RMSEP = 69–79 mg of CE/L; R= 0.93–0.94) as compared to a calibration model developed using all variables (RMSEP = 115 mg of CE/L; R = 0.87) [36].

Spectral analysis based on UV-VIS-NIR was reported as a method to analyse wine [37]. In this study, the authors determined trans-resveratrol, oenin, malvin, catechin, epicatechin, quercetin and syringic acid in commercial red wines from two Spanish regions, namely DO Rias Baixas and DO Ribeira Sacra [37]. Calibration models were developed using principal component regression (PCR) or PLS regression and HPLC as a reference method [37]. The results from this study showed that good calibration statistics using PLS regression models were obtained to quantify all polyphenol compounds from the Rias Baixas wines [37]. Intermediate PLS calibration statistics were obtained for the measurement of quercetin, epicatechin, oenin and syringic acid in wines sourced from Ribeira Sacra, and for catechin and oenin in red wines obtained using Mencia grapes [37].

The composition of phenolic compounds plays an important role in food science and nutrition; thus, there is need for a new method of analysis that is able to speed up the monitoring of product quality parameters [38]. The use of FT-MIR spectroscopy was also evaluated to measure the total antioxidant activity (TAC) of red wines. The PLS regression models showed a good predictive ability (R = 0.85) of the antioxidant activity of red wines in cross validation [38]. These authors concluded that FT-MIR spectroscopy is a promising technique to rapidly provide information on TAC of red wines and has a high potential to be implemented for the rapid screening of TAC during winemaking [38]. Tannin content and composition are critical quality components of red wines [39]. However, few spectroscopic methods assessing these phenols in wine have been described [39]. The use of FT-MIR combined with chemometric techniques was evaluated for the quantitative analysis of red wine tannins [39]. Calibration models were developed using protein precipitation and phloroglucinolysis as analytical reference methods, and after spectra preprocessing, six different predictive PLS models were evaluated by the authors [39]. The best calibration models were obtained for tannin concentration (RMSEC = 2.6%, RMSEP = 9.4%, R = 0.995) and for the prediction of the mean degree of polymerization (mDP) of the tannins (RMSEC = 6.7%, RMSEP = 10.3%, R = 0.958) [39]. Figure 1summarises the main factors or variables affecting the ability of IR spectroscopy to analyse phenolic compounds in grape and wine samples.

CONCLUDING REMARKS

The major advances in the application of spectroscopic techniques for the analysis of phenolic compounds in grapes and wines has been related to the development of powerful mathematical techniques known collectively as chemometrics. These data analysis methods allow the extraction of valuable information from large and complex data sets, underpinning the application of methods based on spectroscopy (e.g., NIR, MIR, UV, VIS). As fast and easy-to-operate techniques, spectroscopy has already gained wide industrial acceptance for routine analysis in the grape and wine industries. However, several critical aspects and limitations still exist, associated with instrument availability, type of application (e.g., grape, juice, wine) and the overall understanding of the technology. For example, one of the main factors that determine the type of instrument to be selected by the industry is the type of sample and parameter to be measured. This simple issue plays an important role in the success of a given application in determining the accuracy of the results obtained.

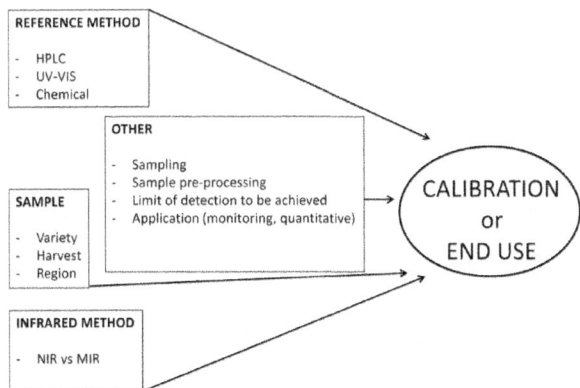

Figure 1: Factors that affect the end use of infrared methods for the analysis of phenolic compounds in grapes and wine samples.

Although, spectroscopy generally cannot measure molecules with low concentration, the indirect effects of such differences in the whole matrix of grape and wine samples can be observed or assessed in the spectrum of a given sample (e.g., fingerprint). This fingerprint, with the application of chemometric techniques (e.g., principal component analysis or discriminant analysis, PLS), can be used to elucidate particular compositional characteristics associated with phenolic compounds in grapes and wine samples not easily detected by traditional targeted chemical analysis.

ACKNOWLEDGMENTS

The financial support of GRDC is acknowledged.

REFERENCES

1.   Crozier, A.; Clifford, M.N.; Ashihara, H. Plant Secondary Metabolites. Occurrence, Structure and Role in the Human Diet; Blacwell Publishing: Oxford, UK, 2006.

2.   Kennedy, J. Grape and wine phenolics: Observations and recent findings. Cienc. Investig. Agrar.2008, 35, 107–120.

3.   Versari, A.; du Toit, W.; Parpinello, G.P. Oenological tannins: A review. Aust. J. Grape Wine Res.2013, 19, 1–10.

4.   Ignat, I.; Volf, I.; Popa, V.I. A critical review of methods for characterisation of polyphenolic compounds in fruits and vegetables. Food Chem. 2011, 126, 1821–1835.

5.   Lorrain, B.; Ky, I.; Pechamat, L.; Teissedre, P.L. Evolution of analysis of polyhenols from grapes, wines, and extracts. Molecules 2013, 18, 1076–1100.

6.   Cozzolino, D.; Dambergs, R.G. Instrumental analysis of grape, must and wine. In Managing Wine Quality: Volume One, Viticulture and Wine Quality; Reynolds, A.G., Ed.; Woodhead Publishing Ltd: Cambridge, UK, 2010.

7.   Gishen, M.; Cozzolino, D.; Dambergs, R.G. The Analysis of Grapes, Wine, and other Alcoholic Beverages by Infrared Spectroscopy. Handb. Vib. Spectrosc. 2010.

8.   Cozzolino, D.; Cynkar, W.; Janik, L.; Dambergs, R.G.; Gishen, M. Analysis of grape and wine by near infrared spectroscopy—A review. J. Near Infrared Spectrosc. 2006, 14, 279–289.

9.   Gishen, M.; Dambergs, R.G.; Cozzolino, D. Grape and wine analysis in the Australian wine industry—Enhancing the power of spectroscopy

with chemometrics. Aust. J. Grape Wine Res.2005, 11, 296–305.

10. Bauer, R.; Nieuwoudt, H.H.; Bauer, F.F.; Kossmann, J.; Koch, K.R.; Esbensen, K.H. FTIR spectroscopy for grape and wine analysis. Anal. Chem. 2008, 80, 1371–1379.

11. Cozzolino, D.; Shah, N.; Cynkar, W.U.; Smith, P. Technical solutions for analysis of grape juice, must and wine: The role of infrared spectroscopy and chemometrics. Anal. Bioanal. Chem. 2011,401, 1479–1488.

12. De Villiers, A.; Alberts, P.; Tredoux, A.G.; Nieuwoudt, H.H. Analytical techniques for wine analysis: An African perspective; a review. Anal. Chim. Acta 2012, 730, 2–23.

13. Guidetti, R.; Beghi, R.; Bodrial, L. Evaluation of Chiavennasca grapefruit technological and phenolic ripening indexes by a portable Vis-NIR device. In Agricultural and Biosystems Engineering for a Sustainable World, International Conference on Agricultural Engineering, Hersonissos, Crete, Greece, 23–25 June 2008.

14. Larrain, M.; Guesalaga, A.R.; Agosin, E. A multipurpose portable instrument for determining ripeness in wine grapes using NIR spectroscopy. IEEE Trans. Inst. Meas. 2008, 57, 294–302.

15. Beghi, R.; Mena, A.; Giovenzana, V. Quick quality evaluation of Chilean grape by a portable vis/NIR device. In Post Harvest, Food and Process Engineering, International Conference of Agricultural Engineering— CIGR-AgEng 2012: Agriculture and Engineering for a Healthier Life, Valencia, Spain, 8–12 July 2012.

16. Giovenzana, V.; Beghi, R.; Mena, A.; Civelli, R.; Guidetti, R.; Best, S.; Leon Gutierrez, L.F. Quick quality evaluation of Chilean grapes by a portable vis/NIR device. Acta Hortic. 2013, 978, 93–100.

17. Cozzolino, D.; Cynkar, W.; Dambergs, R.; Mercurio, M.; Smith, P. Measurement of condensed tannins and dry matter in red grape homogenates using near infrared spectroscopy and partial least squares. J. Agric. Food Chem. 2008, 56, 7631–7636.

18. Ferrer-Gallego, R.; Hernández-Hierro, J.M.; Rivas-Gonzalo, J.C.; Escribano-Bailón, M.T. Determination of phenolic compounds of grape skins during ripening by NIR spectroscopy. LWT Food Sci. Technol. 2011, 44, 847–853.

19. Torchio, F.; Río Segade, S.; Giacosa, S.; Gerbi, V.; Rolle, L. Effect of Growing Zone and Vintage on the Prediction of Extractable Flavanols in Winegrape Seeds by a FT-NIR Method. J. Agric. Food Chem. 2013, 61, 9076–9088.

20. Rolle, L.; Torchio, F.; Lorrain, B.; Giacosa, S.; Río Segade, S.; Cagnasso, F.; Gerbi, V.; Teissedre, P.L. Rapid methods for the evaluation of total phenol content and extractability in intact grape seeds of Cabernet-Sauvignon: Instrumental mechanical properties and FT-NIR spectrum. J. Int. Sci. Vigne Vin 2012, 46, 29–40.

21. Muganu, M.; Paolocci, M.; Gnisci, D.; Barnaba, F.E.; Bellincontro, A.; Mencarelli, F.; Grosu, I. Effect of different soil management practices on grapevine growth and on berry quality assessed by NIR-AOTF spectroscopy. Acta Hortic. 2013, 978, 117–125.

22. Fragoso, S.; Acena, L.; Guasch, J.; Busto, O.; Mestres, M. Application of FT-MIR spectroscopy for fast control of red grape phenolic ripening. J. Agric. Food Chem. 2011, 59, 2175–2183.

23. Passos, C.P.; Cardoso, S.M.; Barros, A.S.; Silva, C.M.; Coimbra, M.A. Application of Fourier transform infrared spectroscopy and orthogonal projections to latent structures/partial least squares regression for estimation of procyanidins average degree of polymerization. Anal. Chim. Acta 2010, 661, 143–149.

24. Nogales-Bueno, J.; Hernández-Hierro, J.M.; Rodríguez-Pulido, F.J.; Heredia, F.J. Determination of technological maturity of grapes and total phenolic compounds of grape skins in red and white cultivars during ripening by near infrared hyperspectral image: A preliminary approach. Food Chem. 2014, 152, 586–591.

25. Liu, X.; Wu, D.; Liang, M.; Yang, S.; Zhang, Z.; Ning, J. Multiple regression analysis of anthocyanin content of winegrape skins using hyper-spectral image technology. Trans. Chin. Soc. Agric. Mach. 2013, 44, 180–186.

26. Rodriguez-Pulido, F.J.; Hernandez-Hierro, J.M.; Nogales-Bueno, J.; Gordillo, J.; Gonzalez-Miret, L.; Heredia, F.J. A novel method for evaluating flavanols in grape seeds by near infrared hyperspectral imaging. Talanta 2014, 122, 145–150.

27. Rodríguez-Pulido, F.J.; Barbin, D.F.; Sun, D.W.; Gordillo, B.; González-Miret, M.L.; Heredia, F.J. Grape seed characterization by NIR hyperspectral imaging. Post. Biol. Technol. 2013, 76, 74–82.

28. Hernández-Hierro, J.M.; Nogales-Bueno, J.; Rodríguez-Pulido, F.J.; Heredia, F.J. Feasibility study on the use of near-infrared hyperspectral imaging for the screening of anthocyanins in intact grapes during ripening. J. Agric. Food Chem. 2013, 61, 9804–9809.

29. Silva, S.D.; Feliciano, R.P.; Boas, L.V.; Bronze, M.R. Application of FTIR-ATR to Moscatel dessert wines for prediction of total phenolic and flavonoid contents and antioxidant capacity. Food Chem. 2014, 150,

489–493.

30. Versari, A.; Parpinello, G.; Laghi, L. Application of Infrared Spectroscopy for the Prediction of Color Components of Red Wines. Spectroscopy 2012, 27, 36–39.

31. Romera-Fernandez, M.; Berrueta, L.A.; Garmon-Lobato, S.; Gallo, B.; Vicente, F.; Moreda, J.M. Feasibility study of FT-MIR spectroscopy and PLS-R for the fast determination of anthocyanins in wine. Talanta 2012, 88, 303–310.

32. Laghi, L.; Versari, A.; Parpinello, G.P.; Nakaji, D.Y.; Boulton, R.B. FTIR Spectroscopy and Direct Orthogonal Signal Correction Preprocessing Applied to Selected Phenolic Compounds in Red Wines. Food Anal. Methods 2011, 4, 619–625.

33. Fragoso, S.; Aceña, L.; Guasch, J.; Mestres, M.; Busto, O. Quantification of Phenolic Compounds during Red Winemaking Using FT-MIR pectroscopy and PLS-Regression. J. Agric. Food Chem.2011, 59, 10795–10802.

34. Vera, L.; Acena, L.; Boque, R.; Guasch, J.; Mestres, M.; Busto, O. Application of an electronic tongue based on FT-MIR to emulate the gustative mouthfeel "tannin amount" in red wines. Anal. Bioanal. Chem. 2010, 397, 3043–3049.

35. Pereira, A.C.; Reis, M.S.; Saraiva, P.M.; Marques, J.C. Madeira wine ageing prediction based on different analytical techniques: UV-vis, GC-MS, HPLC-DAD. Chemom. Intell. Lab. Syst. 2011,105, 43–55.

36. Jensen, J.S.; Egebo, M.; Meyer, A.S. Identification of spectral regions for the quantification of red wine tannins with Fourier transform mid-infrared spectroscopy. J. Agric. Food Chem. 2008,56, 3493–3499.

37. Martelo-Vidal, M.J.; Vazquez, M. Determination of polyphenolic compounds of red wines by UV-VIS-NIR spectroscopy and chemometrics tools. Food Chem. 2014, 158, 28–34.

38. Versari, A.; Parpinello, G.P.; Scazzina, F.; Del Rio, D. Prediction of total antioxidant capacity of red wine by Fourier transform infrared spectroscopy. Food Control. 2010, 21, 786–789.

39. Fernandez, K.; Agosin, E. Quantitative analysis of red wine tannins using Fourier-transform mid-infrared spectrometry. J. Agric. Food Chem. 2007, 55, 7294–7300.

Chapter 9

EXPLORATORY DATA ANALYSIS WITH LATENT SUBSPACE MODELS

José Camacho

Department of Signal Theory, Telematics and Communication, University of Granada, Granada Spain

INTRODUCTION

Exploratory Data Analysis (EDA) has been employed for decades in many research fields, including social sciences, psychology, education, medicine, chemometrics and related fields (1) (2). EDA is both a data analysis philosophy and a set of tools (3). Nevertheless, while the philosophy has essentially remained the same, the tools are in constant evolution. The application of EDA to current problems is challenging due to the large scale of the data sets involved. For instance, genomics data sets can have up to a million of variables (5). There is a clear interest in developing EDA methods to manage these scales of data while taking advantage of the basic importance of simply looking at data (3). In data sets with a large number of variables, collinear data and missing values, projection models based on latent structures, such as Principal Component Analysis (PCA) (6) (7) (1) and Partial Least Squares (PLS) (8) (9) (10), are valuable tools within EDA. Projection models and the set of tools used in combination simplify the analysis of complex data sets, pointing out to special observations (outliers), clusters of similar observations, groups of related variables, and crossed relationships between specific observations and variables.

All this information is of paramount importance to improve data knowledge. EDA based on projection models has been successfully applied in the area of chemometrics and industrial process analysis. In this chapter, several standard tools for EDA with projection models, namely score plots, loading plots and biplots, are revised and their limitations are elucidated. Two recently proposed tools are introduced to overcome these limitations. The first of them, named Missing-data methods for Exploratory Data Analysis or MEDA for short (11), is used to investigate the relationships between variables in projection subspaces. The second one is an extension of MEDA, named observation-based MEDA or oMEDA (33), to discover the relationships between observations and variables. The EDA approach based on PCA/PLS with scores and loading plots, MEDA and oMEDA is illustrated with several real examples from the chemometrics field. This chapter is organized as follows. Section 2 briefly discusses the importance of subspace models and score plots to explore the data distribution. Section 3 is devoted to the investigation of the relationship among variables in a data set. Section 4 studies the relationship between observations and variables in latent subspaces. Section 5 presents a EDA case study of Quantitative Structure-Activity Relationship (QSAR) modelling and section 6 proposes some concluding remarks. Examples and Figure were computed using the MATLAB programming environment, with the PLS-Toolbox (32) and home-made software. A MATLAB toolbox with the tools employed in this chapter is available at http://wdb.ugr.es/ josecamacho/.

PATTERNS IN THE DATA DISTRIBUTION

The distribution of the observations in a data set contains relevant information for data understanding. For instance, in an industrial process, one outlier may represent an abnormal situation which affects the process variables to a large extent. Studying this observation with more detail, one may be able to identify if it is the result of a process upset or, very commonly, a sensor failure. Also, clusters of observations may represent different operation points. Outliers, clusters and trends in the data may be indicative of the degree of control in the process and of assignable sources of variation. The identification of these sources of variation may lead to the reduction of the variance in the process with the consequent reduction of costs.

The distribution of the observations can be visualized using scatter plots. For obvious reasons, scatter plots are limited to three dimensions at most, and typically to two dimensions.

Therefore, the direct observation of the data distribution in data sets with several tens, hundreds or even thousands of variables is not possible. One can always construct scatter plots for selective pairs or thirds of variables, but this is an overwhelming and often misleading approach. Projection models overcome this problem. PCA and PLS can be used straightforwardly to visualize the distribution of the data in the latent subspace, considering only a few latent variables (LVs) which contain most of the variability of interest. Scatter plots of the scores corresponding to the LVs, the so-called score plots, are used for this purpose. Score plots are well known and accepted in the chemometric field. Although simple to understand, score plots are paramount for EDA. The following example may be illustrative of this. In Figureure 1, three simulated data sets of the same size (100 × 100) are compared. Data simulation was performed using the technique named Approximation of a DIstribution for a given COVariance matrix (15), or ADICOV for short. Using this technique, the same covariance structure was simulated for the three data sets but with different distributions: the first data set presents a multi-normal distribution in the latent subspace, the second one presents a severe outlier and the third one presents a pair of clusters. If the scatter plot of the observations in the plane spanned by the first two variables is depicted (first row of Figureure 1), the data sets seem to be almost identical. Therefore, unless an extensive exploration is performed, the three data sets may be though to come from a similar data generation procedure.

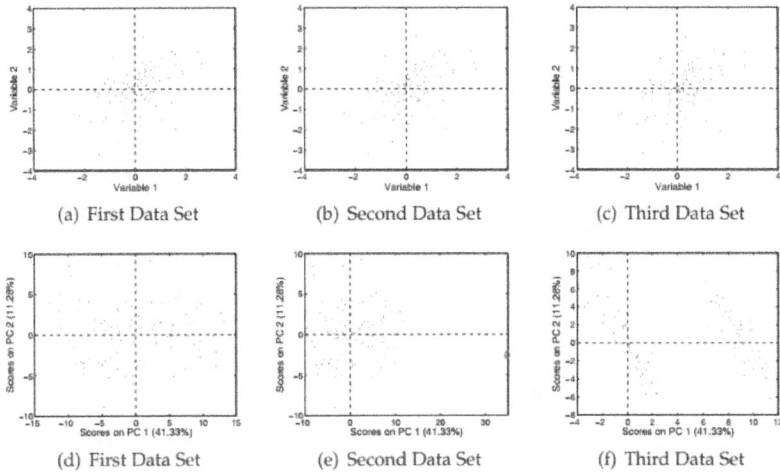

Figure 1: Experiment with three simulated data sets of dimension 100 × 100. Data simulation was performed using the ADICOV technique (15). In the first row of Figureures, the scatter plots corresponding to the first two variables in the data sets are shown. In the second row of Figureures, the scatter plots (score plots) corresponding to the first two PCs in the data sets are shown.

However, if a PCA model for each data set is fitted and the score plots corresponding to the first 2 PCs are shown (second row of Figure 1), differences among the three data sets are made apparent: in the second data set there is one outlier (right side of Figure 1(e)) and in the third data set there are two clusters of observations. As already discussed, the capability to find these details is paramount for data understanding, since outliers and clusters are very informative of the underlaying phenomena. Most of the times these details are also apparent in the original variables, but finding them may be a tedious work. Score plots after PCA modelling are perfectly suited to discover large deviations among the observations, avoiding the overwhelming task of visualizing each possible pair of original variables. Also, score plots in regression models such as PLS are paramount for model interpretation prior to prediction.

RELATIONSHIPS AMONG VARIABLES

PCA has been often employed to explore the relationships among variables in a data set (19; 20). Nevertheless, it is generally accepted that Factor Analysis (FA) is better suited than PCA to study these relationships (1; 7). This is because FA algorithms are designed to distinguish between shared and unique variability. The shared variability, the so-called communalities in the FA community, reflect the common factors–common variability–among observable variables. The unique variability is only present in one observable variable.

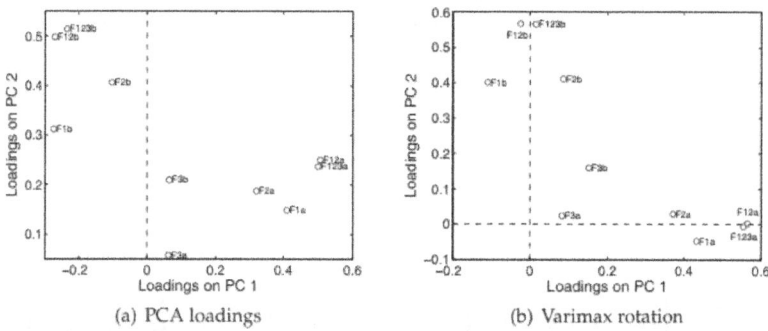

(a) PCA loadings   (b) Varimax rotation

Figure 2: Loading plots of the PCA model fitted from the data in the two pipelines example: (a) original loadings and (b) loadings after varimax rotation.

The common factors make up the relationship structure in the data. PCA makes no distinction between shared and unique variability and therefore it is not suited to find the structure in the data. When either PCA or FA are used for data understanding, a two step procedure is typically followed (1; 7). Firstly, the model is calibrated from the available data. Secondly, the model is rotated to obtain a so-called simple structure. The second step is aimed at obtaining loading vectors with as much loadings close to 0 as possible. That way, the loading vectors are easier to interpret. It is generally accepted that oblique transformations are preferred to the more simple orthogonal transformations (19; 20), although in many situations the results are similar (1). The limitation of PCA to detect common factors and the application of rotation methods will be illustrated using the pipelines artificial examples (14). Data for each pipeline are simulated according to the following equalities:

$$F12 = F1 + F2$$
$$F123 = F1 + F2 + F3$$

where F1, F2 and F3 represent liquid flows which are generated independently at random following a normal distribution of 0 mean and standard deviation 1. A 30% of measurement noise is generated for each of the five variables in a pipeline at random, following a normal distribution of 0 mean:

$$\tilde{x}_i = (x_i + \sqrt{0.3} \cdot n)/(\sqrt{1.3})$$

Where $\tilde{x}_i$ is the contaminated variable, x_1 the noise-free variable and n the noise generated. This simulation structure generates blocks of five variables with three common factors: the common factor F1, present in the observed variables F1, F12 and F123; the common factor F2, present in the observed variables F2, F12 and F123; and the common factor F3, present in F3 and F123. Data sets of any size can be obtained by combining the variables from different pipelines. In this present example, a data set with 100 observations from two pipelines for which data are independently generated is considered. Thus, the size of the data set is 100 × 10 and the variability is built from 6 common factors.

Figure 2 shows the loading plots of the PCA model of the data before and after rotation. Loading plots are interpreted so that close variables, provided they are far enough from the origin of coordinates, are considered to be correlated. This interpretation is not always correct. In Figureure 2(a), the first component separates the variables corresponding to the two pipelines. The second component captures variance of most variables, specially of those in the second pipeline. The two PCs capture variability corresponding to most common factors in the data at the same time, which complicates the interpretation. As already discussed, PCA is focused on variance, without making the distinction between unique and shared variance. The result is that the common factors are not aligned with the PCs. Thus, one single component reflects several common factors and the same common factor may be reflected in several components. As a consequence, variables with high and similar loadings in the same subset of components do not necessarily need to be correlated, since they may present very different loadings in others components. Because of this, inspecting only a pair of components may lead to incorrect conclusions. A good interpretation would require inspecting and interrelating all pairs of components with relevant information, something which may be challenging in many situations. This problem affects the interpretation and it is the reason why FA is generally preferred to PCA.

Figure 2(b) shows the resulting loadings after applying one of the most used rotation methods: the varimax rotation. Now, the variables corresponding to each pipeline are grouped towards one of the loading vectors. This highlights the fact that there are two main and orthogonal sources of variability, each one representing the variability in a pipeline. Also, in the first component variables collected from pipeline 2 present low loadings whereas in the second component variables collected from pipeline 1 present low loadings. This is the result of applying the notion of simple structure, with most of the loadings rotated towards 0. The interpretation is simplified as a consequence of improving the alignment of components with common factors. This is especially useful in data sets with many variables. Although FA and rotation methods may improve the interpretation, they still present severe limitations.

The derivation of the structure in the data from a loading plot is not straightforward. On the other hand, the rotated model depends greatly on the normalization of the data and the number of PCs (1; 21). To avoid this, several alternative approaches to rotation have been suggested. The point in common of these approaches is that they find a trade-off between variance explained and model simplicity (1). Nevertheless, imposing a simple structure has also drawbacks. Reference (11) shows that, when simplicity is pursued, there is a potential risk of simplifying even the true relationships in the data set, missing part of the data structure. Thus, the indirect improvement of data interpretation by imposing a simple structure may also report misleading results in certain situations.

MEDA

MEDA is designed to find the true relationships in the data. Therefore, it is an alternative to rotation methods or in general to the simple structure approach. A main advantage of MEDA is that, unlike rotation or FA methods, it is applied over any projection subspace without actually modifying it. The benefit is twofold. Firstly, MEDA is straightforwardly applied in any subspace of interest: PCA (maximizing variance), PLS (maximizing correlations) and any other. On the contrary, FA methods are typically based on complicated algorithms, several of which have not been extended to regression. Secondly, MEDA is also useful for model interpretation, since common factors and components are easily interrelated. This is quite useful, for instance, in the selection of the number of components. MEDA is based on the capability of missing values estimation of projection models (22–27).

The MEDA approach is depicted in Figureure 3. Firstly, a projection model is fitted from the calibration $N \times M$ matrix X (and optionally Y). Then, for each variable m, matrix X_m is constructed, which is a $N \times M$ matrix full with zeros except in the m-th column where it contains the m-th column of matrix X. Using X_m and the model, the scores are estimated with a missing data method. The known data regression (KDR) method (22; 25) is suggested at this point. From the scores, the original data is reconstructed and the estimation error computed. The variability of the estimation error is compared to that of the original data according to the following index of goodness of prediction:

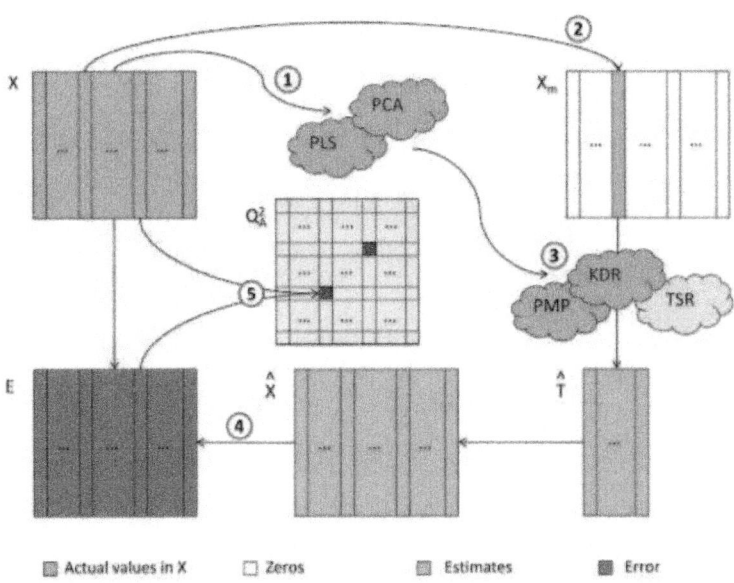

Figure 3: MEDA technique: (1) model calibration, (2) introduction of missing data, (3) missing data imputation, (4) error computation, (5) computation of matrix Q^2 A.

$$q^2_{A,(m,l)} = 1 - \frac{\|\hat{e}_{A,(l)}\|^2}{\|x_{(l)}\|^2}, \quad \forall l \neq m.$$

where $\hat{e}_{A,(l)}$ corresponds to the estimation error for the l-th variable and $x_{(l)}$ is its actual value. The closer the value of the index is to 1, the more related variables m and l are. After all the indices corresponding to each pair of variables are computed, matrix Q^2 A is formed so that $q^2_{A,(m,l)}$ is located at row m and column l. For interpretation, when the number of variables is large, matrix Q^2 A can be shown as a color map. Also, a threshold can be applied to Q^2 A so that elements over this threshold are set to 1 and elements below the threshold are set to 0.

The procedure depicted in Figureure 3 is the original and more general MEDA algorithm. Nevertheless, provided KDR is the missing data estimation technique, matrix Q^2A can be computed from cross-product matrices following a more direct procedure. The value corresponding to the element in the i-th row and j-th column of matrix $Q^2 A$ in MEDA is equal to:

$$q^2_{A,(m,l)} = \frac{2 \cdot S_{ml} \cdot S_{mlA} - (S_{mlA})^2}{S_{mm} \cdot S_{ll}}.$$
(2)

where S_{lmA} stands for the cross-product of variables x_l and x_m, i.e. $S_{lm} = x_l^T \cdot x_m$, and S_{lmA} stands for the cross-product of variables x_l and x_m^A, being x_m^A the projection of x_m in the model sub-space in coordinates of the original space.

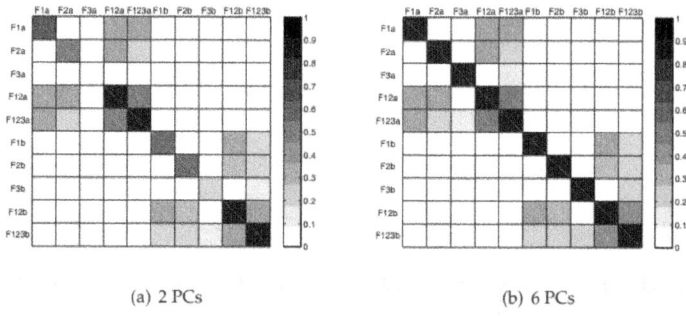

(a) 2 PCs   (b) 6 PCs

Figure. 4: MEDA matrix from the PCA model fitted from the data in the two pipelines example.

Thus, S_{lm} is the element in the l-th row and m-th column of the cross-product matrix $XX = X^T \cdot X$ and S_{lmA} corresponds to the element in the l-th row and m-th column of matrix $XX \cdot P_A \cdot P_A^t$ in PCA and of matrix $XX \cdot R_A \cdot P_A^t$ in PLS, with:

$$R_A = W_A \cdot (P_A^T \cdot W_A)^{-1}.$$
(3)

The relationship of the MEDA algorithm and cross-product matrices was firstly pointed out by Arteaga (28) and it can also be derived from the original MEDA paper (11). Equation (2) represents a direct and fast procedure to compute MEDA, similar in nature to the algorithms for model fitting from cross-product matrices, namely the eigendecomposition (ED) for PCA and the kernel algorithms (29) (30) (31) for PLS.

In Figure 4(a), the MEDA matrix corresponding to the 2 PCs PCA model of the example in the previous section, the two independent pipelines, is shown. The structure in the data is elucidated from this matrix. The separation between the two pipelines is shown in the fact that upper-right and lower-left quadrants are close to zero. The relationship among variables corresponding to factors F1 and F2 are also apparent in both pipelines. Since the variability corresponding to factors F3 is barely captured by the first 2 PCs, these are not reflected in the matrix. Nevertheless, if 6 PCs are selected, (Figureure 4(b)) the complete structure in the data is clearly found. MEDA improves the interpretation of both the data set and the model fitted without actually pursuing a simple structure. The result is that MEDA has better properties than rotation methods: it is more accurate and its performance is not deteriorated when the number of PCs is overestimated. Also, the output of MEDA does not depend on the normalization of the loadings, like rotated models do, an it is not limited to subspaces with two or three components at most. A comparison of MEDA with rotation methods is out of the scope of this chapter. Please refer to (11) for it and also for a more algorithmic description of MEDA.

Loading plots and MEDA

The limitations of loading plots and the application of MEDA were introduced with the pipelines artificial data set. This is further illustrated in this section with two examples provided with the PLS-toolbox (32): the Wine data set, which is used in the documentation of the cited software to show the capability of PCA for improving data understanding, and the PLSdata data set, which is used to introduce regression models, including PLS.

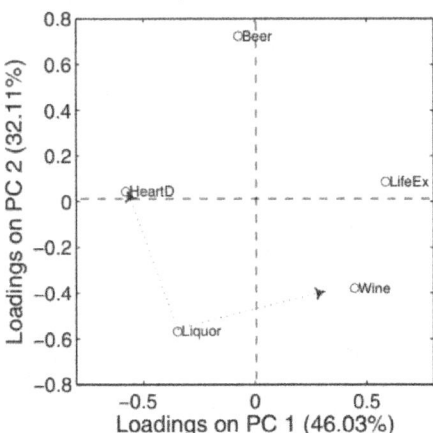

Figure. 5: Loading plot of the first 2 PCs from the Wine Data set provided with the PLS-toolbox (32).

The reading of the analysis and discussion of both data sets in (32) is recommended.As suggested in (32), two PCs are selected for the PCA model of the Wine data set. The corresponding loading plot is shown in Figureure 5. According to the reference, this plot shows that variables HeartD and LifeEx are negatively correlated, being this correlation captured in the first component. Also, "wine is somewhat positively correlated with life expectancy, likewise, liquor is somewhat positively correlated with heart disease". Finally, bear, wine and liquor form a triangle in the Figure, which "suggests that countries tend to trade one of these vices for others, but the sum of the three tends to remain constant". Notice that although these conclusions are interesting, some of them are not so evidently shown by the plot. For instance, Liquor is almost as close to HeartD than to Wine. Is Liquor correlated to Wine as it is to HeartD?

MEDA can be used to improve the interpretation of loading plots. In Figureure 6(a), the MEDA matrix for the first PC is shown. It confirms the negative correlation between HeartD and LifeEx, and the–lower–positive correlation between HeartD and Liqour and LifeEx and Wine. Notice that these three relationships are three different common factors. Nevertheless, they all manifest in the same component, making the interpretation with loading plots more complex. The MEDA matrix for the second PC in Figureure 6(b) shows the relationship between the three types of drinks. The fact that the second PC captures this relationship was not clear in the loading plot. Furthermore, the MEDA matrix shows that Wine and Liquor are not correlated, answering to the question in the previous paragraph. Finally, this absence of correlation refutes that countries tend to trade wine for liquor or viceversa, although this effect may be true for bear. In the PLSdata data set, the aim is to obtain a regression model that relates 20 temperatures measured in a Slurry-Fed Ceramic Melter (SFCM) with the level of molten glass. The x-block contains 20 variables which correspond to temperatures collected in two vertical thermowells. Variables 1 to 10 are taken from the bottom to the top in thermowell 1, and variables 11 to 20 from the bottom to the top in thermowell 2. The data set includes 300 training observations and 200 test observations.

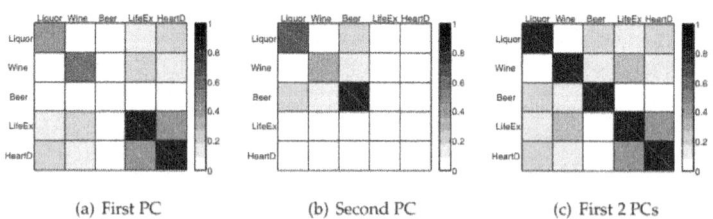

(a) First PC   (b) Second PC   (c) First 2 PCs

Figure. 6: MEDA matrices of the first PCs from the Wine data set provided with the PLS-toolbox (32).

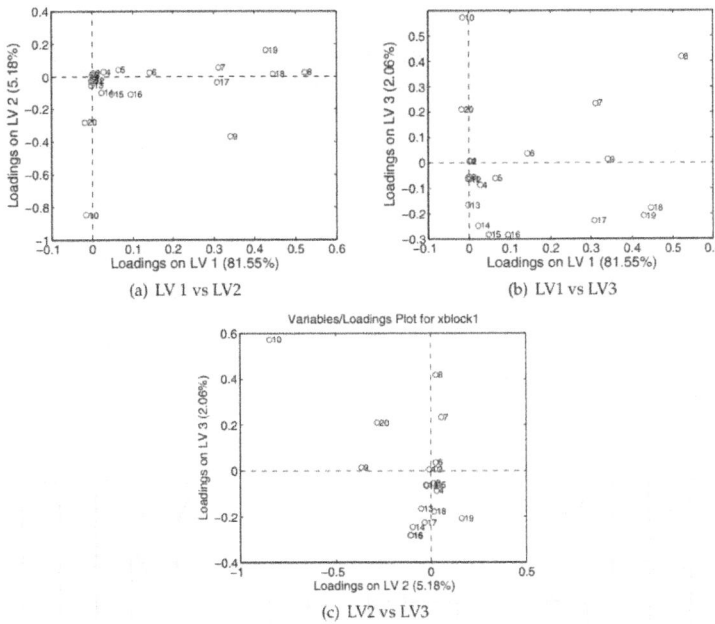

Figure. 7: Loading plots from the PLS model in the Slurry-Fed Ceramic Melter data set.

This same data set was used to illustrate MEDA with PCA in (11) with the temperatures and the level of molten glass together in the same block of data [1].

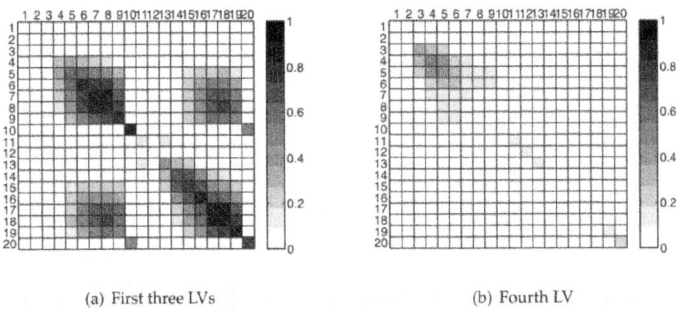

Figure. 8: MEDA matrices from the PLS model in the Slurry-Fed Ceramic Melter data set.

Following recommendations in (32), a 3 LVs PLS model from a mean-centered x-block was fitted. The loading plots corresponding to the three possible pairs of LVs are shown in Figure 7. The first LV captures the predictive variance in the temperatures, with higher loadings for higher indices of the temperatures in each thermowell. This holds exception made on temperatures 10 and 20, which present their predictive variance in the second and third LVs, being the variance in 10 between three and four times higher than that in 20. The third LVs also seems to discriminate between both thermowells. For instance, in Figure 7(c) most temperatures of the first thermowell are in the upper middle of the Figure, whereas most temperatures of the second thermowell are in the lower middle. Nevertheless, it should be noted that the third LV only captures 2% of the variability in the x-block and 1% of the variability in the y-block. The information gained with the loading plots can be complemented with that in MEDA, which brings a much clearer picture of the structure in the data. The MEDA matrix for the PLS model with 3 LVs is shown in Figureure 8(a). There is a clear auto-correlation effect in the temperatures, so that closer sensors are more correlated. This holds exception made on temperatures 10 and 20. Also, the corresponding temperatures in both thermowells are correlated, including 10 and 20. Finally, the temperatures at the bottom do not contain almost any predictive information of the level of molten glass. In (32), the predictive error by cross-validation is used to identify the number of LVs. Four LVs attain the minimum predictive error, but 3 LVs are selected since the fourth LV does not contribute much to the reduction of this error. In Figureure 8(b), the contribution of this fourth LV is shown. It is capturing the predictive variability from the third to the fifth temperature sensors in the first thermowell, which are correlated. The corresponding variability in the second thermowell is already captured by the first 3 LVs. This is an example of the capability of MEDA for model interpretation, which can be very useful in the determination of the number of LVs. In this case and depending on the application of the model, the fourth LV may be added to the model in order to compensate the variance captured in both thermowells, even if the improvement in prediction performance is not high.

The information provided by MEDA can also be useful for variable selection. In this example, temperatures 1-2 and 11-12 do not contribute to a relevant degree to the regression model. As shown in Table 1, if those variables are not used in the model, its prediction performance remains the same.

Also, considering the correlation among thermowells, one may be tempted to use only one of the thermowells for prediction, reducing the associated costs of maintaining two thermowells. If this is done, only 8 predictor variables are used and the prediction performance is reduced, but not to a large extent. Correlated variables in a prediction model help to better discriminate between true structure and noise. For instance, in this example, when only the sensors of one thermowell are used, the PLS model captures more x-block variance and less y-block variance. This is showing that more specific–noisy–variance in the x-block is being captured. Using both thermowells reduces this effect. Another example of variable selection with MEDA will be presented in Section 5.

Table 1: Comparison of three PLS models in the Slurry-Fed Ceramic Melter data set. The variance in both blocks of data and the Root Mean Square Error of Calibration (RMSEC), Cross-validation (RMSECV) and Prediction (RMSEP) are compared.

Variables	Complete model	[3 : 10 13 : 20]	[3 : 10]	[13 : 20]
LVs	3	3	3	3
X-block variance	88.79	89.84	96.36	96.93
Y-block variance	87.89	87.78	84.61	83.76
RMSEC	0.1035	0.1039	0.1166	0.1198
RMSECV	0.1098	0.1098	0.1253	0.1271
RMSEP	0.1396	0.1394	0.1522	0.1631

Correlation matrices and MEDA

There is a close similarity between MEDA and correlation matrices. To this regard, equation (2) simplifies the interpretation of the MEDA procedure. The MEDA index combines the original variance with the model subspace variance in S_{ml} and $S_{ml}{}^A$. Also, the denominator of the index in eq. (2) is the original variance. Thus, those pairs of variables where a high amount of the total variance of one of them can be recovered from the other are highlighted. This is convenient for data interpretation, since only factors of high variance are highlighted. On the other hand, it is easy to see that when the number of LVs, A, equals the rank of X, then Q^2 A is equal to the element-wise squared correlation matrix of X, C^2 (11). This can be observed in the following element-wise equality:

$$q^2_{Rank(X),(m,l)} = \frac{S^2_{ml}}{S_{mm} \cdot S_{ll}} = c^2_{(m,l)}.$$

(4)

This equivalence shows that matrix Q^2 A has a similar structure than the–element-wise squared–correlation matrix. To elaborate this similarity, a correlation matrix can be easily extended to the notion of latent subspace. The correlation matrix in the latent subspace, CA, can de defined as the correlation matrix of the reconstruction of X with the first A LVs. Thus, $C_A = P_A \cdot P_A^t \cdot C \cdot P_A \cdot P_A^t$ in PCA and $C_A = P_A \cdot R_A^t \cdot C \cdot R_A \cdot P_A^t$ i. If the elements of CA are then squared, the element-wise squared correlation in the latent subspace, noted as C^2 A, is obtained. Strictly speaking, each element of C^2 A is defined as

$$c^2_{A,(m,l)} = \frac{S^2_{m^A l^A}}{S_{m^A m^A} \cdot S_{l^A l^A}}.$$

(5)

However, for the same reason explained before, if C^2A is aimed at data interpretation, the denominator should be original variance:

$$c^2_{A,(m,l)} = \frac{S^2_{m^A l^A}}{S_{mm} \cdot S_{ll}}.$$

(6)

If this equation is compared to equation (2), we can see that the main difference between MEDA and the–projected and element-wise squared–correlation matrix is the combination of original and projected variance in the numerator of the former. This combination is paramount for interpretation. Figureure 9 illustrates this. The example of the pipelines is used again, but in this case ten pipelines and only 20 observations are considered, yielding a dimension of 20×50 in the data. Two data sets are simulated. In the first one, the pipelines are correlated. As a consequence, the data present three common factors represented by the three biggest eigenvalues in Figure 9(a).

In the second one, each pipeline is independently generated, yielding a more distributed variance in the eigenvalues (Figure 9(b)). For matrices $Q^2 A$ and C^2A to infer the structure in the data, they should have large values in the elements which represent real structural information (common factors) and low values in the rest of the elements. Since in both data sets it is known a-priori which elements in the matrices represent actual common factors and which not, the mean values for the two groups of elements in matrices $Q^2 A$ and C^2A can be computed. The ratio of these means, computed by dividing the mean of the elements with common factors by the mean of the elements without common factors, is a measure of the discrimination capability between structure and noise of each matrix. The higher this index is, the better the discrimination capability is. This ratio is shown in Figureures 9(c) and 9(d) for different numbers of PCs. Q^2A outperforms C^2A until all relevant eigenvalues are incorporated to the model. Also, $Q^2 A$ presents maximum discrimination capability for a reduced number of components. Notice that both alternative definitions of C^2A in equations (5) and (6) give exactly the same result, though equation (6) is preferred for visual interpretation.

CONNECTION BETWEEN OBSERVATIONS AND VARIABLES

The most relevant issue for data understanding is probably the connection between observations and variables. It is almost useless to detect certain details in the data distribution, such as outliers or clusters, if the set of variables related to these details are not identified. Traditionally, biplots (12) have been used for this purpose. In biplots, the scatter plots of loadings and scores are combined in a single plot. Apart from relevant considerations regarding the comparability of the axes in the plot, which is also important for any scatter plots, and of the scales in scores and loadings (18), biplots may be misleading just because of the loading plot included. In this point, a variant of MEDA, named observation-based MEDA or oMEDA, can be used to unveil the connection between observations and variables without the limitations of biplots.

oMEDA

oMEDA is a variant of MEDA to connect observations and variables. Basically, oMEDA is a MEDA algorithm applied over a combination of the original data and a dummy variable designed to cover the observations of interest. Take the following example: a number of subsets of observations $\{C_1, ..., C_N\}$ form different clusters in the scores plot which are located far from the bulk of the data, L. One may be interested in identifying, for instance, the variables related to the deviation of C_1 from L without considering the rest of clusters.

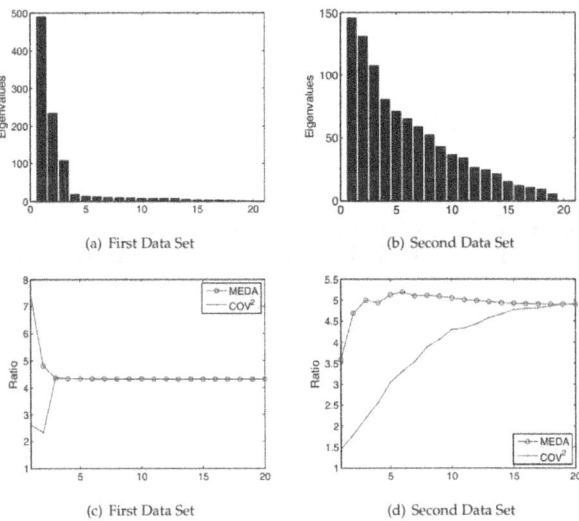

(a) First Data Set

(b) Second Data Set

(c) First Data Set

(d) Second Data Set

Figure. 9: Comparison of MEDA and the projected and element-wise squared co-variance matrix in the identification of the data structure from the PCA model fitted from the data in the ten pipelines example: (a) and (b) show the eigenvalues when the pipelines are correlated and independent, respectively, and (c) and (d) show the ratio between the mean of the elements with common factors and the mean of the elements without common factors in the matrices.

For that, a dummy variable d is created so that observations in C_1 are set to 1, observations in L are set to -1, while the remaining observations are left to 0. Also, values other than 1 and -1 can be included in the dummy variable if desired. oMEDA is then performed using this dummy variable.

The oMEDA technique is illustrated in Figureure 10. Firstly, the dummy variable is designed and combined with the data set. Then, a MEDA run is performed by predicting the original variables from the dummy variable. The result is a single vector, $d^2 A$, of dimension $M \times 1$, being M the number of original variables. In practice, the oMEDA index is slightly different to that used in MEDA. Being d the dummy variable, designed to compare a set of observations with value 1 (or in general positive values) with another set with value -1 (or in general negative values), then the oMEDA index follows:

$$d^2_{A,(l)} = \|x^d_{(l)}\|^2 - \|\hat{e}^d_{A,(l)}\|^2, \quad \forall l.$$

(7)

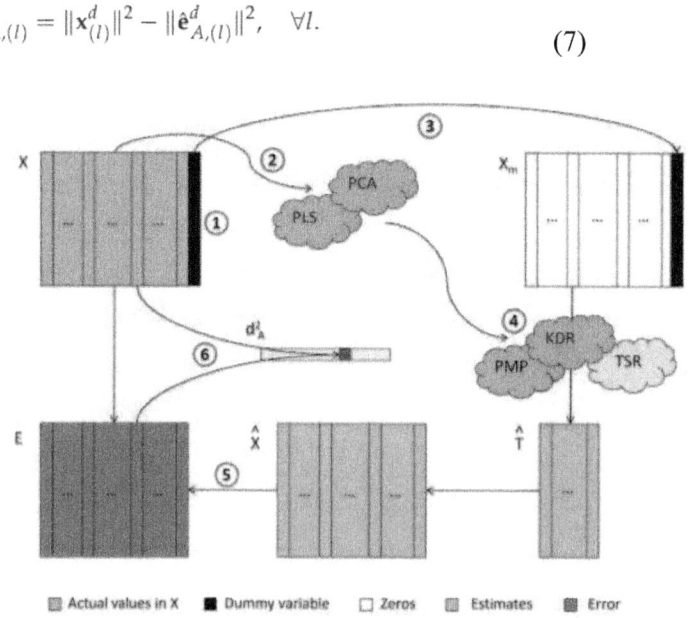

Figure. 10: oMEDA technique: (1) introduction of the dummy variable, (2) model calibration, (3) introduction of missing data, (4) missing data imputation, (5) error computation, (6) computation of vector $d^2 A$.

where $x^d_{(l)}$ represents the values of the l-th variable in the original observations different to 0 in d and 0 in d and $\hat{e}^d_{A,(l)}$ is the corresponding estimation error. The main difference between the computation of index $d^2_{A,(l)}$ in oMEDA and that of MEDA is the absence of the denominator in the former. This modification is convenient to avoid high values in $d^2_{A,(l)}$ when the amount of variance of a variable in the reduced set of observations of interest is very low. Once $d^2 A$ is computed for a given dummy variable, sign information can be added from the mean vectors of the two groups of observations considered (33).

In practice, in order to avoid any modification in the PCA or PLS subspace due to the introduction of the dummy variable, the oMEDA algorithm is slightly more complex than the procedure shown in Figureure 10. For a description of this algorithm refer to (33). However, like in MEDA, the oMEDA vector can be computed in a more direct way by assuming KDR (26) is used as the missing data estimation procedure. If this holds, the oMEDA vector follows:

$$d^2_{A,(l)} = 2 \cdot x^t_{(l)} \cdot D \cdot x_{A,(l)} - x^t_{A,(l)} \cdot D \cdot x_{A,(l)}, \tag{8}$$

where $x_{(l)}$ represents the l-th variable in the–complete–set of original observations and $x_{A,(l)}$ its projection in the latent subspace in coordinates of the original space and:

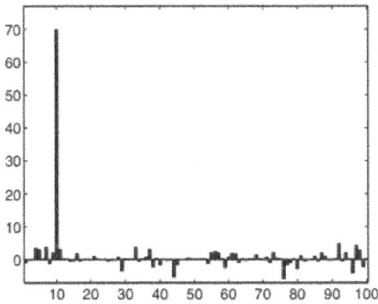

Figure. 11: oMEDA vector of two clusters of data from the 10 PCs PCA model of a simulated data set of dimension 100×100. This model captures 30% of the variability. Data present two clusters in variable 10.

$$D = \frac{d \cdot (d)^T}{\|d\|^2}. \tag{9}$$

Finally, the equation can also be reexpressed as follows:

$$d^2_{A,(l)} = \frac{1}{N} \cdot (2 \cdot \Sigma^d_{(l)} - \Sigma^d_{A,(l)}) \cdot |\Sigma^d_{A,(l)}| \tag{10}$$

with $\Sigma^d_{(l)}$ and $\Sigma^d_{A,(l)}$ being the weighted sum of elements in $x_{(l)}$ and $x_{A,(l)}$ according to the weights in d, respectively. Equation (10) has two advantages. Firstly, it presents the oMEDA vector as a weighted sum of values, which is easier to understand. Secondly, it has the sign computation built in, due to the absolute value in the last element. Notice also that oMEDA inherits the combination of total and projected variance present in MEDA.

In Figure 11 an example of oMEDA is shown. For this, a 100×100 data set with two clusters of data was simulated. The distribution of the observations was designed so that both clusters had significantly different values only in variable 10 and then data was auto-scaled. The oMEDA vector clearly highlights variable 10 as the main difference between both clusters.

Biplots vs oMEDA

Let us return to the discussion regarding the relationship between the common factors and the components. As already commented, several common factors can be captured by the same component in a projection model. As a result, a group of variables may be located close in a loading plot without the need to be correlated. This is also true for the observations. Thus, two observations closely located in a score plot may be quite similar or quite different depending on their scores in the remaining LVs. However, score plots are typically employed to observe a general distribution of the observations. This exploration is more aimed at finding differences among observations rather than similarities. Because of this, the problem described for loading plots is not so relevant for the typical use of score plots. However, this is a problem when interpreting biplots. In biplots, deviations in the observations are related to deviations in the variables. Like loading plots, biplots may be useful to perform a fast view on the data, but any conclusion should be confirmed with another technique. oMEDA is perfectly suited for this.

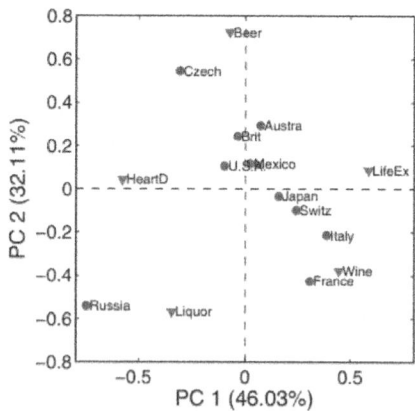

Figure. 12: Biplot of the first 2 PCs from the Wine Data set provided with the PLS-toolbox (32).

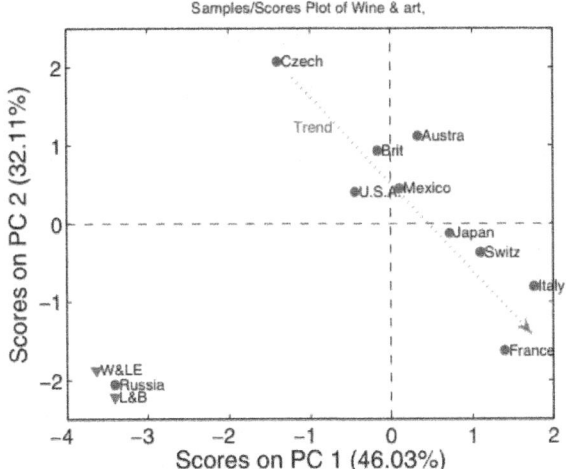

Figure. 13: Score plot of the first 2 PCs from the Wine Data set provided with the PLS-toolbox (32) and two artificial observations.

In Figure 12, the biplot of the Wine data set is presented. The distribution of the scores show that all countries but Russia follow a trend from the Czech Republic to France, while Russia is located far from this trend. The biplot may be useful to make hypothesis on the variables related to the special nature of Russia or to the trend in the rest of countries. Nevertheless, this hypothesis making is not straightforward. To illustrate this, in Figure 13 the scores are shown together with two artificial observations. The artificial observations were designed to lay close to Russia in the score plot of the first two PCs. In both cases, three of the five variables were left to their average value in the Wine data set and only two variables are set to approach Russia. Thus, observation W&LE only uses variables Wine and LifeEx to yield a point close to Russia in the score plot while the other variables are set to the average. Observation L&B only uses Liquor and Beer. With this example, it can be concluded that very little can be said about Russia only by looking at the biplot.

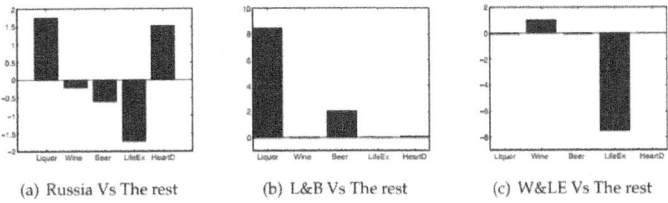

(a) Russia Vs The rest   (b) L&B Vs The rest   (c) W&LE Vs The rest

Figure. 14: oMEDA vectors of the first 2 PCs from the Wine Data set provided with the PLS-toolbox (32). Russia (a), L&B (b) and W&LE (c) compared to the rest of countries

In Figure 14(a), the oMEDA vector to discover the differences between Russia and the rest of countries is shown. For this, a dummy variable is built where all observations except Russia are set to -1 and Russia is set to 1. oMEDA shows that Russia has in general less life expectancy and more heart disease and liquor consumption than the rest of countries. The same experiment is repeated for artificial observations L&B and W&LE in Figure 14(b) and 14(c). oMEDA clearly distinguishes among the three observations, while in the biplot they seem to be very similar.

To analyze the trend shown by all countries except Russia in Figure 12, the simplest approach is to compare the most separated observations, in this case France and the Czech Republic. The oMEDA vector is shown in Figure 15(a). In this case, the dummy variable is built so that France has value 1, the Czech Republic has value -1 and the rest of the countries have 0 value. Thus, positive values in the oMEDA vector identify variables with higher value in France than in the Czech Republic and negative values the opposite. oMEDA shows that the French consume more wine and less beer than Czech people. Also, according to the data, the former seem to be more healthy. Comparing the two most separated observations may be misleading in certain situations. Another choice is to use the capability of oMEDA to unveil the variables related to any direction in a score plot. For instance, let us analyze the trend of the countries incorporating the information in all the countries. For this, different weights are considered in the dummy variable. We can think of these weights as-approximate-projections of the observations in the direction of interest. Following this approach, the weights listed in Table 2 are assigned, which approximate the projection of the countries in the imaginary line depicted by the arrow in Figure 13. Since Russia is not in the trend, it is left to 0. Using these weights, the resulting oMEDA vector is shown in Figure 15(b). In this case, the analysis of the complete set of observations in the trend resembles the conclusions in the analysis of the two most separated observations.

Table 2: Weights used in the dummy variable for the oMEDA vector in Figureure 15(b)

Country	Weight	Country	Weight
France	3	Mexico	-1
Italy	2	U.S.A.	-1
Switz	1	Austra	-1
Japan	1	Brit	-1
Russia	0	Czech	-3

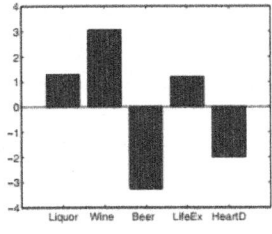
(a) France Vs Czech

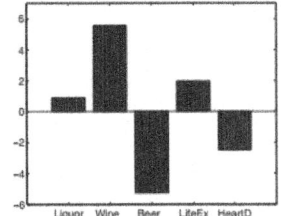
(b) All except Russia

Figure. 15: oMEDA vectors of the first 2 PCs from the Wine Data set provided with the PLS-toolbox (32). In (a), France and Czech Republic are compared. In (b), the trend shown in the score plot by all countries except Russia is analyzed.

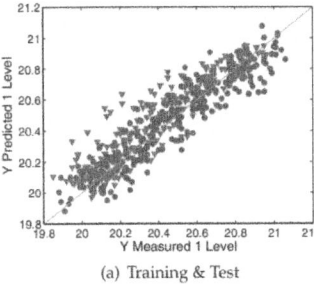
(a) Training & Test

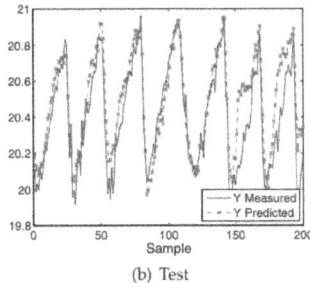
(b) Test

Figure. 16: Measured vs predicted values of molten glass level in the PLS model with 3 LVs fitted from the Slurry-Fed Ceramic Melter (SFCM) data set in (32).

Let us return to the PLSdata data set. A PLS model relating temperatures with the level of molten glass was previously fitted. As already discussed, the data set includes 300 training observations and 200 test observations. The measured and predicted values of both sets of observations are compared in Figureure 16(a). The predicted values in the test observations (inverted triangles) tend to be higher than true values. This is also observed in Figure 16(b). The cause for this seems to be that the process has slightly moved from the operation point where training data was collected. oMEDA can be used to identify this change of operation point by simply comparing training and test observations in the model subspace. Thus, training (value 1) and test observations (value -1) are compared in the subspace spanned by the first 3 LVs of the PLS model fitted only from training data. The resulting oMEDA vector is shown in Figure 17. According to the result, considering the test observations have value -1 in the dummy variable, it can be concluded that the process has moved to a situation in which top temperatures are higher than during model calibration.

CASE STUDY: SELWOOD DATA SET

In this section, an exploratory data analysis of the Selwood data set (34) is carried out. The data set was downloaded from http://michem.disat.unimib.it/chm/download/datasets.htm.

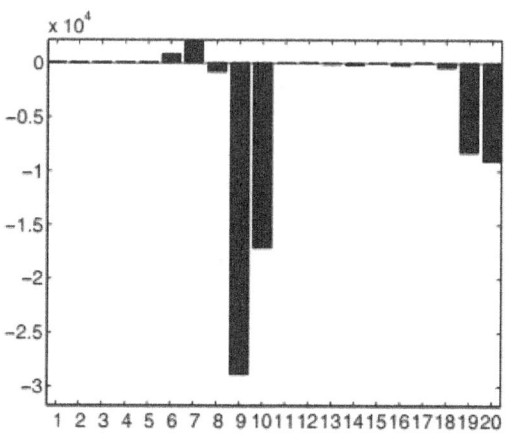

Figure. 17: oMEDA vector comparing training and test observations in the PLS model with 3 LVs fitted from the Slurry-Fed Ceramic Melter (SFCM) data set in (32).

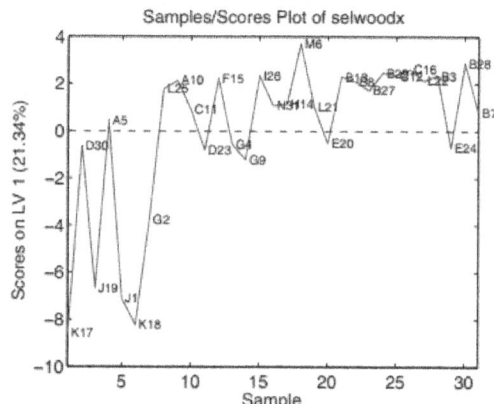

Figure. 18: Scores corresponding to the first LV in the PLS model with the complete set of descriptors of the Selwood dataset.

It consists of 31 antifilarial antimycin A_1 analogues for which 53 physicochemical descriptors were calculated for Quantitative Structure-Activity Relationship (QSAR) modelling. The set of descriptors is listed in

Table 3. These descriptors are used for predicting in vitro antifilarial activity (-LOGEC50). This data set has been employed for testing variables selection methods, for instance in (35; 36), in order to find a reduced number of descriptors with best prediction performance. Generally speaking, these variable selection methods are based on complex optimization algorithms which make use of heuristics to reduce the search space.

Table 3: Physicochemical descriptors of the Selwood dataset

Indices	Descriptors
1:10	ATCH1 ATCH2 ATCH3 ATCH4 ATCH5 ATCH6 ATCH7 ATCH8 ATCH9 ATCH10
11:20	DIPV_X DIPV_Y DIPV_Z DIPMOM ESDL1 ESDL2 ESDL3 ESDL4 ESDL5 ESDL6
21:30	ESDL7 ESDL8 ESDL9 ESDL10 NSDL1 NSDL2 NSDL3 NSDL4 NSDL5 NSDL6
31:40	NSDL7 NSDL8 NSDL9 NSDL10 VDWVOL SURF_A MOFI_X MOFI_Y MOFI_Z PEAX_X
41:50	PEAX_Y PEAX_Z MOL_WT S8_1DX S8_1DY S8_1DZ S8_1CX S8_1CY S8_1CZ LOGP
51:53	M_PNT SUM_F SUM_R

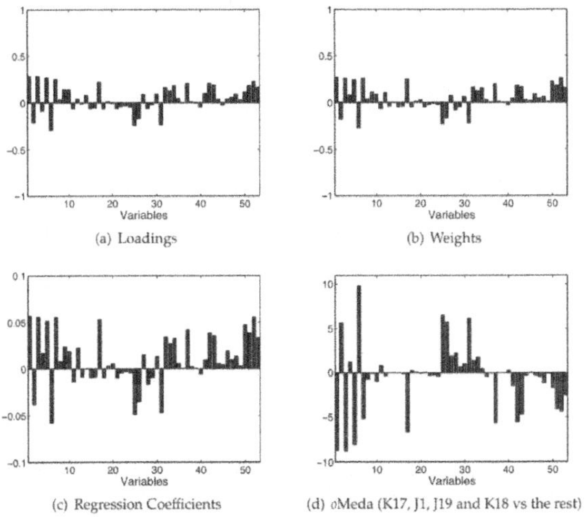

(a) Loadings   (b) Weights

(c) Regression Coefficients   (d) οMeda (K17, J1, J19 and K18 vs the rest)

Figure. 19: Several vectors corresponding to the first LV in the PLS model with the complete set of descriptors of the Selwood dataset.

First of all, a PLS model is calibrated between the complete set of descriptors and -LOGEC50. Leave-one-out cross-validation suggests one LV. The score plot corresponding to the 31 analogues in that LV are shown in Figure 18. Four of the compounds, namely K17, J1, J19 and K18, present an abnormally low score. This deviation is highly contributing to the variance in the first LV and the reason for it should be investigated. Two of these compounds, K17 and K18, were catalogued as outliers by (34), where the authors stated that "Chemically, these compounds are distinct from the bulk of the training set in that they have an n-alkyl side chain as opposed to a side chain of the phenoxy ether type". Since the four compounds present an abnormally low score in the first LV, typically the analyst may interpret the coefficients of that LV to try to explain this abnormality. In Figure 19, the loadings, weights and regression coefficients of the PLS model are presented together with the oMEDA vector. The latter identifies those variables related to the deviation of the four compounds from the rest. The oMEDA vector is similar, but with opposite sign, to the other vectors in several descriptors, but quite different in others. Therefore, the loadings, weights or coefficient vectors should not be used in this case for the investigation of the deviation, or otherwise one may arrive to incorrect conclusions. On the other hand, it may be worth to check whether the oMEDA vector is representative of the deviation in the four compounds. Performing oMEDA individually on each of the compounds confirm this fact (see Figure 20)

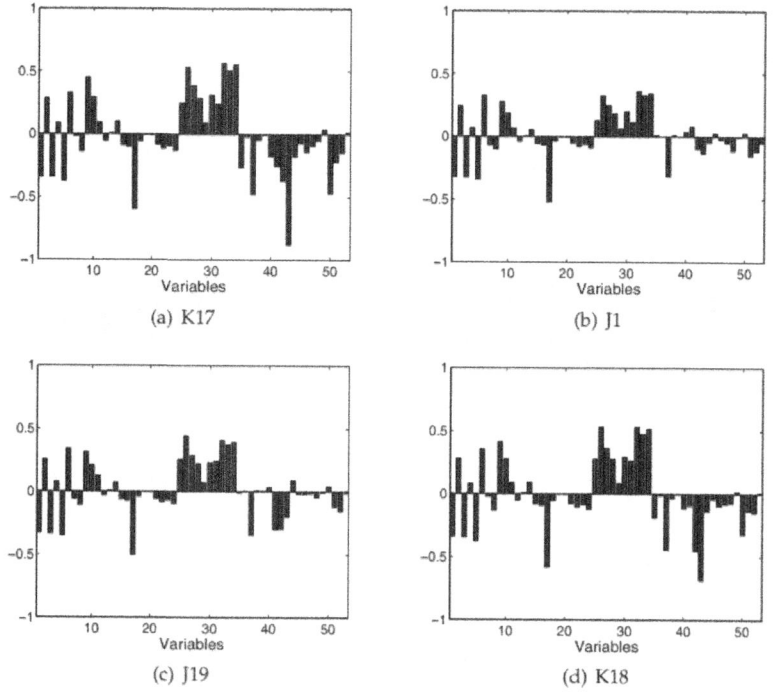

Figure. 20: oMEDA vectors corresponding to the first LV in the PLS model with the complete set of descriptors of the Selwood dataset. To compute each of the vectors, one of the four compounds K17, J1, J19 and K18 are set to 1 in the dummy variable and the other three are set to 0, while the rest of compounds in the data set are set to -1.

The subsequent step is to search for relevant descriptors (variable selection) For this, MEDA will be employed. In this point, there are two choices. The four compounds with low score in the first LV may be treated as outliers and separated from the rest of the data (34) or the complete data set may be modelled with a single QSAR model (35; 36). It should be noted that differences among observations in one model may not be found in a different model, so that the same observation may be an outlier or a normal observation depending on the model. Furthermore, as discussed in (35), the more general the QSAR model is, so that it models a wider set of compounds, the better. Therefore, the complete set of compounds will be considered in the remaining of the example. On the other hand, regarding the analysis tools used, there are different possibilities. MEDA may be applied over the PLS model relating the descriptors in the x-block and -LOGEC50 in the y-block. Alternatively, both blocks may be joined together in a single block of data and MEDA with PCA be applied. The second choice will be generally preferred to avoid over-fitting, but typically both approaches may lead to the same conclusions, like it happens in the present example

The application of MEDA requires to select the number of PCs in the PCA model. Considering that the aim is to understand how the variability in -LOGEC50 is related to the descriptors in the data set, the Structural and Variance Information (SVI) plots are the adequate analysis tool (14).

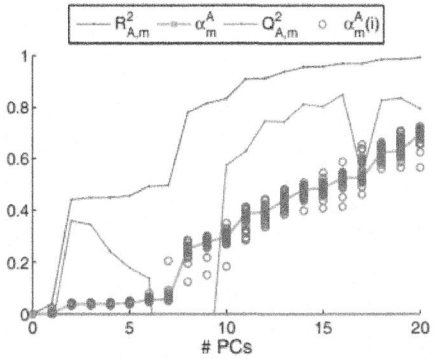

Figure. 21: Structural and Variance Information (SVI) plot of in vitro antifilarial activity (-LOGEC50). The data set considered combines -LOGEC50 with the complete set of descriptors of the Selwood dataset.

The SVI plots combine variance information with structural information to elucidate how a PCA model captures the variance of a single variable. The SVI plot of a variable v reveals how the following indices evolve with the addition of PCs in the PCA model:

- The R^2 statistic, which measures the variance of v.
- The Q^2 statistic, which measures the performance of the missing data

imputation of v, or otherwise stated its prediction performance.

- The α statistic, which measures the portion of variance of v which is identified as unique variance, i.e. variance not shared with other variables.

- The stability of α, as an indicator of the stability of the model calibration.

Figure 21 shows the SVI plot of -LOGEC50 in the PCA model with the complete set of descriptors. The plot shows that the model remains quite stable until 5-6 PCs are included. This is seen in the closeness of the circles which represents the different instances of α computed on a leave-one-out cross-validation run. The main portion of variability in -LOGEC50 is captured in the second and eighth PCs. Nevertheless, is not until the tenth PC that the missing data imputation (Q^2) yields a high value. For more PCs, the captured variability is only slightly augmented. Since MEDA makes use of the missing data imputation of a PCA model, Q^2 is a relevant index. At the same time, from equation (2) is clear that MEDA is also influenced by captured variance. Thus, 10 PCs are selected. In any case, it should be noted that MEDA is quite robust to the overestimation in the number of PCs (11) and very similar MEDA matrices are obtained for 3 or more PCs in this example.

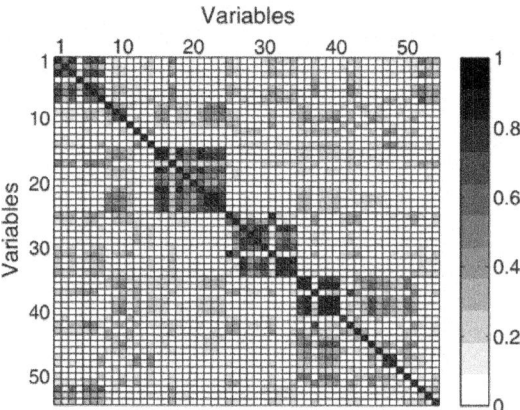

Figure. 22: MEDA matrix of the PCA model with 10 PCs from the data set which combines the in vitro antifilarial activity (-LOGEC50) with the complete set of descriptors of the Selwood dataset.

The MEDA matrix corresponding to the PCA model with 10 PCs from the data set which combines -LOGEC50 with the complete set of descriptors of the Selwood dataset is presented in Figure 22. For variable selection, the most relevant part of this matrix is the last column (or row), which corresponds to -LOGEC50. This vector is shown in Figure 23(a). Those descriptors with high value in this vector are the ones from which -LOGEC50 can be better predicted. Nevertheless, the selection of, say, the first n variables with higher value is not an adequate strategy because the relationship among the descriptors should also be considered. Let us select the descriptor with better prediction performance, in this case ATCH6, though

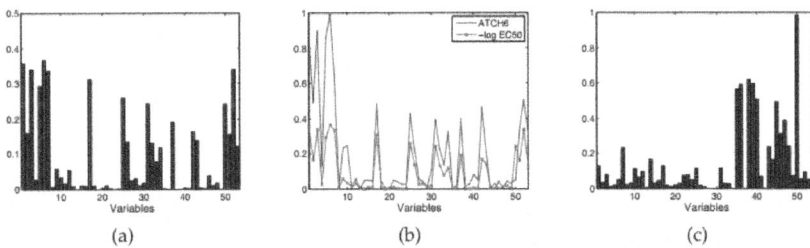

(a)   (b)   (c)

Figure. 23: MEDA vector corresponding to the in vitro antifilarial activity (-LO-GEC50) (a) comparison between this vector and that corresponding to ATCH6 (b) and MEDA vector corresponding to LOGP (c) in the PCA model with 10 PCs from the data set which combines -LOGEC50 with the complete set of descriptors of the Selwood dataset.

ATCH1, ATCH3, ATCH7 or SUM_F have a very similar prediction performance. The least-squares regression model with ATCH6 as regressor attains a Q^2 equal to 0.30, more than the Q^2 attained by any number of LVs in the PLS model with the complete set of descriptors. If for instance ATCH6 and ATCH1 are used as regressors, $Q^2 = 0.26$ is obtained for least squares regression and $Q^2 = 0.31$ for PLS with 1 LV, which means an almost negligible improvement with the addition of ATCH1. The facts that the improvement is low and that the 1 LV PLS model outperforms the least squares model are caused by the correlation between ATCH6 and ATCH1, correlation clearly pointed out in the MEDA matrix (see the element at the sixth column and first row or the first column and sixth row) Clearly, both ATCH6 and ATCH1 are related to the same common factor in -LOGEC50. However, the variability in -LOGEC50 is the result of several sources of variability, which may be common factors with other descriptors.

Therefore, in order to introduce a new common factor in the model other than that in ATCH6, we need to find a descriptor related to -LOGEC50 but not to ATCH6. Also, the model may be improved by introducing a descriptor related to ATCH6 but not to -LOGEC50. For this, Figureure 23(b) compares the columns in the MEDA matrix corresponding to ATCH6 and -LOGEC50. The comparison should not be performed in terms of direct differences between values. For instance, ATCH1 and ATCH6 are much more correlated than ATCH1 and -LOGEC50. It is the difference in shape which is informative. Thus, we find that -LOGEC50 present a high correlation with LOGP (variable 50) which is not found in ATCH6. Thus, LOGP presents a common factor with -LOGEC50 which is not present in ATCH6. Using LOGP and ATCH6 as regressors, the least squares model presents $Q^2 = 0.37$.

If an additional descriptor is to be added to the model, again it should present a different common factor with any of the variables in the model. The MEDA vector corresponding to LOGP is shown in Figureure 23(c). This descriptor is related to a number of variables which are not related to -LOGEC50. This relationship represents a common factor in LOGP but not in -LOGEC50. The inclusion of a descriptor containing this common factor, for instance MOFI_Y (variable 38) may improve prediction because it may help to distinguish the portion of variability in LOGP which is useful to predict -LOGEC50 from the portion which is not. Using LOGP, ATCH6 and MOFI_Y as regressors yields $Q^2 = 0.56$, illustrating that the addition of a descriptor which is not related to the predicted variable may be useful for prediction.

Table 4: Top 10 models obtained after variable selection of the Selwood data set in (35)

Descriptors	Q^2
SUM_F (52) LOGP (50) MOFI_Y (38)	0.647
ESDL3 (17) LOGP (50) SURF_A (36)	0.645
SUM_F (52) LOGP (50) MOFI_Z (39)	0.644
LOGP (50) MOFI_Z (39)	0.534
ESDL3 (17) LOGP (50) MOFI_Y (38)	0.605
ESDL3 (17) LOGP (50) MOFI_Z (39)	0.601
LOGP (50) MOFI_Y (38)	0.524
LOGP (50) PEAX_X (40)	0.518
LOGP (50) SURF_A (36)	0.501
SUM_F (52) LOGP (50) PEAX_X (40)	0.599

In Figure 24, the two common factors described before, the one present in ATCH6 and -LOGEC50 and the one present in LOGP and MOFI_Y, are approximately highlighted in the MEDA matrix. If variables ATCH6 and MOFI_Y are replaced by others with the same common factors, the prediction performance of the model remains similar. However, LOGP is utmost for the model since is the only descriptor which relates the second common factor and -LOGEC50. These results are coherent with findings in the literature. Both (35) and (36) highlight the relevance of LOGP, and justify it with the results in several more publications. Furthermore, the top 10 models found in (35), presented in Table 4, follow the same patter of the solution found here. The models with three descriptors contain LOGP with one descriptor from the first and second common factors. The models with two descriptors contain LOGP and a variable with the second common factor.

Finally, in Figureure 25 the plot of measured vs predicted values of -LOGEC50 in the model resulting from the exploration is shown.

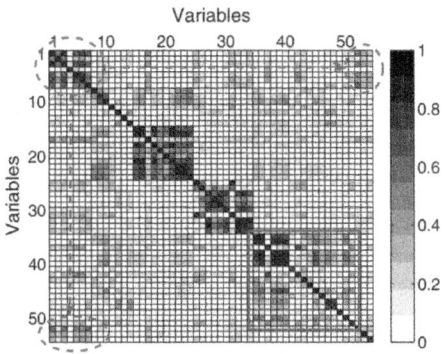

Figure. 24: MEDA matrix of the PCA model with 10 PCs from the data set which combines the in vitro antifilarial activity (-LOGEC50) with the complete set of descriptors of the Selwood dataset. Two common factors are highlighted. The first one is mainly found in descriptors 1 to 3, 5 to 7, 17, 52 and 53. The second one is mainly found in descriptors 35, 36, 38 to 40, 45, 47 and 50. Though the second common factor is not present in -LOGEC50, it is in LOGP.

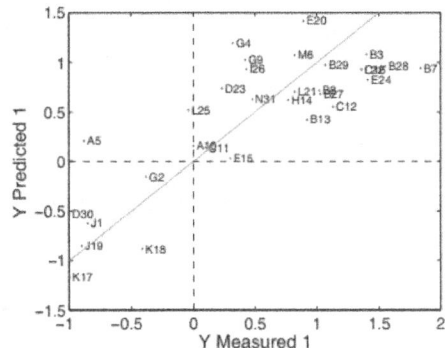

Figure. 25: Plot of measured vs predicted values of -LOGEC50 in the model with regressors LOGP, ATCH6 and MOFI_Y of the Selwood dataset.

No outliers are identified, though the four compounds previously highlighted are found at the bottom left corner. This result support the non-convenience of isolating these compounds. Notice that MEDA is not a variable selection technique per se and therefore other methods may be more powerful for this purpose. Nevertheless, being an exploratory method, the benefit of using MEDA for variable selection is that the general solution can be identified and understood, like in the present example. On the contrary, most variable selection approaches are based on highly complex algorithms which can only report a set of possible alternative solutions (e.g. Table 4).

CONCLUSION

In this chapter, new tools for exploratory data analysis are presented and combined with already well known techniques in the chemometrics field, such as projection models, score and loading plots. The shortcomings and potential pitfalls in the application of common tools are elucidated and illustrated with examples. Then, the new techniques are introduced to overcome these problems. The Missing-data methods for Exploratory Data Analysis technique, MEDA for short, studies the relationships among variables. As it is discussed in the chapter, while chemometric models such as PCA and PLS are quite useful for data understanding, they have a main problem which complicates its interpretation: a single component captures several sources of variability or common factors and at the same time a single common factor is captured in several components. MEDA, like rotation methods or Factor Analysis (FA), is a tool for the identification of the common factors in subspace models, in order to elucidate the structure in the data. The output of MEDA is similar to a correlation matrix but with better properties associated. MEDA is the perfect complement of loading plots. It gives a different picture of the relationships among variables which is especially useful to find groups of related variables. Using a Quantitative Structure-Activity Relationship (QSAR) example, it was shown that the understanding of the relationships among variables in the data may lead to perform variable selection with similar performance of highly sophisticated algorithms, with the extra benefit that the global solution is not only found but also understood. The second technique introduced in this chapter is a variant of MEDA, named observation-based MEDA or oMEDA. oMEDA was designed to identify the variables which differ between two groups of observations in a latent subspace, but it can be used for the more general problem of identifying the variables related to a given direction in the score plot. Thus, when a number of observations are located in a specific direction in the score plot, oMEDA gives the variables related to that distribution. oMEDA is the perfect complement of score plots and much more reliable than biplots. It can also be seen as an extension of contribution plots to groups of observations. It may be especially useful to check whether the distribution of a new set of observations agree with a calibration model. Though MEDA and oMEDA are grounded on missing-data imputation methods and their original algorithms are complex to a certain extent, both tools can be computed with very simple equations. A MATLAB toolbox with the tools employed in this chapter, including MEDA, oMEDA, ADICOV and SVI plots, is available at http://wdb.ugr.es/ josecamacho/.

ACKNOWLEDGEMENT

Research in this work is partially supported by the Spanish Ministry of Science and Technology through grant CEI BioTIC GENIL (CEB09-0010).

REFERENCES

1.   Jolliffe I.T.. Principal component analysis. EEUU: Springer Verlag Inc. 2002.

2.   Han J., Kamber M.. Data Mining: Concepts and Techniques. agora. cs.illinois.edu: Morgan Kaufmann Publishers, Elsevier 2006.

3.   Keren Gideon, Lewis Charles. A Handbook for data analysis in the behavioral sciences: statistical issues. Hillsdale, NJ: Lawrence Erlbaum Associates 1993.

4.   Tukey John W. Exploratory data analysis. Addison-Wesley Series in Behavioral ScienceReading, MA: Addison-Wesley 1977.

5.   Teo Yik Y.. Exploratory data analysis in large-scale genetic studies Biostatistics. 2010;11:70-81.

6.   Pearson K.. On Lines and Planes of Closest Fit to Systems of Points in Space Philosophical Magazine. 1901;2:559–572.

7.   Jackson J.E.. A User's Guide to Principal Components. England: Wiley-Interscience 2003.

8.   Wold H., Lyttkens E.. Nonlinear iterative partial least squares (NIPALS) estimation procedures in Bull. Intern. Statist. Inst. Proc., 37th session, London:1–15 1969.

9.   Geladi P., Kowalski B.R.. Partial Least-Squares Regression: a tutorial Analytica Chimica Acta. 1986;185:1–17.

10.   Wold S., om M. Sj Eriksson L.. PLS-regression: a basic tool of chemometrics Chemometrics and Intelligent Laboratory Systems. 2001;58:109–130.

11.   Camacho J.. Missing-data theory in the context of exploratory data analysis Chemometrics and Intelligent Laboratory Systems. 2010;103:8–18.

12.   Gabriel K.R.. The biplot graphic display of matrices with application to

principal component analysis Biometrika. 1971;58:453–467.

13. Westerhuis J.A., Gurden S.P., Smilde A.K.. Generalized contribution plots in multivariate statistical process monitoring Chemometrics and Intelligent Laboratory Systems. 2000;51:95–114.

14. Camacho J., Picó J., Ferrer A.. Data understanding with PCA: Structural and Variance Information plots Chemometrics and Intelligent Laboratory Systems. 2010;100:48–56.

15. Camacho J., Padilla P., Díaz-Verdejo J., Smith K., Lovett D.. Least-squares approximation of a space distribution for a given covariance and latent sub-space Chemometrics and Intelligent Laboratory Systems. 2011;105:171–180.

16. Kosanovich K.A., Dahl K.S., Piovoso M.J.. Improved Process Understanding Using Multiway Principal Component Analysis Engineering Chemical Research. 1996;35:138–146.

17. Ferrer A.. Multivariate Statistical Process Control based on Principal Component Analysis (MSPC-PCA): Some Reflections and a Case Study in an Autobody Assembly Process Quality Engineering. 2007;19:311–325.

18. Kjeldahl K., Bro R.. Some common misunderstandings in chemometrics Journal of Chemometrics. 2010;24:558–564.

19. L. Fabrigar, D. Wegener, R. MacCallum, E. Strahan, Evaluating the use of exploratory factor analysis in psychological research, Psycological Methods 4 (3) (1999) 272–299.

20. A. Costello, J. Osborne, Best practices in exploratory factor analysis: Four recommendations for getting the most from your analysis, Practical Assessment, Research & Evaluation 10 (7) (2005) 1–9.

21. I. Jolliffe, Rotation of principal components: choice of normalization constraints, Journal of Applied Statistics 22 (1) (1995) 29–35.

22. P. Nelson, P. Taylor, J. MacGregor, Missing data methods in pca and pls: score calculations with incomplete observations, Chemometrics and Intelligent Laboratory Systems 35 (1996) 45–65.

23. D. Andrews, P. Wentzell, Applications of maximum likelihood principal component analysis: incomplete data sets and calibration transfer, Analytica Chimica Acta 350 (1997) 341U–352. "

24. B. Walczak, D. Massart, Dealing with missing data: Part i, Chemometrics and Intelligent Laboratoy Systems 58 (2001) 15–27.

25. 25F. Arteaga, A. Ferrer, Dealing with missing data in mspc: several methods, different interpretations, some examples, Journal of

Chemometrics 16 (2002) 408–418.

26. F. Arteaga, A. Ferrer, Framework for regression-based missing data imputation methods in on-line mspc, Journal of Chemometrics 19 (2005) 439–447.

27. M. Reis, P. Saraiva, Heteroscedastic latent variable modelling with applications to multivariate statistical process control, Chemometrics and Intelligent Laboratoy Systems 80 (2006) 57–66.

28. F. Arteaga, Unpublished results.

29. F. Lindgren, P. Geladi, S. Wold, The kernel algorithm for pls, Journal of Chemometrics 7 (1993) 45–59.

30. S. de Jong, C. ter Braak, Comments on the pls kernel algorithm, Journal of Chemometrics 8 (1994) 169–174.

31. B. Dayal, J. MacGregor, Improved pls algorithms, Journal of Chemometrics 11 (1997) 73–85.

32. B. Wise, N. Gallagher, R. Bro, J. Shaver, W. Windig, R. Koch, PLSToolbox 3.5 for use with Matlab, Eigenvector Research Inc., 2005.

33. Camacho J. Observation-based missing data methods for exploratory data analysis to unveil the connection between observations and variables in latent subspace models Journal of Chemometrics 25 (2011) 592 - 600.

34. D.L. Selwood, D.J. Livingstone, J.C.W. Comley, A.B. O'Dowd, A.T. Hudson, P. Jackson, K.S. Jandu, V.S. Rose, J.N. Stables Structure-Activity Relationships of Antifiral Antimycin Analogues: A Multivariate Pattern Recognition Study, Journal of Medical Chemistry 33 (1990) 136–142.

35. S.J. Cho, M.A. Hermsmeier, Genetic Algorithm Guided Selection: Variable Selection and Subset Selection, J. Chem. Inf. Comput. Sci. 42 (2002) 927–936.

36. S.S. Liu, H.L. Liu, C.S. Yin, L.S. Wang, VSMP: A Novel Variable Selection and Modelling Method Based on the Prediction, J. Chem. Inf. Comput. Sci. 43 (2003) 964–969.

Chapter 10

METABOLOMICS AND CHEMOMETRICS AS TOOLS FOR CHEMO(BIO)DIVERSITY ANALYSIS - MAIZE LANDRACES AND PROPOLIS

Marcelo Maraschin et al

Plant Morphogenesis and Biochemistry Laboratory, Federal University of Santa Catarina, Florianopolis, SC, Brazil

INTRODUCTION

Developments in analytical techniques (GC-MS, LC-MS, 1H-, 13C-NMR, FT-MS, e.g.) are progressing rapidly and have been driven mostly by the requirements in the healthcare and food sectors. Simultaneous high-throughput measurements of several analytes at the level of the transcript (transcriptomics), proteins, (proteomics), and metabolites (metabolomics) are currently performed, producing a prodigious amount of data. Thus, the advent of omic studies has created an information explosion, resulting in a paradigm shift in the emphasis of analytical research of biological systems. The traditional approaches of biochemistry and molecular cell biology, where the cellular processes have been investigated individually and often independent of each other, are giving way to a wider approach of analyzing the cellular composition in its entirety, allowing achieving a quasi-complete metabolic picture.

The exponential growth of data, largely from genomics and genomic technologies, has changed the way biologists think about and handle data. In order to derive meaning from these large data sets, tools are required to analyze and identify patterns in the data, and allow data to be placed into a biological context. In this scenario, biologists have a continuous need for tools to manage and analyze the ever-increasing data supply. Optimal use of the data set, primarily of chemical nature, requires effective methods to analyze and manage them. It is obvious that all omic approaches will rely heavily upon bioinformatics for the storage, retrieval, and analysis of large data sets. Thus, and taking into account the multivariate nature of analysis in omic technologies, there is an increase emphasis in research on the application of chemometric techniques for extracting relevant information.

Metabolomics* and chemometrics† have been used in a number of areas to provide biological information beyond the simple identification of cell constituents. These areas include:

- Fingerprinting of species, genotypes or ecotypes for taxonomic or biochemical (genediscovery) purposes;

- Monitoring the behavior of specific classes of metabolites in relation to applied exogenous chemical and/or physical stimuli;

- Studying developmental processes such as establishment of symbiotic associations orfruit ripening;

- Comparing and contrasting the metabolite content of mutant or transgenic plants with that of their wild-type counterparts.

In general sense, strategies to obtain biological information in the above mentioned areas have focused on the analysis of metabolic differences that evidence responses to a range of extrinsic (ambient) and intrinsic (genetic) stimuli. Since no single analytical method has been found to obtain a complete picture of the metabolome of an organism, an association of advanced analytical techniques (GC-MS, LC-MS, FTIR, 1H-, 13C-NMR, FT-MS, e.g.) coupled to chemometrics, e.g., univariate (ANOVA, correlation analysis, regression analysis) or multivariate (PCA, HA, PLS) statistical techniques, has been performed in order to rapidly identify up- or down-regulated endogenous metabolites in complex matrices such as plant extracts, flours, starches, and biofluids, for instance. Plant extracts are recognized to be a complex matrix containing a wide range of primary and secondary metabolites that vary according to the environmental condition, genotype, developmental stage, and agronomic traits, for example. Such a complex matrix has long been used to characterize plant genotypes growing in a given geographic region and/or subjected to external stimuli, giving rise to additional information of interest, e.g., plant genetic breeding programs, local biodiversity conservation, food industry, and quality control in drug development/production processes. In the former case, programs for genetic breeding of plants have often focused on the analysis of landraces‡ genotypes (i.e., creole and local varieties), aiming at to identify individuals well adapted to specific local environmental conditions (soil and climate) and with superior agronomic performance and biomass yield. Indeed, the analysis and exploitation of the local genotypes' diversity has long been used as a strategy to improve agronomic traits by conventional breeding methods in plant crops of economical interest, as well as for stimulating the preservation of plant genetic resources. Taking into consideration that a series of primary (e.g., proteins and starch) and secondary metabolites (alkaloids, phenolic acids, and carotenoids, for instance) are well recognized compounds associated to the plants' adaptation mechanisms to their surroundings ecological factors, metabolomics and chemometrics have emerged as an interesting approach for helping the selection of superior genotypes, as further described in the first part of this chapter for maize landraces developed and cultured in southern regions of Brazil.

In a second part of this chapter is described the adoption of a typical metabolomic platform, i.e., FTIR and UV-visible spectroscopies coupled to chemometrics, for discriminating propolis samples produced in southern Brazil, a region of huge plant biodiversity. Propolis is typically a complex matrix and has been recognized for its broad pharmacological activities (anti-inflammatory, antibacterial, antifungal, anticancer, and antioxidant, e.g.) since ancient times. Propolis (registration number chemical abstracts service - CAS 9009-62- 5) is a beekeeping resinous and complex product, with a variable physical appearance, collected and transformed by honey bees, Apis mellifera, from the vegetation they visit. It may be ochre, red, brown, light brown or green; some are friable and steady, while the others are gummy and elastic.

Phenolics such as flavonoids and phenol-carboxylic acids are strategic components in propolis to render it bioactive against several pathogenic microorganisms, for instance as bacteriostatic and/or bactericidal agents. The flora (buds, twigs, bark, and less importantly, flowers) surrounding the hive is the basic source for the phenolics stuff and thus exerts an outstanding importance on the propolis final composition and on its physical, chemical, and biological properties. Although the wax component is an unquestionable supplement provided by the bee secretory apparatus by far less is known about the degree of intensity that these laborious insects play changing all the other chemical constituents collected in the Nature including minor ingredients like essential oils (10%), most of them responsible for the delicate and pleasant odor. All this flora contribution to propolis and the exact wax content may then explain physical properties such as color, taste, texture, melting point, and more importantly, from the health standpoint, a lot of pharmaceutical applications. However, for purpose of industrial applications, the propolis hydroalcoholic extract needs to meet specific composition in order to guarantee any claimed pharmacological activity.

One common method used by the industry for quality control is analyzing the propolis sample for the presence of chemical markers known to be present in the specific propolis product they market. Even though this has been the acceptable method for quality control, the presence of the chemical markers do not always guarantee an individual is getting the actual propolis stated by the product label, especially if the product has been spiked with the chemical markers. The quantitation method for the chemical markers will confirm the compounds presence, but it may not confirm the presence of the propolis known to contain the chemical markers. Authentication of the propolis material may be possible by a chemical fingerprint of it and, if possible, of its botanical sources. Thus, chemical fingerprinting, i.e., metabolomics and chemometrics, is an additional method that has been claimed to be included in the quality control process in order to confirm or deny the propolis sample quality being used for manufacturing of a derived product of that resinous and complex matrix. The second part of this chapter aims to demonstrate the possibility of a FTIR and UV-vis metabolomic-chemometrics approach to identify and classify propolis samples originating from nineteen geographic regions (Santa Catarina State, southern Brazil) in different classes, on the basis of the concerted variation in metabolite levels detected by those spectroscopic techniques. Exploratory data analysis and patterns of chemical composition based on, for instance, principal component analysis, as well as discriminating models will be described in order to unravel propolis chemotypes produced in southern Brazil.

MAIZE: METABOLOMIC AND CHEMOMETRIC ANALYSES FOR THE STUDY OF LANDRACES

Maize (Zea mays L.) was chosen as a model for metabolomic analysis because although most of this cereal produced worldwide is used for animal feeding, an important amount is also used in human diet and for industrial purposes, providing raw material for food, pharmaceuticals, and cosmetics production. The maize grain is composed of several chemicals of commercial value and the diversity of its applications depends on the differences in relative chemical composition, e.g. protein, oil, and starch contents, traits that show prominent genetic components (Baye et al., 2006; White, 2001). Over the last centuries, farmers have created thousands of maize varieties suitable for cultivation in numerous environments. Accordingly, it seems consensual that the maize landraces' phenotypes, e.g., morphological and agronomic traits and special chemical characteristics of grains are resultant of the domestication process. Thus, high throughput metabolomic analysis of maize genotypes could improve metabolic singularities knowledge about landraces, helping their characterization and evaluation, and indicating new alternatives for their use. In this context, to distinguish metabolic profiles it is necessary to consider the use of diverse analytical tools, such as spectroscopic and chromatographic techniques for instance. Techniques that are reproducible, stable with time, and do not require complex sample preparation such as infrared vibrational spectroscopy and nuclear magnetic resonance spectroscopy are desirable for metabolic profiling.

Metabolic profiling of maize landraces through FTIR-PCA – integral and degermed flours

Vibrational spectroscopy, and particularly Fourier transform infrared spectroscopy (FTIR) is thought to be interesting as one aims at discriminating and classifying maize landraces according to their chemical traits. FTIR is a physicochemical method that measures the vibrations of bonds within functional groups and generates a spectrum that can be regarded as a metabolic fingerprint. It is a flexible method that can quickly provide qualitative and quantitative information with minimal or no sample preparation of complex biological matrices (Ferreira et al., 2001). By other hand, a FTIR spectrum is complex, containing many variables per sample and making visual analysis very difficult. Hence, to extract useful information from the whole spectra, multivariate data analysis is needed, particularly through the determination of the principal components (PCA - Fukusaki & Kobayashi, 2005). Such a multivariate analysis technique could allow the characterization of the sample relationships (scores plans or axis) and the recovery of their subspectral profiles (loadings). This approach was applied to classify flour samples from whole (integral) and degermed maize grains of twenty-six landraces developed and cultivated by small farmers in the farwest region of Santa Catarina State, southern Brazil (Anchieta County – 26°31'11"S, 53°20'26"W).

Previously to multivariate analysis, FTIR spectra were normalized, baseline-corrected in the region of interest by drawing a straight line before resolution enhancement (k factor of 1.7) was applied using Fourier self deconvolution (Opus v. 5.0, Bruker Biospin, GmbH, Rheinstetten, Germany). Chemometric analysis used normalized, baseline-corrected (3000– 600 cm⁻¹. 1700 data points) and deconvoluted spectra, which were transferred via a JCAMP format (OPUS v. 5.0, Bruker Biospin GmbH, Rheinstetten, Germany) into the data analysis software for PCA (The Unscramble v. 9.1, CAMO Software Inc., Woodbridge, USA)

Previously to PCA analysis each spectrum within the (3000–600 cm⁻¹) region was standard normal deviates corrected. Figureure 1 shows a PCA scores scatter plot for flour samples from whole and degermed grains using the whole FTIR spectral window data set (3000–600 cm⁻¹). The scores scatter plot (PC1 vs. PC2) that contains 93% of the data set variability shows a clear discrimination among flour samples of whole and degermed grains.

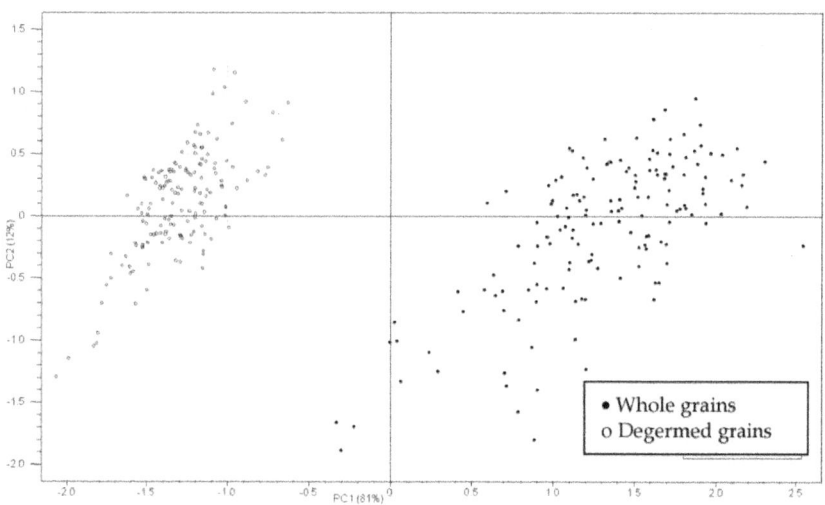

Figure. 1: Principal component analysis scores scatter plot of the FTIR data set in the spectral window of 3000–600 cm⁻¹ wavenumber of landrace maize flours of whole and degermed grain cultivated in the southern Brazil.

The samples of whole grains grouped in PC1+ axis seemed to be more discrepant in their chemical composition, appearing more scattered through the quadrants of the PCA representation. Figure 2 shows the loadings plot of PC1, revealing the most important wavenumbers which explain the distinction of the samples previously found (scores scatter plot). The loadings indicated a prominent effect of the lipid components (2924, 2850, and 1743 cm⁻¹) for the segregation observed. The two major structures of the grains are the endosperm and the germ (embryo) that constitute approximately 80 and 10% of the mature kernel dry weight, respectively. The endosperm is largely starch (approaching 90%) and the germ contains high levels of oil (30%) and protein (18% - Boyer & Hannah, 2001). The greatest chemical diversity observed in whole grains can be explained by genetic variation of embryos resulting from sexual reproduction. Some authors suggest that the high level of genetic and epigenetic diversity observed in maize could be responsible for its great adaptation capacity to a wide range of ecological factors. Lemos (2010) analyzing the metabolic profile of maize landraces' leaf tissues from Anchieta County (southern Brazil) found a prominent chemical variability among individuals of same variety, although intervariety variability has also been observed.

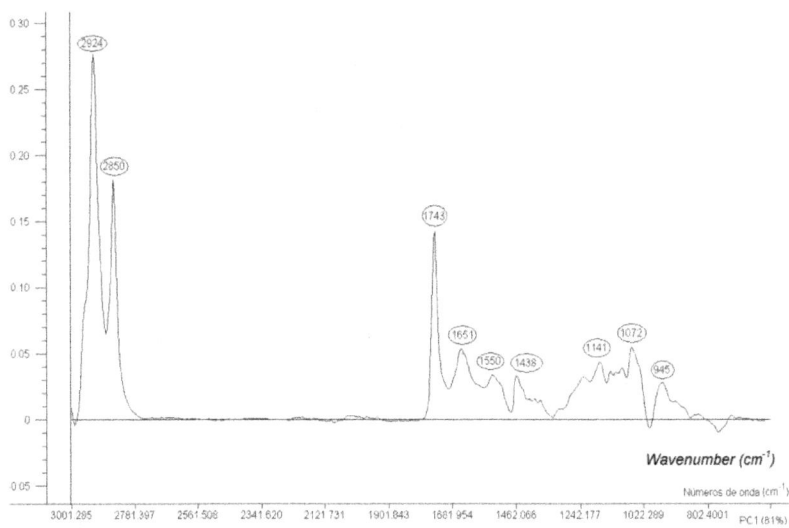

Figure. 2: PC1 loadings plot of the FTIR spectra of maize flours of whole and de-germed grains in the 3000–600 cm⁻¹ wavenumber region.

Starch recognition pattern of maize landraces by NMR spectroscopy and PCA

The composition of maize grains can be heterogeneous for both the quantity and quality of compounds from endosperm as starch, protein, and oil. In this context, metabolomics coupled to chemometrics approach was successfully applied to the discrimination of starches from the studied twenty-six maize landraces. The starches were extracted from flours with distilled water (1: 70, w/v) under reflux (80°C, ¹ h), precipitated with ethyl alcohol (12 h, 4°C), and oven-dried (55°C until constant weight). Samples (50 mg) were dissolved in DMSO-d6 (0.6 mL) and ¹H-NMR spectra obtained under standard conditions. Sodium-3-trimethylsilylpropionate (TMSP-2, 2, 3, 3-d₄) was used as internal reference (δppm 0.0). Spectra were processed using 32768 data points by applying an exponential line broadening of 0.3 Hz for sensitivity enhancement before Fourier transformation and were accurately phased, baseline adjusted, and converted into JCAMP format to build the data matrix. All calculations were carried out using the Pirouette software (v. 3.11, InfoMetrix, Woodinville, Washington, USA). The PCA analysis of the whole ¹H-NMR data set (32.000 data points) was performed including the spectra of amylose and amylopectin standards. The chemical structures and the purity of the standards of amylose and amylopectin were confirmed by ¹³C-NMR spectroscopy. The PC1 (32%) vs. PC2 (28%) scores scatter plot allowed a clear segregation of the amylopectin standard and a discrimination for the maize flours samples in two groups (Figure. 3) by PC1 axis.

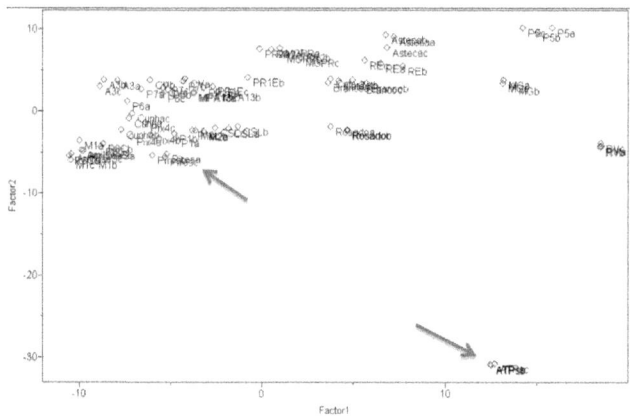

Figure. 3: Classification plot of starch fractions of maize landraces for ^{1}H-NMR after PCA analysis. The arrows shows the amylose and amylopectin stardands.

The Roxo 41 variety located closer to the amylopectin standard suggesting the predominance of that polysaccharide in relation to amylose in its starchy fraction. This result is in accordance to PCA analysis of the IR data set of the fingerprint region of carbohydrates that diagnosed the starch granules from Roxo 41 with superior amylopectin amount in respect to their amylose content (data not shown). By other hand, the amylose standard was grouped with MPA1 and Rajado 8 Carreiras varieties, suggesting that the starch granules contain superior amount of that polysaccharide in its starchy fraction.

PROPOLIS: ATR-FTIR AND UV-VISIBLE SPECTROPHO-TOMETRY COUPLED TO CHEMOMETRICS AS AN ANA-LYTICAL APPROACH FOR PATTERN RECOGNITION

Attenuated total reflectance-Fourier transform infrared spectroscopy

Propolis (registration number chemical abstracts service - CAS 9009-62-5) is a sticky colored material, which honeybees collect from different plants exudates and modify in its hypopharyngeal glands, being used in the hive to fill gaps and to protect against invaders as insects and microorganisms. Raw propolis usually contains 50% resin and balsam, 30% wax,10% aromatic oils, 5% pollen, and 5% other substances as inorganic salts and amino acids. This resin has been used by humanity since ancient civilizations like Egyptian,

Assyrian, Greek, Roman, and Inca. In these days, a number of studies have confirmed important biological activities such as antibacterial, antifungal, antiviral, antioxidant, antiinflammatory, hepatoprotective, and antitumoral (for review see Bankova, 2009; Banksota et al., 2001; Castaldo & Capasso, 2002). The aspect, texture and the chemical composition of propolis is highly variable and depends on the climate, season, bee species and mainly the local flora which is visited by bees to collect resin (Markham et al., 1996). For this reason, comparing propolis samples from distinct regions might be the same as to compare extracts of two plants that belong to different taxonomical families (Bankova, 2005).

Propolis from Europe is the best known type of propolis. In European regions with temperate climate bees obtain resin mainly from the buds of Poplus species and the main bioactive components are flavonoids (Greenaway et al., 1990). In tropical countries, the botanical resources are much more variable in respect to temperate zones, so bees find much more possibilities of collecting resins and hence the chemical composition of tropical propolis are more variable and distinct from European ones (Sawaya et al., 2011). Different compounds have been reported in tropical propolis such as terpenoids and prenylated derivatives of p-coumaric acids in Brazilian propolis (Marcucci, 1995), lignans in Chilean samples (Valcic et al., 1998), and polyisoprenylated benzophenones in Venezuelan, Brazilian, and Cuban propolis (Cuesta-Rubio et al., 1999; Marcucci, 1995).

In order to be accepted officially into the main stream of the healthcare system and for industrial applications, propolis needs chemical standardization that guarantees its quality, safety, efficacy, and provenance. The chemical diversity mainly caused by the botanical origin makes the standardization difficult. Since the chemistry and biological activity of propolis depends on its geographical origin, a proper method to discriminate its origin is needed (Bankova, 2005). Chromatographic methods (HPLC, TLC, GC, e.g.) are largely used to identification and quantification of propolis compounds, but it its becoming clear that to separate and evaluate all constituents of propolis is an almost impossible task (Sarbu & Mot, 2011). Even thought the presence of the chemical markers are considered an acceptable method for quality control, not always is guarantee about what is stated by the product label, especially if the product has been spiked with the chemical markers. Besides, literature has demonstrated that is not possible to ascribe the pharmacological activity solely to a unique compound and until now no single propolis component has shown to possess anti-bacterial activity higher than total extract (Kujumgiev et al., 1999; Popova et al., 2004). Thus, a possibility is offered by the fingerprinting methods that can analyze in a non-selective way the propolis samples as a whole.

Poplar propolis, for example, can be distinguished by UV-visible spectrophotometric determination of all three important components (flavones and flavonols, flavonones and dihydroflavonols, and total phenolics - Popova et al., 2004), but some constraints regarding such an analytical approach has been claimed for propolis from tropical regions (Bankova & Marcucci, 2000)

The search for faster screening methods capable of characterizing propolis samples of different geographic origins and composition has lead to the use of direct insertion mass spectrometric fingerprinting techniques (ESI-MS and EASI-MS), which has proven to be a fast and robust method for propolis characterization (Sawaya et al., 2011), although this analytical approach can only detect compounds that ionize under the experimental conditions. Similarly, Fourier transform infrared vibrational spectroscopy (FTIR) has also demonstrated to be valuable to chemically characterize complex matrices such as propolis (Wu et al., 2008). In order to achieve the goal of treat propolis sample as a whole than just be focused only in marker compounds, chemometric methods are being considered an important tool to analyze the huge data sets generated by non-selective analytical techniques such as UV-vis, MS, NMR, and FT-IR, generating information not only about chemical composition of propolis but also discriminating its geographical origin.

Authentication of propolis material may be possible by a chemical fingerprint of it and, if possible, of its botanical sources. Thus, chemical fingerprinting, i.e., metabolomics and chemometrics, is an additional method that has been claimed to be included as a quality control method in order to confirm or deny the propolis sample being used for the manufacturing of a derived product of that resinous and complex matrix. Over the last decades, infrared (IR) vibrational spectroscopy has been well established as a useful tool for structure elucidation and quality control in several industrial applications. Indeed, the development of Fourier transform (FT) IR and attenuated total reflectance (ATR) techniques have also evolved allowing rapid IR measurements of organosolvent extracts of plant tissues, edible oils, and essential oils, for example (Damm et al., 2005; Lai et al., 1994; Schulz & Baranska, 2007). In consequence of the strong dipole moment of water, IR spectroscopy applications have mostly focused on the analysis of dried or non-aqueous plant matrices and currently IR methods are widely used as a fast analytical technique for the authentication and detection of adulteration of vegetable oils.

ATR-FTIR spectroscopy was applied to propolis samples collected in the autumn-2010 and originated from nineteen geographic regions of Santa Catarina State (southern Brazil) in order to gain insights as to the chemical profile of those complex matrices. FTIR spectroscopy measures the vibrations of bonds within functional groups and generates a spectrum that can be regarded as a metabolic fingerprint. Similar IR spectral profiles (3000 – 600 cm⁻¹, Figureure 4) were found by a preliminary visual analysis for purpose of an exploratory overview of data, revealing typical signals of e.g., lipids (2910 – 2845 cm⁻¹), monoterpenes (1732, 1592, 1114, 1022, 972 cm⁻¹), sesquiterpenes (1472 cm⁻¹), and sucrose (1122 cm⁻¹ - Schulz & Baranska, 2007) for all the studied samples. However, we were not able to identify by visual inspection of the spectra a clear picture regarding a discriminating effect of any primary or secondary metabolites among the propolis samples.

A FTIR spectrum is complex, containing many variables per sample and making visual analysis very difficult. Hence, to extract extra useful information, i.e., latent variables, from the whole spectra chemometric analysis was performed considering the whole FTIR data set using principal components analysis (PCA) for an exploratory overview of data. This method could reveal similarity/dissimilarity patterns among propolis samples, simplifying data dimensions and results interpretation, without missing the more relevant information associated to them (Fukusaki & Kobayashi, 2005; Leardi, 2003).

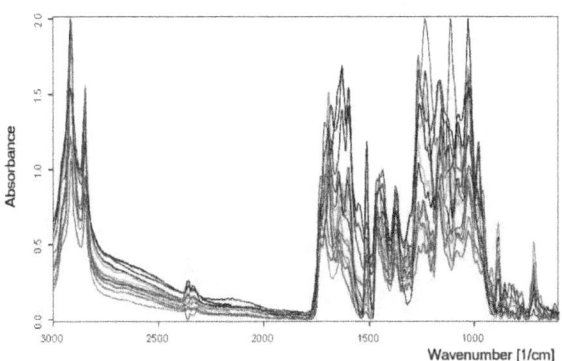

Figure. 4: IR spectral profile of propolis samples (autumn, 2010) produced in southern Brazil, according to the geographic regions of origin in Santa Catarina State. IR spectra are shown from top to bottom following the geographic precedence, i.e. 19 counties, of the propolis samples: Angelina (ANG), Balneário Gaivotas (BG), Bom Retiro (BR₁ and BR₂), Caçador (Cç), Campo-Erê (CE), Canoinhas (CA), Campos Novos (CN), Descanso (DS), José Boiteux (JB), Porto União (PU), Serra Alta (SA), São Joaquim (SJ₁ and SJ₂), São José do Cerrito (SJC), Urupema (URU), Vidal Ramos (VR), Florianópolis (FLN), and Xaxim (XX)

The covariance was choose for matrix construction in PCA calculation, since all variables considered were expressed in the same unit. By doing so, the magnitude differences were maintained, i.e., data were not standardized and variables contribution to samples distribution along axes was direct proportional to their magnitude. For the purpose of the propolis chemical profile analysis this kind of information is thought to be very useful, because wavenumber with higher absorbances (higher metabolites concentration) contribute more significantly with objects distribution into PCA, introducing quantitative information beside the compositional information of the sample data.

The principal component analysis (PCA) of the whole spectral data (3000 – 600 cm^{-1}, 1700 data points) revealed that PC1 and PC2 defined 88% of the variability from the original IR data and a peculiar pattern of lipids (2914 cm^{-1} and 2848 cm^{-1} - C-H stretching vibrations) for the samples from the northern region (NR) of Santa Catarina State. The climate in that region is typically mesothermic, humid subtropical with a mild summer and an annual temperature average of 17.2°C – 19.4°C. On the other hand, the propolis produced in the highlands (1360m altitude, annual maximum and minimum temperatures average of 18.9°C and 9.3°C, respectively) were discrepant regarding their monoterpene (1114 cm^{-1} and 972 cm^{-1} - v-CH$_2$) and sesquiterpene (1472 cm^{-1} - d CH$_2$) compounds (Schulz & Baranska, 2007) – Figureure 5. In despite of NR$_1$ and NR$_2$ propolis samples have grouped in PC1- and PC2+, they differ somewhat in respect to their chemical composition, an effect attributed to the flora composition found in those regions, e.g., mostly Atlantic Rainforest in NR$_1$ as NR$_2$ shows extensive areas covered by artificial reforestations i.e., Eucalyptus spp and Pinus spp, furnishing distinct raw materials for propolis production.

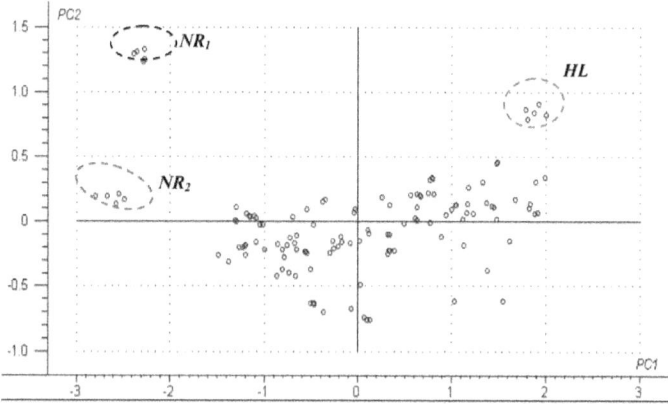

Figure. 5: Principal component analysis scores scatter plot of the FTIR data set in the spectral window of 3000–600 cm^{-1} wavenumber (1700 data points) of propolis samples produced in the southern Brazil (Santa Catarina State). NR$_1$, NR$_2$ and HL refer to propolis samples originated from northern and highland regions of Santa Catarina

State. The calculations were carried out using The Unscrambler software (v. 9.1, Oslo - Norway). PC1 and PC_2 accounts for 88% of the variance preserved.

Further chemometric analysis took into consideration the fact that propolis is a very well known source of phenolic compounds, e.g., phenolics acids and flavonoids. Indeed, phenolic compounds occur ubiquitously in most plant species and take part of the chemical constitution of propolis worldwide. IR spectroscopy allows to identify phenolic compounds since they demonstrate strong IR bands due to C-H wagging vibrations between $1260 - 1180$ cm^{-1} and $900 - 820$ cm^{-1} (Schulz & Baranska, 2007). The principal

component calculations were performed for both $1260 - 1180$ cm^{-1} and $900 - 820$ cm^{-1} spectral windows and PC1 and PC_2 resolved about 96% of the spectral data variability. An interesting discrimination profile was detected where samples from the far-west (FW) region grouped distinctly in respect to northern (NR_1) ones, which also differed from the highlands (HL) propolis samples. Such findings can be explained in any extension based on the flora composition of the studied geographic regions. In the northern and far-west regions of Santa Catarina State the Atlantic Rainforest is typically found, but the floristic composition varies according to the altitude, e.g., 240 m altitude – NR1 and 830 m – FW. Besides, as a mesothermic humid subtropical climate is found in NR_1, the FW region is characterized by a temperate climate that determines a discrepant composition of plant species. Finally, the HL region (1360m altitude, temperate climate) is covered by the Araucaria Forest, where parana pine (Araucaria angustifolia, Gymnospermae, Araucariaceae) occurs as a dominant plant species. A. angustifolia produces a resinous exudate rich in guaiacyl type lignans, fatty acids, sterols (Anderegg & Rowe, 2009), phenolics, and terpenic acids that is thought to be used by honey bee (Apis mellifera) for the propolis production. Since the plant species populations influence the propolis chemical composition, the discrimination profile detected by ATR-FTIR coupled to chemometrics seems to be an interesting analytical approach to gain insights as to the effect of the climatic factors and floristic composition on the chemical traits of that complex matrix.

Ultraviolet-visible scanning spectrophotometry

Combination of UV-visible spectrophotometric wavelength scans and chemometric (PCA) analysis seems to be a simple and fast way to prospect plant extracts. This analytical strategy revealed to be fruitful for discrimination of habanero peppers according to their content of capsaicinoids, substances responsible for the pungency of their fruits (Davis et al., 2007). Chemometric analysis was performed considering the absorbance values of the total UV visible data set (200 ηm to 700 ηm, 450 data points) for the propolis samples in study, by using principal components analysis (PCA) for an exploratory overview of data.

In a first approach, principal components analysis (PCA) was tested by both correlation and covariance matrices of calculations. If correlation is used, the data set is standardized (meancentered and columns scaled to the unit of variance), decreasing the effect of differences in magnitude between variables and leading to a distribution of objects (eigenvalues) with equal influence from all variables. On the other hand, if covariance is used, data is only meancentered; retaining its original scale. The resulting distribution is then determined either by composition and magnitude of variables, leading to a PCA representation more influenced by larger observed values (Manetti et al., 2004). A similar distribution of objects was found for both correlation and covariance matrices in PC calculations, as PC1 and PC2 resolved 91.2% and 96.3%, respectively of the variability of the spectrophotometric data set. Thus, the covariance matrix was chosen for PCA calculations, since all variables considered were expressed in the same unit. By doing so, the magnitude differences were maintained, i.e., data were not standardized and variables contribution to samples distribution along axes was direct proportional to their magnitude. For the purpose of the chemical profile analysis of the propolis samples this kind of information is thought to be very useful, because wavelengths with higher absorbances (higher metabolites concentration) contribute more significantly with objects distribution into PCA, introducing quantitative information beside the compositional information of the sample data.

Principal component analysis was performed using The Unscrambler software (v. 9.1, Oslo - Norway) and revealed mostly a distribution of the propolis samples along the PC1 axis (91% sample total variability), as PC2 (5% sample total variability) seemed to be lesser discriminator of the objects. A clear separation of the samples according to the east-west axis of Santa Catarina State could be found, where propolis produced near coastal regions (CR_1 and CR_2) grouped in PC1+/PC2-, as the sample from the far-west region (FW) was detected in the opposite side of PC1 axis, along with the samples from the northern region (PU, Cç, and CA - Figureure 6). Interestingly, propolis samples from the counties SJ, URU, BR (highlands counties), and ANG, which shown geographic proximities and a certain common floral composition, seemed to be similar in their chemical profiles as determined by UV-visible scanning spectrophotometry, grouping in PC1+/PC2+.

High loadings associated to the wavelengths 394 ηm, 360 ηm, 440 ηm, and 310 ηm seemed to influence the observed distribution of the propolis samples and could be associated to the presence of (poly)phenolic compounds. In fact, the λ_{max} for the cinnamic acid and its derivatives is near 310-320 ηm as for the flavonols is usually around 360 ηm (Tsao & Deng, 2004). Further chemical analysis of the total content of phenolics ad flavonoids in the propolis originated from the counties SJ, URU, BR, and ANG revealed similar contents, with average concentrations of 1411.52 μg/ml and 4.61 μg/ml of those secondary metabolites, respectively, in the hydroalcoholic (70: 30, v/v) extract. Such findings differed (P

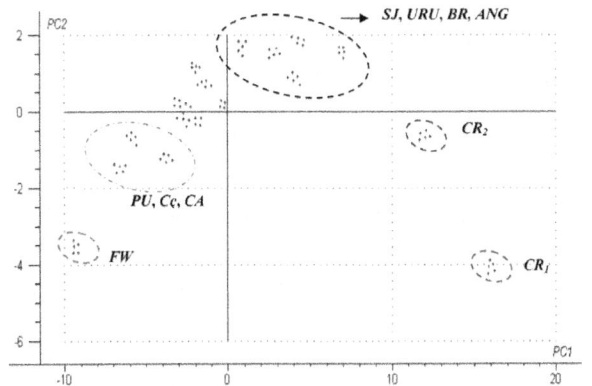

Figure. 6: Principal component analysis scores scatter plot of the UV-visible data set in the spectral window of 200 ηm to 700 ηm (450 data points) of propolis samples produced in the southern Brazil (Santa Catarina State). CR_1, CR_2, and FW refer to propolis samples originated from coastal (BG and FLN Counties) and far-west (CE County) regions, respectively, of Santa Catarina State. The sample grouping of propolis with similar UV-visivel scanning profiles regarding their (poly)phenolic composition is detached in the PC1+/PC2+ quadrant. PC1 and PC2 resolved 96% of the total variability of the spectral data set.

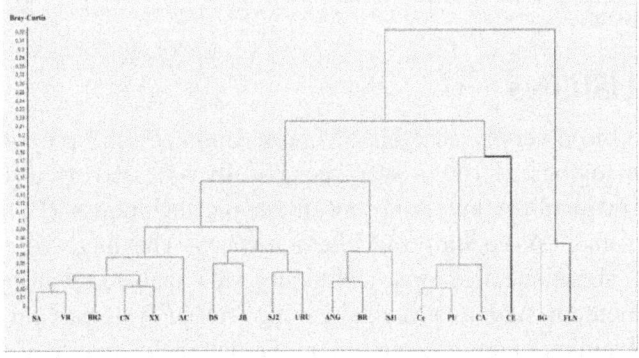

Figure. 7: Dendrogram of propolis samples using average linkage with Bray-Curtis dissimilarity measure. Data calculations were based on the absorbance values for the

UVvisible spectral window of 200 ηm to 700 ηm of propolis samples produced in Santa Catarina State - southern Brazil, autumn-2010.

In order to check the chemical similarity pattern of propolis samples detected by PCA, further cluster analysis of the whole absorbance UV-vis data set, i.e., absorbance values of 200 ηm to 700 ηm (450 data points), was performed by using the Unweighted Pair Group Method with Arithmatic Mean (UPGMA) based on Bray-Curtis dissimilarity coefficient. UPGMA is a simple agglomerative or hierachical clustering method used in for the creation of phonetic trees, i.e., phenograms, hierarchical trees or dendrograms that indicate the similarity degree among samples/objects of interest, so that observations in the same cluster are similar in some sense. In UPGMA method after two objects with the least dissimilarity fuse together an arithmetic average of the dissimilarity of this new cluster and the rest of the objects is calculated. This leads to a reduction in the size of the original dissimilarity matrix. The procedure continues with the dissimilarity matrix being correspondingly reduced. When the average between an object and a cluster is calculated, the method gives equal weights to the members of the clusters when averaging, i.e., unweighted. Thus, in the progressive reduction of the dissimilarity matrix, only relationships between groups are considered, which are given equal weighting and this leads to loss of information about the relationships between pairs of objects (Legendre, 1998; Singh, 2008).

The hierarchical tree of the similarity of chemical profiles of the propolis samples is shown in Figureure 7. The findings suggest a resemblance of grouping as found by PCA calculations in respect to the SJ, URU, BR, and ANG samples, as well as for the propolis originated from the coastal (BG and FLN) and northern regions (CA, PU, and Cç). Additionally, UPGMA analysis also discriminate the propolis produced in the western (AC, XX, and CN) and farwest regions.

CONCLUSIONS

The chemo(bio)diversity analysis of maize landraces and propolis produced in southern regions of Brazil was successfully assessed by using a typical metabolomic platform involving spectroscopic techniques (FTIR, 1H- and 13C-NMR, and UV-visible) and chemometrics. The huge amount of data afforded by those spectroscopic techniques was analyzed using multivariate statistical methods such as principal component analysis and cluster analysis allowing obtaining extra information on the metabolic profile of the complex matrices in study. The analytical approach described showed to be suitable when ones aim to discriminate maize flour samples from whole and degermed maize, an issue thought to be important for the food, cosmetic, and pharmaceutical industries regarding the usage and quality control process of that raw material. Similarly, the classification of maize landraces according to their starch traits is considered technologically relevant in order to optimize the usage of non-chemically modified starches in industrial process, for instance. The classification of Brazilian propolis as to their chemical profiles and geographic regions seems to be relevant because that biomass is typically quite complex, making difficult and expensive to perform a complete characterization in that sense. By doing so, the propolis produced in southern Brazil might be better evaluated as to their potential usage in cosmetic and pharmaceutical industry, taking into consideration their secondary metabolite constituents, e.g., mono/sesquiterpenes and phenolics. The coupling of chemometricsspectroscopic techniques used is thought to be essential to allow detecting peculiar chemical traits of the propolis samples according to their geographic regions in a simple and fast way.

ACKNOWLEDGMENT

Authors are indebt to FAPESC, CNPq, and CAPES for financial support and fellowships.

REFERENCES

1.   Anderegg, RJ & Rowe, JW. (2009). Lignans, the major component of resin from Araucaria angustifolia knots. International Journal of the Biology, Chemistry, Physics and Technology of Wood, Vol. 28, pp.171–175. ISSN 0018-3830

2.   Bankova, V. (2005). Chemical diversity of propolis and the problem of standardization. Journal of Ethnopharmacology, Vol. 100, pp.114-117. ISSN: 0378-8741

3.   Bankova, V & Marcucci, MC. (2000). Standardization of propolis: present status and perspectives. Bee World, Vol. 81, pp.182-188. ISSN: 0005-772X

4.   Banksota, AH., Tezuka, Y & Kadota, S. (2001). Recent progress in pharmacological research of propolis. Phytotherapy Research, Vol. 15, pp.561-571. ISSN: 1099-1573

5.   Baye, TM., Pearson, TC & Settles, AM. (2006). Development of a calibration to predict maize seed composition using single kernel near infrared spectroscopy. Journal of Cereal Science, Vol. 43, pp.236–243. ISSN: 0733-5210

6.   Boyer, CD & Hannah, C. (2001). Kernel mutants of corn. In: Specialty corns. HALLAUER, AR. (Ed.). 2nd ed. pp. 153, CRC Press, London.

7. Castaldo, S & Capasso, F. (2002). Propolis, an old remedy used in modern medicine. Fitoterapia, Vol. 73, pp.S1-S6. ISSN: 0367-326X

8. Cuesta-Rubio, O., Cuellar, AC., Rojas, N., Velez, HC., Rastrelli, L & Aquino, R. (1999). A polyisoprenylated benzophenone from Cuban propolis. Journal of Natural Products, Vol. 62, pp.1013-1015. ISSN: 0974-5211 Damm, U., Lampen, P., Heise, HM., Davies, AN & Mcintyre, PS. (2005). Spectral variable selection for partial least squares calibration applied to authentication and quantification of extra virgin olive oils using Fourier transform Raman spectroscopy. Applied Spectroscopy, Vol. 59, pp.1286-1294. ISSN: 0003-7028

9. Davis, CB., Markey, CE., Busch, MA & Busch, KW. (2007). Determination of capsaicinoids in habanero peppers by chemometric analysis of UV spectral data. Journal of Agricultural and Food Chemistry, Vol. 55, pp. 5925-5933. ISSN: 0021-8561

10. Ferreira, D., Barros, A., Coimbra, MA & Delgadillo, I. (2001). Use of FTIR spectroscopy to follow the effect of processing in cell wall polysaccharide extracts of a sun-dried pear. Carbohydrate Polymers, Vol. 45, pp.175–182. ISSN: 0144-8617 Fukusaki, E & Kobayashi, A. (2005). Plant metabolomics: potential for practical operation. Journal of Bioscience and Bioengineering, Vol. 100, pp.347–354. ISSN: 1389-1723

11. Greenaway, W., Scaysbrook, T & Whately, FR. (1990). The composition and plant origin of propolis: a report of work at Oxford. Bee World, Vol. 71, pp. 107-118. ISSN: 0005- 772X

12. Kujumgiev, A., Tsvetkova, I., Serkedjieva, YU., Bankova, V., Christov, R & Popov, S. (1999).

13. Antibacterial, antifungal and antiviral activity of propolis of different geographical origins. Journal of Ethnopharmacology, Vol. 64, pp. 235-240. ISSN: 0378-8741

14. Lai, YW., Kemsley, EK & Wilson, J. (1994). Potential of Fourier transform infrared spectroscopy for the authentication of vegetable oils. Journal of Agricultural and Food Chemistry, Vol. 42, pp.1154-1159. ISSN: 0021-8561

15. Leardi, R. (2003). Chemometrics in data analysis. In: A user-friendly guide to multivariate calibration and classification. Naes, T., Isaksson, T., Fearn, T & Davies, T (eds). NIR Publications, West Sussex.

16. Legendre, P. (1998). Numerical Ecology. Elsevier Science, New York.. Lemos PMM (2010). Análise do metaboloma foliar parcial de variedades locais de milho (Zea mays L.) e dos efeitos anti-tumoral in vitro e na morfogênese embrionária de Gallus domesticus. PhD thesis, Federal

University of Santa Catarina, Brazil.

17. Manetti, C., Bianchetti, C., Bizarri, M., Casciani, L., Castro, C., D´Ascenzo, G., Delfini, M., DI Cocco, ME., Laganà, A., Miccheli, A., Motto, M & Conti, F. (2004). NMR-based metabonomic study of transgenic maize. Phytochemistry, Vol. 65, pp.3187-3198. ISSN: 0031-9422

18. Marcucci, MC. (1995). Propolis: chemical composition, biological properties and therapeutic activity. Apidologie, Vol. 26, pp.83-99. ISSN: 1297-9678 Markham, KR., Mitchell, KA., Wilkins, AL., Daldy, JA & Lu, Y. (1996). HPLC and CG-MS identification of the major organic constituents in New Zealand propolis.

19. Phytochemistry Vol. 42, pp.205-211. ISSN: 0031-9422 Popova, M., Bankova, V., Butovska, D., Petkov, V., Damynova, BN., Sabatini, AG., Marcazzan, GL & Bogdanov, S (2004). Validated methods for the quantifications of biologically active constituents of poplar-type propolis. Phytochemical Analysis, Vol. 15, pp.235-240. ISSN: 1099-1565

20. Sarbu, C & Mot, AC. (2011). Ecosystem discrimination and fingerprinting of Romain propolis by hierarchical fuzzy clustering and image analysis of TLC patterns. Talanta, Vol. 85, pp.1112-1117. ISSN: 0039-9140

21. Sawaya, ACHF., Silva, IB & Marcucci, MC. (2011). Analytical methods applied to diverse types of Brazilian propolis. Chemistry Central Journal, Vol. 5, pp.1-10. ISSN: 1752- 153X

22. Schulz, H & Baranska, M. (2007). Identification and quantification of valuable plant substances by IR and Raman spectroscopy. Vibrational Spectroscopy, Vol. 43, pp.13- 25. ISSN: 0924-2031

23. Singh, W. (2008). Robustness of three hierarchical agglomerative clustering techniques for ecological data. M.Sc. thesis, University of Iceland, Iceland.

24. Tsao, R & Deng, Z. (2004). Separation procedures for naturally occurring antioxidant phytochemicals. Journal of Chromatography B, Vol. 812, pp.85-99. ISSN: 1570- 0232

25. Valcic, S., Montenegro, G & Timmermann, BN. (1998). Lignans from Chilean propolis. Journal of Natural Products, Vol. 61, pp.771-775. ISSN: 0974-5211

26. White, PJ. (2001). Properties of corn starch. In: Specialty corns. HALLAUER, AR. (Ed.). 2nd ed. pp. 189, CRC Press, London.

27. Wu, YW., Sun, SQ., Zhao, Y., Li, Q & Zhou, J. (2008). Rapid discrimination of extracts of Chinese propolis and poplar buds by FT-IR

and 2D IR correlation spectroscopy. Journal of Molecular Structure, Vol. 884, pp.48-54. ISSN: 0022-2860

Chapter 11

PARAFAC ANALYSIS FOR TEMPERATURE-DEPENDENT NMR SPECTRA OF POLY(LACTIC ACID) NANOCOMPOSITE

Hideyuki Shinzawa[1], Masakazu Nishida[1], Toshiyuki Tanaka[2], Kenzi Suzuki[3] and Wataru Kanematsu[1]

[1]Research Institute of Instrumentation Frontier, Advanced Industrial Science and Technology (AIST)

[2]Mikawa Textile Research Center, Aichi Industrial Technology Institute (AITEC)

[3]Department of Chemical Engineering, Graduate School of Engineering, Nagoya University Japan

INTRODUCTION

This chapter provides a tutorial on the fundamental concept of Parallel factor (PARAFAC) analysis and a practical example of its application. PARAFAC, which attains clarity and simplicity in sorting out convoluted information of highly complex chemical systems, is a powerful and versatile tool for the detailed analysis of multi-way data, which is a dataset represented as a multidimensional array. Its intriguing idea to condense the essence of the information present in the multi-way data into a very compact matrix representation referred to as scores and loadings has gained considerable popularity among scientists in many different areas of research activities. The basic idea of PARAFAC is so flexible and general that its application is not limited to a particular field of spectroscopy confined to a specific electromagnetic probe. Examples of the application include fluorescence (Christensen et al., 2005; Rinnan et al., 2005), IR (Wu et al., 2003), NMR (Bro et al., 2010), UV (Ebrahimi et al., 2008; Van Benthem et al., 2011) and mass spectroscopy (Amigo et al., 2008). The first part of this chapter covers the theoretical background of trilinear decomposition of three-way data by PARAFAC with comparison to bilinear decomposition of two-way data by Principal component analysis (PCA). In the second part of this chapter, an illustrative example of PARAFAC analysis for threeway data obtained in an actual laboratory experiment is presented to show

how PARAFAC trilinear model can be constructed and analyzed to derive in-depth understanding of the system from the data. Thermal deformation of several types of poly lactic acid (PLA) nanocomposites undergoing grass-to-rubber transition is probed by cross-polarization magic-angle (CP-MAS) NMR spectroscopy. Namely, sets of temperature-dependent NMR spectra are measured under varying clay content in the PLA nanocomposite samples. While temperature strongly affects molecular dynamics of PLA, the clay content in the samples also influences the molecular mobility. Thus, NMR spectra in this study become a three-way dataset described as a function of both temperature and clay content. Details of the effects of the temperature and clay content on the physical state of nanocomposite are elucidated by using PARAFAC trilinear model.

PARAFAC

Multi-way data

So, what does a multi-way data look like? It is insightful first to note the data structures of two-way and three-way data. Schematic descriptions of two-way and three-way data based on external perturbation(s) are shown in Figure . 1. In a general spectroscopic measurement, external perturbations are applied to the system of interest to induce the response to the stimuli. Characteristic response of the system is presented in the form of spectrum. For example, when the thermal behaviour of a sample is studied by a spectroscopic method, such as IR, Raman and NMR, the sample is heated up to undergo thermal deformation and its molecular level variation induced by the stimulus is captured at each spectral variable, e.g. wavenumber. The spectral dataset thus obtained will be represented as a two-way array with the (i,j)th element denoting the spectral intensity value at the ith temperature and the jth wavenumber.

Now, let us consider another experiment with one more perturbation. As described above, stimulation of a single sample ends up with two-way data array. But what if we still have some more samples, whose properties (e.g. concentration) are different? We will repeat a similar experiment for every single sample. This generates multiple two-way data. Thus, the entire dataset eventually becomes a stuck of the multiple two-way data like a cube, which contains two dimensions concerning applied two perturbations. Such spectral dataset is described as a three-way array with the (i,j,k)th element denoting the spectral intensity value at the ith concentration, the jth temperature, and the kth wavenumber. For example, the samples will show the variation of their molecular structure depending on the temperature. This may be also influenced by the change in the concentration. Thus the spectral intensities of the samples are potentially influenced by the temperature as well as concentration.

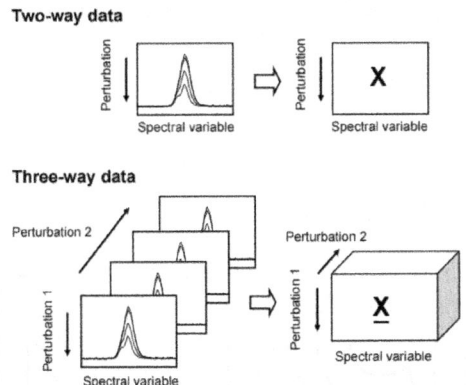

Figure. 1: Schematic illustration of two-way and three-way data.

PARAFAC model

It is possible to condense the essence of the information present in multi-way data into a very compact matrix representation referred to as scores and loadings. The basic hypothesis of factor analysis techniques is that the improved proxy of the original data matrix can be reconstructed from only a limited number of significant factors. Thus, while the score and loading matrices contain only a small number of factors, it effectively carries all the necessary information about spectral features and, eventually, it becomes possible to sorting out the convoluted information content of highly complex chemical systems. The detailed analysis of such matrices potentially brings useful insight into building a mechanistic model for understanding complex phenomena studied by spectroscopic method.

Principal component analysis (PCA) is mathematical decomposition of two-way data in terms of the orthogonal set of dominant factors, i.e., eigenvectors (Smilde et al., 2004; Shinzawa et al., 2010). Two-way data decomposition by PCA results in yielding two matrices called scores and loadings which complementarily represent the entire features broadly distributed in the two-way data as follows,

$$X = TP^t + E_{PCA} \tag{1}$$

where T and P are PCA score and loading matrices consisting of r vectors, respectively. The rank r corresponds to the number of principal components representing the significant portion of the information contained within the data matrix X. The selection of r is somewhat arbitrary. It is usually set to be a number, as small as possible but sufficiently large enough such that there are no obvious spectral features found in the residual matrix E_{PCA}. The residual matrix E_{PCA} is the portion of the original data, which is not accounted for by the first r principal components used for the data representation. The two matrices T and P complementally represent the entire features broadly distributed in X. Namely, T holds abstract information concerning the relationship among the samples and P contains summary of variable, e.g. wavenumber which provides chemically or physically meaningful interpretation to the pattern observed in T. For example, PCA of the two-way data based on temperature-dependent spectra provides T describing similar or dissimilar thermal behaviour of the sample during the perturbation period and corresponding P represent information on key molecular structure associated with such similar or dissimilar thermal behaviour of the sample.

For even more data, PARAFAC is used to decompose a multi-way data and Figure 2 illustrates graphical representation of PARAFAC operation to decompose a three-way data into score and loading vectors. PARAFAC is utilized to decompose the multi-way data into a linear combination of score and loading matrices (Smilde et al., 2004; Bro, 2004). The information on behavior induced by the perturbations is effectively described by score vectors and corresponding loading vectors provide chemically or physically meaningful interpretation to the patterns observed in the scores of the PARAFAC trilinear model.

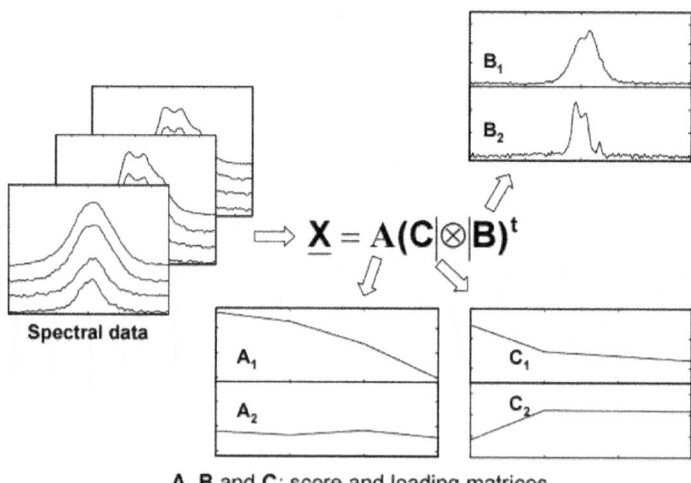

Figure. 2: Schematic illustration of PARAFAC trilinear model.

Namely, by using PARAFAC operation, $I \times J \times K$ array matrix $\underline{X}$ can be expressed in terms of a product of score and loading matrices, A, B, and C, and a residual matrix E as follows

$$X^{(I \times JK)} = A(C|\otimes|B)^t + E^{(I \times JK)}$$

(2)

where $(I \times JK)$ refers to the way that the multi-way array is unfolded. The notation $|\otimes|$ means Khatri-Rao product which operate Kronecker product $|\otimes|$ on partitioned matrices defined as

$$C|\otimes|B = [c_1 \otimes b_1 \quad c_2 \otimes b_2 \quad \cdots \quad c_F \otimes b_F]$$

(3)

In PARAFAC analysis, the set of matrices A, B and C are usually obtained by iteratively solving alternating least-squares (ALS) problems $\min_{A,B,C} \|X^{(I \times JK)} - A(C|\otimes|B)^t\|$ over A for fixed B and C, as well as the minimization over B or C in the similar matrix operation manner under appropriate model constraints, such as the non-negativity of concentration and spectral intensity (Bro & de Jong, 1997; Bro & Sidiropoulos, 1998). General procedure of PARAFAC becomes as follows, Initialize B and C to obtain Z as

$$Z = C \otimes B$$

(4)

A is given by

$$A = X^{(I \times JK)}Z(Z^tZ)^+$$

(5)

where the superscript + means the Moore-Penrose inverse. Then update Z as

$$Z = C \otimes A$$

(6)

B is obtained as

$$B = X^{(J \times IK)}Z(Z^tZ)^+$$

(7)

Update Z as

$$Z = B \otimes A$$

(8)

C is given by

$$C = X^{(K \times IJ)}Z(Z^tZ)^+$$

(9)

If the residual between the original $\underline{X}$ and reconstructed $\underline{X}$ by Eq. 2 is greater than error criteria, one repeats Eqs. (4)-(9) until convergence.

The initial estimates for B and C is important to obtain sufficient minimization of the error criteria (Shinzawa et al., 2007, 2008a & 2008b).

Although ALS algorithm usually offers an eventual convergence to the optimal solution with a sufficiently large number of iterations, it sometimes reaches the suboptimal local minimum (Jiang et al., 2003 & 2004). Unfortunately, such local convergence does not usually offer a global minimum, but it may just be stuck in a local minimum, producing insufficient solution. The major cause of the suboptimal local convergence may be a poor initial estimation. One possible solution for this problem is to select proper initial estimate which is less sensitive to the presence of a local minimum, e.g. via signaler value decomposition (Bro & de Jong, 1997; Bro & Sidiropoulos, 1998; Wang et al., 2006; Awa et al., 2008).

EXAMPLE

PLA nanocomposite

A pertinent example for PARAFAC analysis based on NMR spectra of PLA nanocomposites is provided here to show how certain useful information can be effectively extracted from an actual laboratory experiment. Figure . 3 shows the molecular structure of PLA. PLA polymer is made up of many long chains consisting of the repeat unit shown in the Figure ure. PLA is derived from renewable resources, such as corn starch via fermentation and it is biodegradable under the right conditions, such as the presence of oxygen (Tsuji et al., 2010). Thus, PLA is a possible candidate of a new class of renewable polymers as a substitute for the petrochemical polymers. However, the physical properties of PLA are inadequate for the replacement of conventional commodity plastics in many applications. Nanocomposite is a technique to improve the physical strength, thermal resistance and gas barrier by the dispersion of nanoclay into the polymer (Katti et al., 2006). The improvement of such polymer properties by using nanocomposite is one of the primary areas of interest due to its potential applications. The polymer nanocomposites are generally formed by the addition of a small amount of nanoclay dispersion.

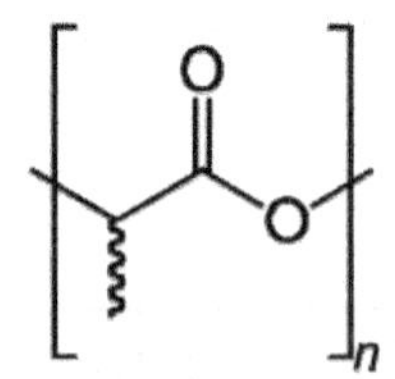

Figure. 3: Molecular structure of PLA.

Figure. 4 shows a schematic illustration of polymer nanocomposite. A typical form of the nanocomposite is intercalated nanocomposite, in which the unit cells of clay structure are expanded by the insertion of polymer into the interlayer spacing, while the periodicity of clay crystal structure is maintained. Most commonly, montmorillonite (MMT) is used as clay due to its highly expansive characteristic (Suguna Lakshmi et al., 2008; Cervantes-Uc et al., 2009). The MMT unit cell is composed of aluminum octahedra sandwiched between two silica tetrahedra with the unit cell dimension of about 1 nm in thickness. For facilitating better miscibility of hydrophobic polymer with the clay and increasing the spacing of the interlayer clay gallery, it is often treated with organic modifiers which are generally long carbon chain compounds with alkylammonium or alkylphosphonium cations.

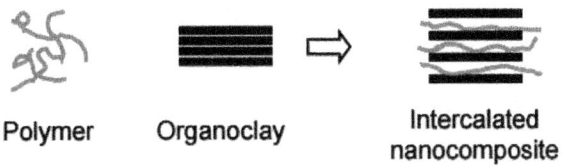

**Polymer   Organoclay   Intercalated
                                        nanocomposite**

Figure. 4: Schematic illustration of polymer nanocomposite.

PLA nanocomposite samples used in this study were prepared with PLA (Teramac®, Unitika) and organically modified clay (S-BEN W®, Hojun). The samples were put into a Labo-plastomill consisting of a 30C150 kneader and an R100 mixer (Toyo Seiki Seisaku-sho, Ltd., Tokyo) to melt-blend at 190 °C and 50 rpm for about 10 minutes. Pellets thus obtained were pressed into 0.2 mm thick sheet sandwiched between two thick Teflon® films by a hot press at 190 °C. Figure . 5 represents the effect of nanocomposite on PLA probed by Thermomechanical analysis (TMA). TMA is a technique to monitor the physical deformation of object under a constant load, while varying the temperature. For example, in this case, the elongation of the PLA nanocomposite samples (clay content = 0, 5 and 15 wt%) were measured by imposing a 9.8 mN load, while varying the temperature from 35 to 140 °C at a rate of 10 °C per a minute. The elongation of the samples starts when the temperature reaches glass transition temperature (T_g) of PLA, i.e. approximately 60 °C (Zhang et al., 2010). Then it gradually increases with the increase of temperature and it finally reached constant levels at the close of the observation period, indicating that the observed plastic deformation is closely related to glass-to-rubber transition of the amorphous component of PLA. It is also noted that the samples results in the different levels of elongation depending on the clay content. For example, the neat PLA sample shows 14.4 % of elongation. In contrast, the PLAnanocomposite including 15 wt % of clay ends up with 9.1 % of elongation. The leveling off of the elongation indicates the formation of a network structure due to the presence of physical crosslinks created by the crystalline domain. Although such observation effectively detects the macroscopic changes in the mechanical properties caused by the presence of clay particles, additional fundamental molecular level understanding of the reinforcement mechanism is also desired. Spectroscopic method should become an important tool to probe the phenomena at the molecular level.

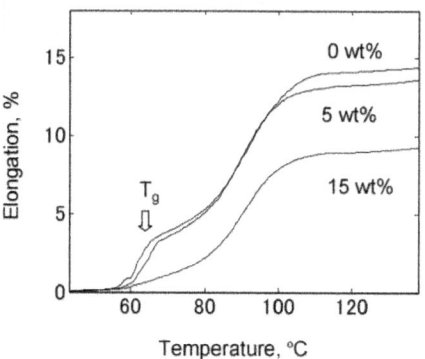

Figure. 5: Physical property of PLA samples proved by TMA.

PALAFAC analysis of NMR spectra of PLA nomocomposites

The temperature-dependent NMR spectra of the PLA samples collected under the varying temperature from 20 to 80 °C are shown in Figure . 6. Cross polarization-magic angle spinning (CP-MAS) NMR experiments were carried out on a Varian 400 NMR system spectrometer operated at 100.56 MHz for 13C resonance with a cross polarization contact time of 2 ms (Fawcett, 1996). A zirconium oxide rotor of 4 mm diameter was used to acquire the NMR spectra at a spinning rate of 15 kHz. Each sample was packed into a 4 mm cylinder-type MAS rotor. A set of temperature-dependent NMR spectra were obtained under varying ambient temperature from 20 to 80 °C at every 20 °C step. The heating rate was approximately 10 °C per an hour. Samples of semicrystalline polymers prepared from their melt possess complex supermolecular structure consisting of crystalline lamellae embedded in an amorphous matrix (Wunderlich, 1980). PLA essentially undergoes highly convoluted transition process, when temperature and its constitution are altered. These transitions include the melting of ordered molecular segments, as well as the grass-to-rubber transition and other relaxation of process of the amorphous component (Zhang et al., 2005; Meaurio et al., 2006).

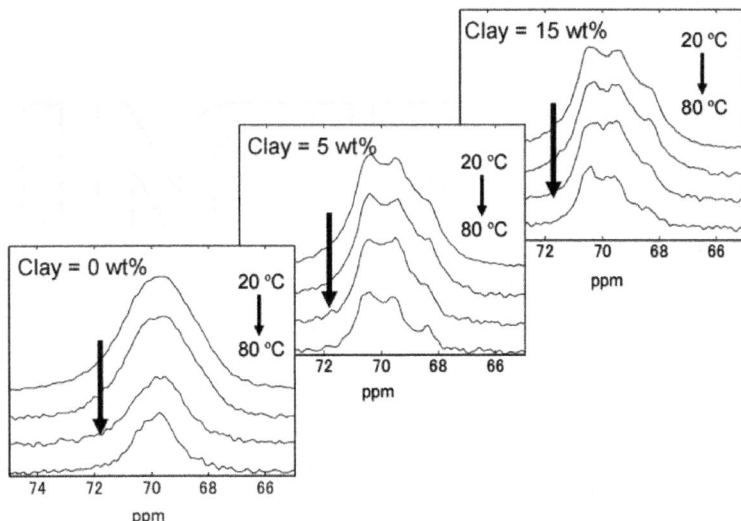

Figure. 6: Temperature-dependent CP-MAS NMR spectra of neat PLA and PLA nanocomposite samples.

The CP-MAS technique is ideal for the observation of ^{13}C spectra of solid samples. Since the local environment of a chemical group in solids are generally rigid, this leads to further considerations for crystallography or, more generally, molecular packing (Fawcett, 1996). The CPMAS NMR study of semicrystalline PLA samples is often complicated by the presence of overlapped contributions from coexisting crystalline and amorphous. For example, the unimodal peak observed around 69.5 ppm is assignable to CH structure which represents mobility of the main chain of the PLA (Tsuji et al., 2010; Kister et al., 1998). It is noted that the peak intensity gradually decreases with the increase of the temperature. This may be explained as the decrease in the cross polarization efficiency by the change in the molecular dynamics during the heating. Thus, the variation of the spectral intensity here reflects the structural alternation of PLA induced by the temperature. More importantly, careful comparison of the samples reveals that the main feature of the NMR spectra of the three samples looks somewhat different. For example, the temperaturedependent NMR spectra of the PLA nanocomposite including 15 wt% clay provides specific three peaks at 70.5, 69.5 and 68.4, indicating the presence of the crystalline structure in the sample (Tsuji et al., 2010; Kister et al., 1998).

When the sample has no clay in the system,these crystalline peaks are disappeared and compensated by the development of seemingly unimodal peak probably assigned to the amorphous of PLA (Tsuji et al., 2010; Kister et al., 1998). This indicates that the presence of the clay substantially influences supermolecular structure of the PLA. Consequently, it is very likely that the change in the spectral feature of the three-way data is closely related to temperature and clay content of the system. Thus, in turn, the fully detailed analysis of the data provides an interesting opportunity to probe the nature of the PLA nanocomposite by elucidating the variation of the NMR spectral intensity induced by the each perturbation with PARAFAC trilinear decomposition.

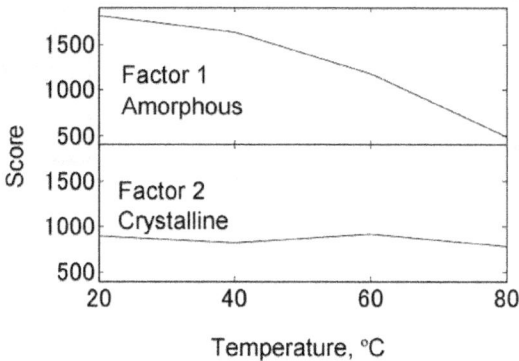

Figure. 7: Score vectors in score matrix A representing thermal behaviours of amorphous and crystalline components in PLA samples.

Figure. 7, 8 and 9 show results obtained from A, B and C matrices derived from PARAFAC analysis of the three-way NMR spectral data collected under varying temperature and clay content, respectively. Two major factors are indicated here, reflecting the fact that there are two species present in the system. One of the important benefits derived from PARAFAC decomposition of the multi-way data is the ability to rationally clarify the effect of the applied perturbations. For example, the matrix A represents abstract information on the temperature-induced behavior of the PLA under the influence of the clay content. In contrast, the matrix C holds essential information on the spectral intensity variation induced by the addition of the clay under the influence of the temperature. The matrix B contains loading vectors which provides chemical or physical interpretation to the patterns observed in the score matrices A and C.

It is noted that the loading vector of the first component of the matrix B (Figure . 8) resembles the spectral feature of the amorphous component of PLA. The loading vector of the second component of the matrix B shows characteristic three peaks assignable to crystalline component of the PLA. Thus it is most likely that the second factor represents thermal behaviours of the crystalline components in PLA samples. Once the assignments for the loading vectors are established, it becomes possible to provide chemically meaningful interpretation to the score matrices A and C representing the dynamic behaviour of the components induced by the perturbations. For example, the score vector of the first factor in the matrix A represents the temperature-induced behaviour of the amorphous component of the PLA. On the other hand the score vector of the second factor means that of the crystalline component of the PLA. It is noted the score vector of the amorphous components exhibits obvious decrease with the temperature and such decrease becomes significant when the temperature exceeds its T_g. In contrast, the change in the score value of the crystalline component is small, indicating no major variation during the heating process.

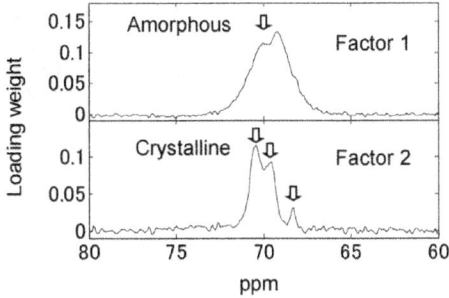

Figure. 8: Loading vectors in score matrix B representing thermal behaviours of amorphous and crystalline components in PLA samples.

The predominant variation of the amorphous component in the temperature region is explained as its glass-to-rubber transition. The change induced in the temperature region is associated with the Micro-Brownian motion of the PLA polymer segment. At a low temperature the amorphous regions of a polymer are in the glassy state. In this state the molecules are frozen on place. They may be able to vibrate slightly, but do not have any segmental motion. When the polymer is heated up to reach its T_g, the molecules can start to wiggle around to become rubbery state. Such segmental motion predominantly occurs in amorphous region of PLA while such motion is strongly restricted in systematically folded crystalline lamellae structure. Thus, it is very likely the observed change of the amorphous is related to glass-to-rubber transition of the amorphous component. Now it is important to point out again that the predominant elongation in the TMA occurred around T_g. This elongation behaviour agrees well with the thermal behaviour of the amorphous component observed in the score matrix A. It thus suggests the physical elongation of the samples is essentially associated with the glass-to-rubber transition mainly occurred in the amorphous region. It also becomes possible to provide the detailed interpretation to the pattern observed in the matrix C representing the clay-induced behaviours of amorphous and crystalline components in the PLA samples. The gradual decrease of the score of the first factor can be explained as the decrease of the amorphous component and the change in the sore of the first factor corresponds to the increase in the crystalline component by the addition of the clay.

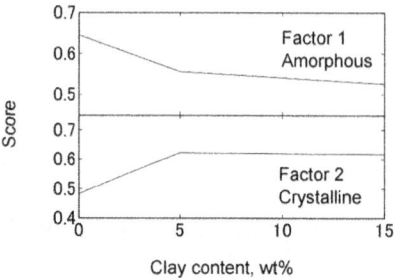

Figure. 9: Score vectors in score matrix C representing clay-induced behaviours of amorphous and crystalline components in PLA samples.

It seems that the decrease in the amorphous is compensated by the development of the crystalline structure. In other words, the clay increases the frequency of the spontaneous nucleation of the PLA crystals. PARAFAC trilinear model of the three-way NMR data of the PLA nanocomposites reveals that the crystalline and amorphous structures of the PLA nanocomposites undergo different transition under the heating. Namely, the change in the micro-Brownian motion of the polymer segments mainly occurs in the amorphous region. In addition, the different variations between the crystalline and amorphous component suggest the different effects of the presence of clay particles on them, i.e. nucleating effect of the clay. The decrease in the amorphous portion should result in the reduction of the structure undergoing the glass-torubber transition. Such variation of the crystallinity agrees well with the decreased elongation observed in the TMA. For example, in Figure . 5, the level of the elongation starting around T_g clearly decreases with the inclusion of the clay. This hypothesis is also clearly supported with corresponding transmission electron microscope (TEM) images and differential scanning calorimetry (DSC) results of the PLA nanocomposite sample. Figure . 10 represents the TEM images of the PLA nanocomposite sample including 15 wt% clay. For example, in Figure . 10(a), one can see that the clay is broadly distributed over the PLA matrix. On the other hand, Figure . 10(b) reveals that some parts of the interlayer gallery is obviously extended, suggesting the insertion of the PLA polymer into the clay layers, namely intercalation. DSC curves of the PLA samples, represented in Figure . 11, clearly show the presence of glass transition temperature around 60 °C. It is important to point out that this glass-to-rubber transition of amorphous component agrees well with the change in the elongation observed in the TMA. More importantly, the samples also provide obvious crystallization peak around 110 °C. The crystallization peak shows gradual increase by the inclusion of the clay content, suggesting quantitative increase in the amount of the crystalline structure. Thus, it is very likely that the clay works as the nucleating agent to increase the frequency of the spontaneous nucleation of the PLA crystals.

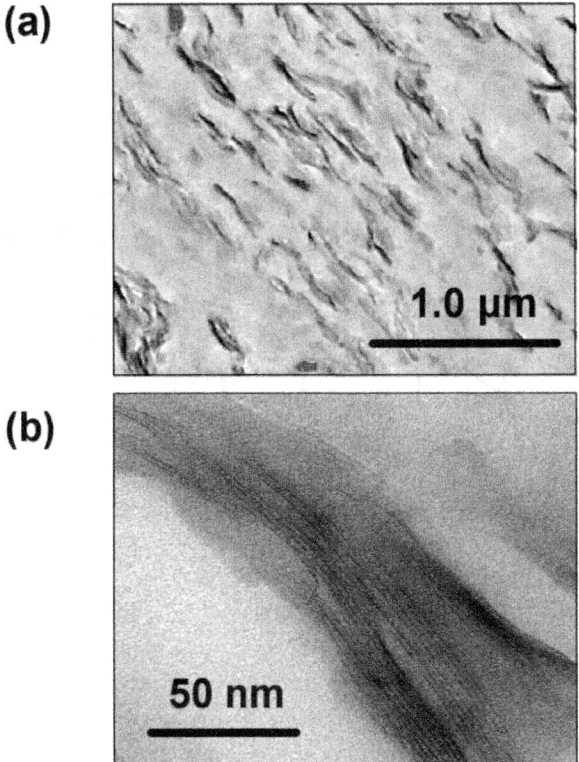

Figure. 10: TEM images of PLA nanocomposite sample.

All the results put together, it provides overall picture of the system. When the clay is dispersed in the PLA matrix, the PLA polymer located at interlayer or around surface layer of the clay develops crystalline structure more frequently. The generation of the crystalline structure of PLA is compensated by the decrease of the amorphous content. This should decrease the structural portion substantially undergoing glass-to-rubber transition above T_g. Thus, in turn, it restricts the elongation of the samples during the heating process under a certain level of load. Consequently, it is demonstrated that PARAFAC analysis of the three way data of the PLA nanocomposite samples effectively elucidates the mechanisms of the improvement of the mechanical property by the clay. By carrying out detailed band position shift analysis of the three way data of the temperature- and clay- dependent NMR spectra of the PLA samples, it becomes possible to extract chemically meaningful information concerning the variation of the crystalline structure closely associated with the nanocomposite system.

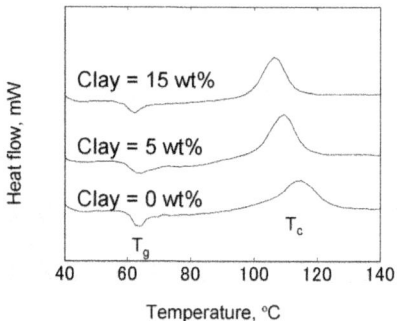

Figure. 11: DSC curves of neat PLA and PLA nanocomposite samples.

CONCLUSION

The basic background of PARAFAC and its practical example based on the temperaturedependent NMR spectra of the PLA nanocomposite samples are presented. The central concept of PARAFA decomposition of multi-way data lies in the fact that it can condense the essence of the information present in the multi-way data into a very compact matrix representation referred to as scores and loadings. Thus, while the score and loading matrices contain only a small number of factors, it effectively carries all the necessary information about spectral features and leads to sorting out the convoluted information content of highly complex chemical systems. The effect of PLA nanocomposite is studied by the PARAFAC analysis of the temperaturedependent NMR spectra of several PLA nanocomposite samples including different clay contents. The PARAFAC analysis for the three-way data of the PLA nanocomposites revealed that the crystalline and amorphous structures of the PLA nanocomposites substantially undergo different transition under the heating. Namely, the change in the micro-Brownian motion of the polymer segments mainly occurs in the amorphous region when the PLA samples are heated up to their Tg. It also revealed that clay potentially works as nucleating effect of the clay. Namely, it increases the frequency of the spontaneous nucleation of the PLA crystals. Thus, in turn, the change in the population of the rigid crystalline and rubbery amorphous provides the improvement of the physical property. Consequently, it is possible to derive in-depth understanding of the PLA nanocomposites.

ACKNOWLEDGMENT

A part of this work was financially supported by NEDO "Technological Development of Ultra-hybrid Materials" Project.

REFERENCES

1. Amigo, J. M., Skov, T., Coello, J., Maspoch, S. & Bro, R. (2008) Solving GC-MS problems with PARAFAC2. TrAC Trends in Analytical Chemistry, Vo. 27, No. 8, pp. 714-725

2. Awa, K., Okumura, T., Shinzawa, H., Otsuka, M. & Ozaki, Y. (2008). Self-modeling Curve Resolution (SMCR) Analysis of Near-infrared (NIR) Iimaging Data of Pharmaceutical Tablets. Analytica Chimica Acta, Vol. 619, No. 1, pp. 81-86

3. Bro, R. (2004). PARAFAC. Tutorial and applications. Chemometrics and Intelligent Laboratory Systems, Vol. 37, No. 2, pp. 149-171

4. Bro, R. & de Jong, S. (1997). A fast non-negativity constrained linear least squares algorithm for use in multi-way algorithms. Journal of Chemometrics, Vol. 11, No. 5, pp. 393-401

5. Bro, R. & Sidiropoulos, N. (1998). Least squares algorithms under unimodality and nonnegativity constraints. Journal of Chemometrics. Vol. 12, No. 4, pp. 223–247

6. Bro, R., Viereck, N., Toft, M., Toft, H., Hansen, P. I. & Engelsen, S. B. (2010). Mathematical chromatography solves the cocktail party effect in mixtures using 2D spectra and PARAFAC. TrAC Trends in Analytical Chemistry, Vol. 29, No. 4, pp. 281-284

7. Cervantes-Uc, J. M., Espinosa, J. I. M., Cauich-Rodriguez, J. V., Avila-Ortega, A., VazquezTorres, H., Marcos-Fernandez. A. & San Roman, J. (2009). TGA/FTIR studies of segmented aliphatic polyurethanes and their nanocomposites prepared with commercial montmorillonites. Polymer Degradation and Stability, Vol. 94, No. 10, pp. 1666-1677

8. Christensen, J., Miquel Becker, B. & Frederiksen, C. S. (2005). Fluorescence spectroscopy and PARAFAC in the analysis of yogurt. Chemometrics and Intelligent Laboratory Systems, Vol. 75, No. 2, pp. 201-208

9. Ebrahimi, D., Kennedy, D. F., Messerle, B. A. & Hibbert, D. B. (2008).

High throughput screening arrays of rhodium and iridium complexes as catalysts for intramolecular hydroamination using parallel factor analysis. Analyst, Vol. 133,

10. No. 6, pp. 817-822

11. Fawcett, A. H. (1996). Polymer Spectroscopy, John Wiley & Sons, ISBN 0471960292, West Sussex, UK

12. Jiang, J.-H., Šašic, S., Yu, R.-Q. & Ozaki, Y. (2003). Resolution of two-way data fromspectroscopic monitoring of reaction or process systems by parallel vector analysis (PVA) and window factor analysis (WFA): inspection of the effect of mass balance, methods and simulations. Journal of Chemometrics, Vol. 17, No. 3, pp. 186-197

13. Jiang, J.-H., Liang, Y. & Ozaki, Y. (2004). Principles and methodologies in self-modeling curve resolution. Chemometrics and Intelligent Laboratory Systems, Vol. 71, No. 1, pp. -12

14. Katti, K. S., Sikdar, D., Katti, D. R., Ghosh, P. & Verma. D. (2006). Molecular interactions in intercalated organically modified clay andclay–polycaprolactam nanocomposites: Experiments and modeling. Polymer, Vol. 47, No. 1, pp. 403-414

15. Kister, G., Cassanas, G. & Vert, M. (1998). Structure and morphology of solid lactideglycolide copolymers from 13C n.m.r., infra-red and Raman spectroscopy. Polymer,Vol. 39, No. 15, pp. 3335-3340

16. Meaurio, E., Zuza, E., López-Rodríguez. N. & Sarasua, J. R. (2006). Conformational Behavior of Poly(L-lactide) Studied by Infrared Spectroscopy. Journal of Physical Chemistry B, Vol. 110, No. 11 pp. 5790-5800

17. Rinnan, Å. & Andersen, C. M. (2005). Handling of first-order Rayleigh scatter in PARAFAC modelling of fluorescence excitation–emission data. Chemometrics and Intelligent Laboratory Systems, Vol. 76, No. 1, pp. 91-99

18. Shinzawa, H., Iwahashi, M., Noda, I. & Ozaki. Y. (2008a). Asynchronous Kernel Analysis for Binary Mixture Solutions of Ethanol and Carboxylic Acids. Journal of Molecular Structure, Vol. 883-884, No. 30, pp. 27-30

19. Shinzawa, H., Iwahashi, M., Noda, I. & Ozaki, Y. (2008b). A Convergence Criterion in Alternating Least Squares (ALS) by Global Phase Angle. Journal of Molecular Structure, Vol. 883-884, No. 30, pp. 73-78

20. Shinzawa, H., Jiang, J.-H., Iwahashi, M., Noda, I. & Ozaki, Y. (2007). Self-modeling Curve Resolution (SMCR) by Particle Swarm Optimization (PSO). Analytica Chimica Acta, Vol. 595, No. 1-2, pp. 275-281

21. Shinzawa, H., Awa, K., Kanematsu, W. & Ozaki, Y. (2010). Multivariate Data Analysis for Raman Spectroscopic Imaging. Journal of Raman Spectroscopy, Vol. 40, No. 12, pp.720-1725

22. Smilde, A., Bro, R. & Geladi, P. (November 2004). Multi-way Analysis: Applications in the Chemical Sciences. John Wiley & Sons, ISBN: 0471986911, West Sussex, UK

23. Suguna Lakshmi, M., Narmadha, B. & Reddy, B. S. R. (2008). Enhanced thermal stability and structural characteristics of different MMT-Clay/ epoxy-nanocomposite materials.

24. Polymer Degradation and Stability, Vol. 93, No. 1, pp 201-213

25. Tsuji, H., Kamo, S. & Horii, F. (2010). Solid-state 13C NMR analyses of the structures of crystallized and quenched poly(lactide)s: Effects of crystallinity, water absorption, hydrolytic degradation, and tacticity. Polymer, Vol. 51, No. 10, pp. 2215-2220

26. Van Benthem, M. H., Lane, T. W., Davis, R. W., Lane, R. D. & Keenan, M. R., (2011).

27. PARAFAC modeling of three-way hyperspectral images: Endogenous fluorophores as health biomarkers in aquatic species, Chemometrics and Intelligent Laboratory Systems, Vol. 106, No. 1, pp. 115-124

28. Wang, Z.-G., Jiang, J.-H., Ding, Y.-J., Wu, H.-L. & Yu, Ru-Qin., (2006). Trilinear evolving factor analysis for the resolution of three-way multi-component chromatograms. Analytica Chimica Acta, Vol. 558, No. 1-2, pp. 137-143

29. Wu, Y., Yuan, B., Zhao, J.-G. & Ozaki, Y. (2003). Hybrid Two-Dimensional Correlation and Parallel Factor Studies on the Switching Dynamics of a Surface-stabilized Ferroelectric Liquid Crystal. Journal of Physical Chemistry B, Vol. 107, No. 31, pp. 7706–7715

30. Wunderlich, B. (1980). Macromolecular Physics: Vol. 2 Crystal Nucleation, Growth, Annealing, Academic Press, New York, USA

31. Zhang, J., Li, C., Duan, Y., Domb, A. J. & Ozaki, Y. (2010). Glass transition and disorder-toorder phase transition behavior of poly(l-lacticacid) revealed by infrared spectroscopy. Vibrational Spectroscopy, Vol. 53, No. 2, pp. 307–310

32.   Zhang, J., Sato, H., Tsuji, H., Noda, I. & Ozaki, Y. (2005). Differences in the CH3…O=C interactions among poly(L-lactide), poly(L-lactide)/ poly(D-lactide) stereocomplex, and poly(3-hydroxybutyrate) studied by infrared spectroscopy. Journal of Molecular Structure. Vol. 735–736, No. 14, pp. 249–257

Chapter 12

USING PRINCIPAL COMPONENT SCORES AND ARTIFICIAL NEURAL NETWORKS IN PREDICTING WATER QUALITY INDEX

Rashid Atta Khan[2], Sharifuddin M. Zain[2], Hafizan Juahir[1], Mohd Kamil Yusoff[1] and Tg Hanidza T.I.[1]

[1]Department of Environmental Science, Faculty of Environmental Study, University Putra Malaysia, Serdang

[2]Chemistry Department, Faculty of Science, University of Malaya, Kuala Lumpur Malaysia

INTRODUCTION

The management of river water quality is a major environmental challenge. One of the major challenges is in determining point and non-point sources of pollutants. Industrial and municipal wastewater discharges can be considered as constant polluting sources, unlike surface water runoff which is seasonal and highly affected by climate. According to Aiken et al. (1982), 42 tributaries in Peninsular Malaysia are categorized as very polluted including the Langat River. Until 1999, there were about 13 polluted tributaries and 36 polluted rivers due to human activities such as, industry, construction and agriculture (Department of Environment, Malaysia (DOE), 1999). In 1990, there were 48 clean rivers classified as clean but the number is reduced to 32 rivers in 1999 (Rosnani Ibrahim, 2001).

Surface water pollution is identified as the major problem affecting the Langat River Basin in Malaysia. Increase in developing areas within the river basin has in turn increased pollution loading into the Langat River. To avoid further degradation, the DOE have installed telemetric stations along the river basin to continuously monitor the water quality. As a result, abundant data were collected since 1988. There are 927 monitoring stations located within 120 river basins throughout Malaysia. Water quality data were used to determine the water quality status and to classify the rivers based on water quality index (WQI) and Interim National Water Quality Standards for Malaysia (INWQS).

WQI provides a useful way to predict changes and trends in the water quality by considering multiple parameters. WQI is calculated from six selected water quality variables, namely dissolved oxygen (DO), biochemical oxygen demand (BOD), chemical oxygen demand (COD), suspended solid (SS), ammonical nitrogen (AN) and pH (DOE, 1997). It is a well-known phenomenon that the contribution of pollution loading into river systems from the environment involves a complex interaction of many factors (e.g. chemical, physical and meteorological interaction). These primary pollutants are emitted from land use activities surrounding the river basin (e.g. agriculture, forest, urban, industrial and others) Rapid urbanization along the Langat River plays an important role in the increase of point source (PS) and non-point source (NPS). In view of this complex interaction, use of modelling techniques to solve this problem, is needed. However, the problem of obtaining models that adequately represent the dynamic behaviour of field data is not easy. Lack of good understanding and description of the phenomena involved, the availability of reliable and complete field data set and the estimation of the numerous parameters involved are the major factors contributing to this problem. Beck (1986) noted that, increase in model complexity will undoubtedly increase the number of parameters, leading to the problems of identification.

Applications of ANN (Artificial Neural Networks) to environmental problems are becoming more common (Silverman and Dracup, 2000; Scardi, 2001; Recknagel et al., 2002; Bowden et al., 2005; Muttil and Chau, 2007). The applications of ANN, which are computing systems that were originally designed to simulate the structure and function of the brain (Rumelhart et al, 1986) is a relatively new concept in environmental modeling. If trained properly, a neural network model is capable of 'learning' linear as well as the nonlinear features in the data (Elsner and Tronis, 1992).

ANN consists of a set of simple processing units (neurons) arranged in a defined architecture and connected by weighted channels which act to transform remotely-sensed data into a classification. The classification techniques of ANN are unlike the conventional ones. It is distribution-free, may sometimes use small training sets (Hepner et al., 1990) and, once trained; it is rapid computationally, which will be of value in processing large data sets (Gershon and Miller, 1993). Furthermore, ANNs have been shown to be able to map land cover more accurately compared to many widely used statistical classification techniques (Benediktsson et al., 1990; Foody et al., 1995) and alternatives such as evidential reasoning (Peddle et al., 1994).

It has been proposed that the best tool to model non-linear environmental relationship is ANN (Zhang and Stanley, 1997; Jain and Indurthy, 2003). Research have been undertaken at Imperial College, London which attempts to investigate the capability of ANN approach in modelling spatial and temporal variations in river water quality (Clarici, 1995). ANNs were used as a predictive model to predict cyanobacteria Anabaena spp. in the River Murray, South Australia (Maier et al., 1998). DeSilets et al. (1992), have also used ANN to predict salinity. Ha and Stenstrom (2003), proposed a neural network approach to examine the relationship between storm water quality and various types of land use. Zainudin, 2001; Mohd Ekhwan Toriman and Hafizan Juahir, 2003; Hafizan Juahir et al., 2003a,b; Hafizan et al, 2004a,b; 2005; Ruslan Rainis et al., 2004). An approach for identifying possibilities of water quality improvement could be developed by using this concept. Such information could provide opportunities for better river basin management to control river water pollution in Malaysia. In the Malaysian context, Hafizan Juahir et al. (2003a) showed that the ANN model gives a better performance compared to the autoregressive integrated moving average (ARIMA) model in forecasting DO. The use of ANN for river regulation (Mohd. Ekhwan Toriman and Hafizan Juahir, 2003) and the application of the second order back propagation method (Hafizan Juahir et al., 2004a) on water quality of the Langat River have also been demonstrated.

In natural environment, water quality is a multivariate phenomenon, at least as reflected in the multitude of constituents which are used to characterize the quality of water body. Water quality is very difficult to model because of the different interactions between pollutants and meteorological variables. The principal component analysis (PCA) is one of the approaches to avoid this problem and has received increasing attention as an accepted method in environmental pattern recognition (Simeonov et al., 2003; Wunderline et al., 2001; Helena et al., 2000; Loska and Wiechula, 2003)

In natural environment, water quality is a multivariate phenomenon, at least as reflected in the multitude of constituents which are used to characterize the quality of water body. Water quality is very difficult to model because of the different interactions between pollutants and meteorological variables. The principal component analysis (PCA) is one of the approaches to avoid this problem and has received increasing attention as an accepted method in environmental pattern recognition (Simeonov et al., 2003; Wunderline et al., 2001; Helena et al., 2000; Loska and Wiechula, 2003)

METHODOLOGY

The data and monitoring sites

The water quality data in this study were obtained from seven stations along the main Langat River (Figure . 1).

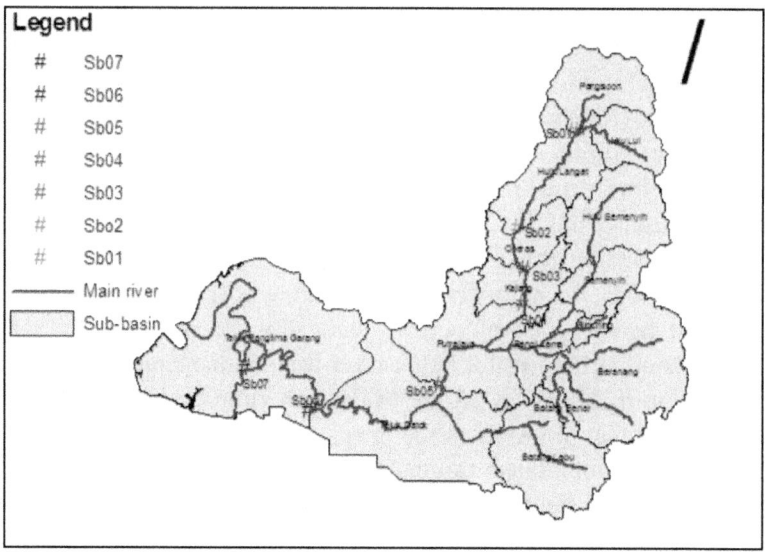

Figure. 1: Data from seven water quality stations (Sb) were selected in this study along the main river.

The water quality monitoring stations are manned by the DOE and Ministry of Natural Resource and Environment of Malaysia. The selected stations are illustrated in Table 1. The data used in the study is from September 1995 to May 2002. Seven sites were chosen, namely, Teluk Panglima Garang (site 7), Teluk Datok (site 6), Putrajaya (site 5), Kajang (site 4), Cheras (site 3), Hulu Langat (site 2), Pangsoon and Ulu Lui (site 1). Sites 3 to site 7 are located in the region of high pollution load as there are a several wastewater drains situated in the middle and downstream of the Langat River basin. Site 2 is partly situated in the middle stream region, designated as moderately polluted. Site 1 and a part of site 2 are located upstream of the Langat River, in an area of relatively low river pollution. It is worth mentioning here that some stations have missing data and not all stations were consistently sampled.

Although there are 30 water quality parameters available, only 23 completely monitored parameters were selected. A total of 254 samples were used for the analysis. The 23 water quality parameters were dissolved oxygen (DO), biological oxygen demand (BOD), electrical conductivity (EC), chemical oxygen demand (COD), ammoniacal nitrogen (AN), pH, suspended solids (SS), temperature (T), salinity (Sal), turbidity (Tur), dissolved solid (DS), total solid (TS), nitrate (NO), chlorine (Cl), phosphate (PO), zinc (Zn), calcium (Ca), iron (Fe), potassium (K), magnesium (Mg), sodium (Na), E.coli and coliform.

Table 1: DOE sampling station at study area

DOE Station No.	Study Code	Distance From Estuary (km)	Grid Reference	Location
2814602	Sb07	4.19	2°. 52.027′N 101° 26.241′E	Kampung Air Tawar (penghujung jalan)
2815603	Sb06	33.49	2° 48.952′N 101° 30.780′E	Telok Datuk, near Banting Town
2817641	Sb05	63.43	2° 51.311′N 101° 40.882′E	Bridge at Kampung Dengkil
2918606	Sb04	81.14	2° 57.835′N 101° 47.030′E	Near West Country Estate
2917642	Sb03	86.94	2° 59.533′N 101° 47.219′E	Kajang bridge
3017612	Sb02	93.38	3° 02.459′N 101° 46.387′E	Junction to Serdang, Cheras at Batu 11
3118647	Sb01	113.99	3° 09.953′N 101° 50.926′E	Bridge at Batu 18

Principal component analysis

In this work, PCA was performed on the above mentioned water quality parameters to rank their relative significance and to describe their interrelation patterns. Chosen PC scores of the 23 water quality parameters were used as input variables in ANN model to predict the WQI. The principal components (PCs) can be expressed as

$$z_{ij} = a_{i1}x_{1j} + a_{i2}x_{2j} + ... + a_{im}x_{mj}$$

(1)

Where z is the component score, a is the component loading, x the measured value of variable, i is the component number, j is the sample number and m is the total number of variables. The PCs generated by PCA are sometimes not readily interpreted; therefore, it is advisable to rotate the PCs by varimax rotation. Varimax rotation ensures that each variable is maximally correlated with only one PC and a near zero association with the other components (Abdul-Wahab et al., 2005; Sousa et al., 2007). Varimax rotations applied on the PCs with eigenvalues more than 1 are considered significant (Kim and Mueller, 1987) where the typical criteria are 75-95% of total variance (Chen and Mynett, 2003). The rotations were carried out, in order to obtain new groups of variables. Variables with communality greater than 0.7 are considered, having significant factor loadings (Stevens, 1986).

Artificial neural networks for WQI prediction

In this work, the back propagation (BP) ANN was used in the development of all the prediction models. The Activation Transfer Function of a back-propagation network is usually a differentiable Sigmoid (S-shape) function, which helps to apply the non-linear mapping from inputs to outputs. A three layer back-propagation ANN is used in this study. The number of input and output neurons is determined by the nature of the problem under study. In this study, the networks were trained, tested and validated with one hidden layer and 1 to 10 hidden neurons. This choice was based on the work of Jiang et al. (2004), who found that the results with one hidden layer was better than that of two hidden layers, and the best performance was obtained using a structure with 3 to 6 neurons in the hidden layer. The output neuron (layer) gives the predicted WQI value. Two different types of ANN models were developed. In the first type, prediction was performed based on the original PCs. In the second type of ANNs developed, scores of rotated (varimax rotation) PCs (ANN-RPCs) with eigenvalues greater than 1 were selected as input. For this model, prediction of WQI was performed using two to six rotated principal components separately. The original PCs and rotated PCs (RPCs) data sets consist of 305 observations (305 rows) and are divided into training, testing and validating phases for WQI prediction. The ANN predicted WQI values are compared to the WQI values calculated using the DOE-WQI formula which is based on 6 water quality parameters, namely the DO, COD, BOD, AN, SS and pH (DOE, 1997). The input data matrix consists of 23 water quality variables (column) and 305 observations (rows) [23×305]. The observed data for each station is arranged according to time of observation from September 13, 1995 to June 7, 2002. Table 2 describes the data structure. The validation data is at least 10% of the whole data set, with 75% training set and 25% testing set data (Kuo et al., 2007).

Table 2: The data structure for ANN prediction model

No. of Observations	Input parameters								Output
	$Input_1$	$Input_2$	$Input_3$	.	.	.	$Input_{23}$	$Output_1$
1	$Obs_{1,1}$	$Obs_{1,2}$	$Obs_{1,3}$	.	.	.	$Obs_{1,23}$	$O_{1,1}$
2	$Obs_{2,1}$	$Obs_{2,2}$	$Obs_{2,3}$	.	.	.	$Obs_{2,23}$	$O_{2,1}$
				.	.	.		
				.	.	.		
...	....	....	...	....	....	....	...	
305	$Obs_{305,1}$	$Obs_{305,2}$	1				$Obs_{305,23}$	$O_{305,1}$

Determination Of Model Performance

The model's behaviour in both learning (training and testing) and validating phase, is evaluated using the following statistical methods; the correlation coefficient (R) at 95% confidence limit, given by equations;

Coefficient of correlation (R)

$$r = \frac{\left[\sum_{i=1}^{n} x_i \hat{x}_i - \frac{1}{n}(\sum x_i)(\sum \hat{x}_i)\right]^2}{\sqrt{\left[\sum x_i^2 - \frac{1}{n}(\sum x_i)^2\right]\left[\sum \hat{x}_i - \frac{1}{n}(\sum \hat{x}_i)^2\right]}}$$

(2)

and the mean bias error or residual error given by;

Mean bias error (MBE),

$$MBE = \frac{1}{n}\sum_{i=1}^{n}(\hat{x}_i - x_i)$$

(3)

Where $\hat{x}_i$ and x_i represent observed values and the corresponding forecast values for i =1,2,.....,n.

The prediction performance evaluated using these two methods are used to evaluate the accuracy of the forecast and for comparing the forecasting ability of each approach. The 95% confidence limit is used to determine that the predicted output lie within the confidence range. It is assumed that a predicted value fall into an interval within which there is an associated uncertainty. According to Wackerly et al. (1996), this uncertainty is derived from the residual errors that have already been calculated within that range of values. If the residual errors are randomly distributed, there is a general rule of thumb which states that they will lie within two standard deviations of their mean with a probability of 0.95. This method was used in the measurements of the ANN prediction performance conducted by some researchers (Bishop, 1995; Tibshirani, 1996; Shao et al., 1997; Zhang et al., 1998; Lowe and Zapart, 1999; Townsend and Tarassenko, 1999) ANN models and statistical analyses were carried out using MATLAB 7.0 and XLSTAT2008 (Excel2003 add-in) for Windows.

RESULTS AND DISCUSSION

Post PCA, out of the 23 principal components generated, only six PCs with eigenvalues higher than 1 (Table 3) were selected for the ANN input parameters. Selected PCs explained 75.1% of the total variation. Furthermore, communality values were high for the selected PCs, for example, the values are 93% for Cond., 95% for Sal, 98% for DS and TS (Table 4). These results further confirm the choice of the selected number of PCs (Stevens, 1986). For the first six rotated PCs (RPCs), the loadings from PCA are given in Table 4. The highest correlations between variables are noted in bold. For instance, Cond., Sal, DS, TS, Cl, Ca, K, Mg and Na, have high correlations with RPC1. Eighteen variables with strong loadings were included in the six selected RPCs. Significant variables in RPC1 are Cond., Sal., DS, TS, Cl, Ca, K, Mg, and Na; in RPC2 they are DO, BOD and AN; in RPC3 they are SS and Tur and in RPC4, NO^{3-} and PO4 $^{3-}$ The only meaningful loads in RPC5 and RPC6 are pH and Zn.

Table 3: Descriptive statistics of selected original PCs with eigenvalues more than 1.

	PC1	PC2	PC3	PC4	PC5	PC6
Eigenvalue	9.074	2.387	2.067	1.492	1.225	1.026
Variability (%)	39.451	10.380	8.987	6.488	5.326	4.459
Cumulative %	39.451	49.830	58.817	65.305	70.631	75.091

Table 4: Rotated factor loadings using six PCs.

Variables	RPC1	RPC2	RPC3	RPC4	RPC5	RPC6	Communalities
DO	-0.205	-0.722	-0.121	0.046	0.485	-0.066	0.82
BOD	0.035	0.740	0.071	0.110	0.110	0.022	0.58
COD	0.340	0.103	0.081	-0.166	0.268	0.326	0.34
SS	-0.042	-0.009	0.920	0.010	-0.025	0.017	0.85
pH	0.189	-0.109	-0.204	0.020	0.792	-0.083	0.72
AN	-0.092	0.797	-0.151	0.161	0.023	-0.032	0.69
T	0.337	0.368	-0.242	-0.298	-0.317	0.208	0.54
Cond.	0.963	0.022	-0.043	0.035	0.013	-0.022	0.93
Sal.	0.974	0.023	-0.038	0.030	0.008	-0.004	0.95
Tur.	-0.031	-0.007	0.863	0.011	-0.140	-0.035	0.77
DS	0.988	0.017	-0.034	0.013	0.009	-0.005	0.98
TS	0.985	0.017	0.069	0.014	0.007	-0.003	0.98
NO_3-	0.018	0.033	0.107	0.688	-0.126	0.300	0.59
Cl	0.986	0.010	-0.029	-0.004	0.020	0.005	0.97
$PO_4$$^{3-}$	0.023	0.312	-0.106	0.700	0.112	-0.073	0.62
Zn	-0.019	0.044	-0.011	0.186	-0.128	0.767	0.64
Ca	0.980	0.028	-0.026	-0.043	-0.024	0.039	0.97
Fe	-0.080	0.043	0.475	0.540	0.066	0.192	0.57
K	0.984	0.004	-0.031	-0.004	0.004	0.010	0.97
Mg	0.974	0.000	-0.022	-0.028	-0.002	0.037	0.95
Na	0.986	0.002	-0.025	-0.020	0.005	0.017	0.97
COLI	-0.254	0.361	0.097	-0.424	0.457	0.056	0.60
COLIFORM	-0.032	0.049	-0.025	0.042	-0.077	-0.517	0.28

Using the original principal component scores as inputs, the best architecture consist of a three layer network with 23 input neurons, 10 neurons in the hidden layer and one neuron in the output layer. Considering RPC scores as inputs, the best architectures were achieved with almost the same number of hidden neurons. The hidden neurons consist of 9 and 10 neurons respectively. Training was carried out for a maximum 10000 iterations. Selection of the network was performed at maximum correlation coefficient (R) and 95% confidence limit.

Table 5 and Figure ure 2 illustrate the prediction performances of ANN models using different combinations of PC scores as input variables. ANN using the first 2 PCs (PC1 and PC2) does not perform very well as far as accuracy is concerned for all the training, testing and validation phases. It is observed that the prediction performance of the validation phase is slightly worse compared to the training and testing phases. It is important to point out that for this model, the cumulative percentage in explaining the variance given by these two RPCs is only 49.8%. None of the strong loading variables contains the variables forming the WQI equation. DO, BOD and pH loadings in PC2 explain only 10.4% of the total variance. Based on the results, it is apparent that the WQI prediction performance increases with the increase in number of input variables. The highest accuracy in predicting WQI is given by model ANN-RPC6, which contains six RPCs with 75.1% variation explained, giving an R2 value of 0.64 (training), 0.87 (testing), and 0.72 (validation) respectively.

Table 5: The prediction performances of the different ANN models

Model	No.of PC	R squared			MBE		
		Training	Testing	Validation	Training	Testing	Validation
ANN-RPC2 (2 inputs)	2	0.43	0.70	0.32	28.01	-167.90	-40.71
ANN-RPC3 (3 inputs)	3	0.60	0.78	0.61	64.95	-109.68	6.60
ANN-RPC4 (4 inputs)	4	0.53	0.79	0.47	0	-165.04	-89.78
ANN-RPC5 (5 inputs)	5	0.53	0.79	0.47	140.12	-143.75	-44.77
ANN-RPC6 (6 inputs)	6	0.64	0.87	0.72	67.93	-58.57	-44.61
ANN-PC23 (23 original PC inputs)	23	0.60	0.85	0.66	-18	-81.59	-49.83

Table 5. The prediction performances of the different ANN models. From table 5, it can be observed that the prediction performance of the ANN model using original PCs (23 input PC scores) is not significantly different from the RPC models. However, as RPC models use fewer variables and is far less complex, the advantage over the ANN-PC23 model is obvious. Comparing the MBE values, it is generally observed that the signs for the validation phases are negative for both the un-rotated and rotated PC models. This is an indication that the predicted WQI values are consistently underestimated in both approaches.

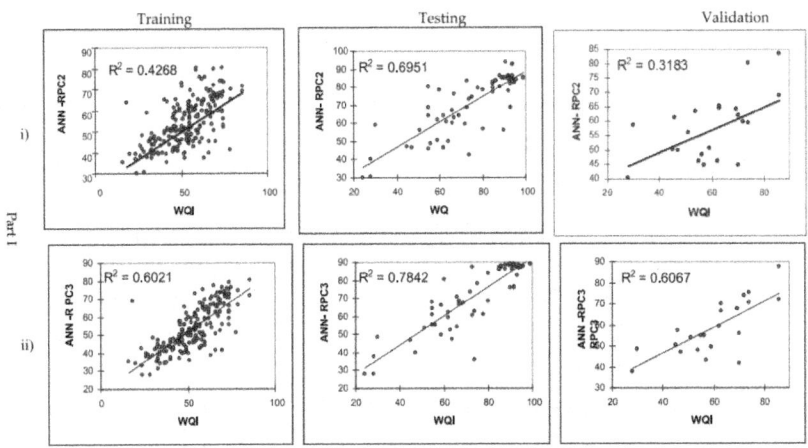

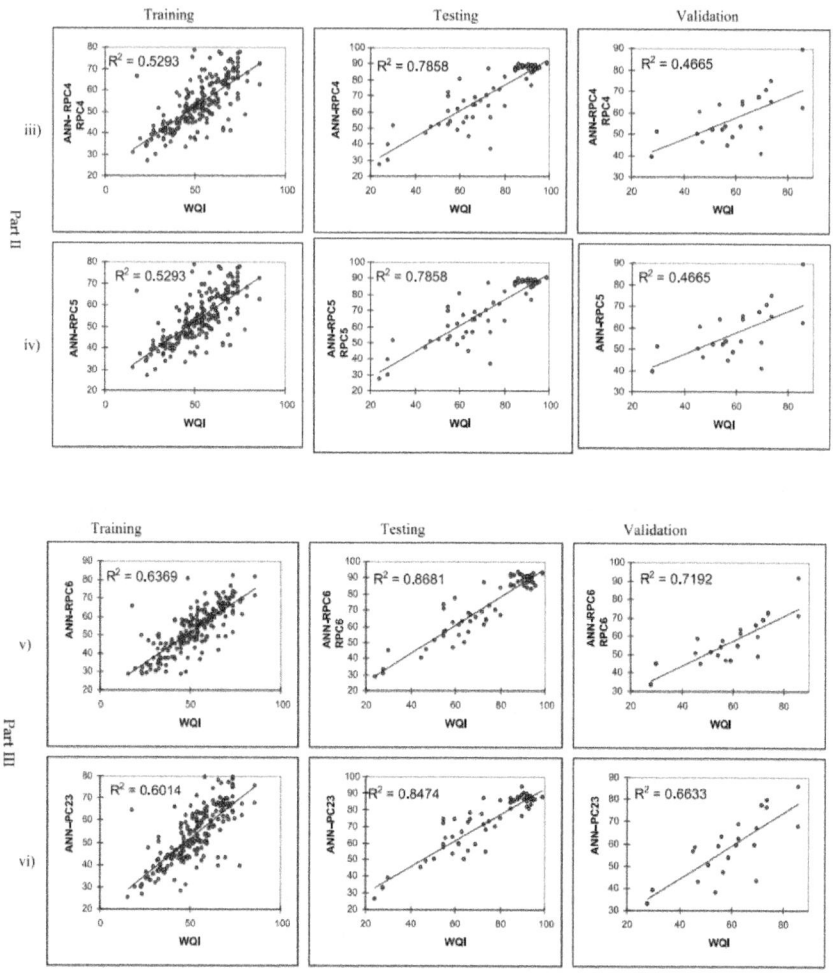

Figure. 2: The prediction performances for different combination of PC scores during training, testing and validation phases : (i) 2 RPCs, (ii) 3 RPCs, (iii) 4 RPCs, (iv) 5 RPCs, (v) 6 RPCs and, (vi) 23 original PCs.

This study also attempts to allocate 95% confidence interval on the WQI prediction produced by the best ANN model. Figure ure 3, 4 and 5 show the comparison between predicted values and the upper (UL) and lower limits (LL) lying within 95% confidence interval. This was carried out for ANN-RPC6 and ANN-PC23 models. It can be seen that only 4.3% out of the 305 predicted values were identified beyond the 95% confidence limit (1% fall below the LL and 3.3% fall beyond the UL) for ANN-RPC6. For ANN-PC23, 25% of the 305 observations fall beyond the upper and lower of 95% confidence interval limit (14% fall below the LL and 11.8% fall beyond the UL). This basically shows that by using reduced rotated PC scores as input, better results can be obtained without losing information. It is thus apparent that ANN prediction using scores of varimax rotated PCs result in a more accurate WQI prediction.

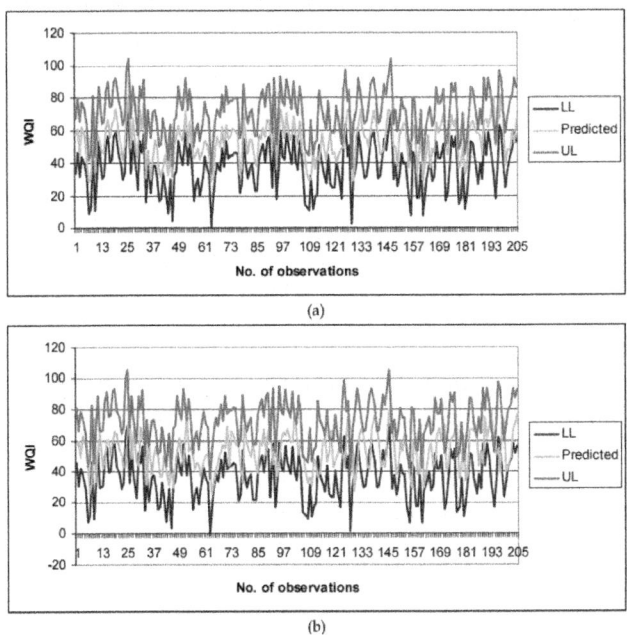

(a)

(b)

Figure. 3: Predicted WQI within the 95% confidence interval during training phase using (a) six rotated PCs, and (b) 23 original PCs.

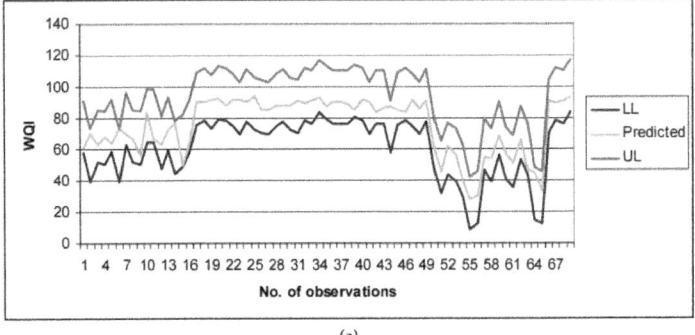

(a)

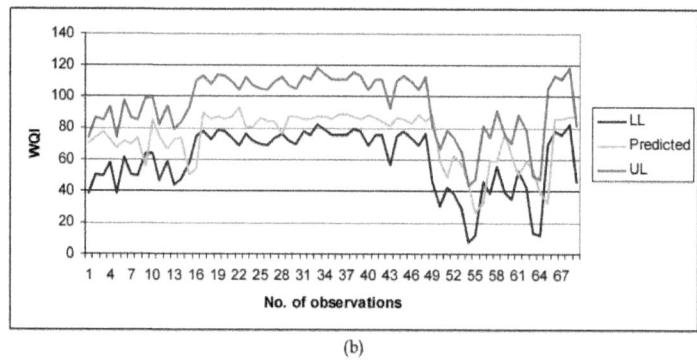

(b)

Figure. 4: Predicted WQI within the 95% confidence interval during testing phase using (a) six rotated PCs, and (b) 23 original PCs.

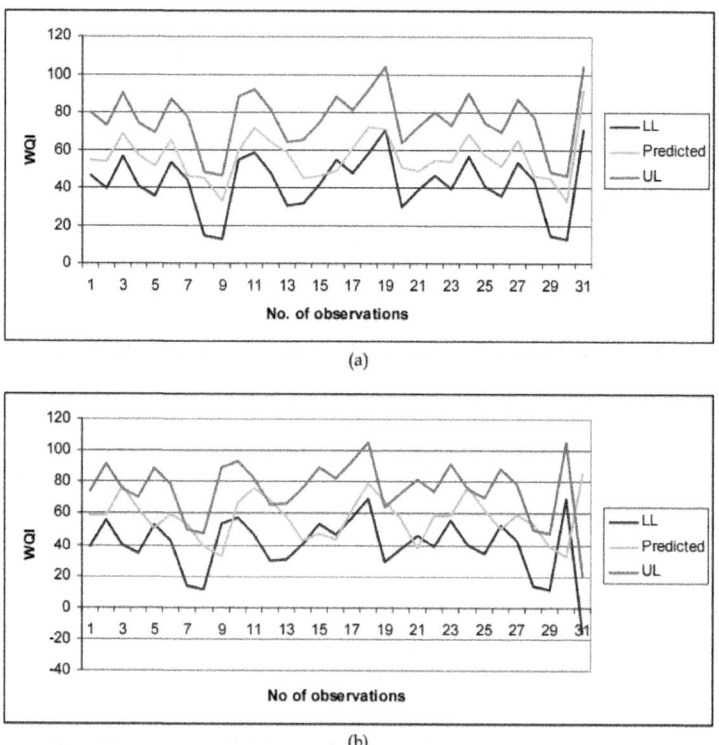

(a)

(b)

Figure. 5: Predicted WQI within the 95% confidence interval during validation phase using (a) six rotated PCs, and (b) 23 original PCs.

CONCLUSION

In this work, a combination of PCA and ANN is used to predict WQI based on 23 historical water quality parameters. The original predictors were selected based on the available Malaysian DOE data. To obtain the latent variables as inputs into the ANN, two different approaches were used; one based on un-rotated original PCs and the other based on varimax rotated PCs. Using six PCs, significant loadings are observed for Cond, Sal, DS, TS, Cl, Ca, K, Mg and Na in PC1, DO, BOD and AN in PC2, SS and Tur in PC3, NO3- and PO43- in PC4, pH in PC5 and Zn in PC6. ANN models based on these 6 PC scores can predict WQI with acceptable accuracy (within 95% confidence limit). Moreover, the ANN model using the 23 original PCs as input, do not render the prediction more accurate, even with a complex network structure. The use of rotated PC scores based models is clearly more effective and efficient due to the elimination of collinearity and reduction of predictor variables without losing important information.

ACKNOWLEDGMENT

The authors acknowledge the financial and technical support for this project provided by the Ministry of Science, Technology and Innovation and Universiti Putra Malaysia under the Science Fund Project no. 01-01-04-SF0733. The authors wish to thank,the Department of Environment, and Department of Irrigation and Drainage, Ministry of Natural Resources and Environment of Malaysia, Institute for Development and Environment (LESTARI), Universiti Kebangsaan Malaysia, Universiti Malaya Consultancy Unit (UPUM) and Chemistry Department of Universiti Malaya, who have provided us with secondary data and valuable advice.

REFERENCES

1.   Abdul-Wahab, S.A., Bakheit, C.S. and Al-Alawi, S.M., 2005. Principal component and multiple regression analysis in modeling of ground-level ozone and factors affecting its concentrations. Environmental Modelling & Software 20, p.1263-1271.

2.   Aiken, R.S., Leigh, C.H., Leinbach, T.R., and Moss, M.R., 1982. Development and Environment in Peninsular Malaysia. McGraw-Hill International Book Company: Singapore.

3.   Beck, M.B., 1986. Identification, estimation and control of biological waste-water treatment processes. IEE Proceeding 133, p.254-264

4.   Benediktsson, J.A., Swain, P.H., and Ersoy, O.K., 1990. Neural network approaches versus statistical methods in classification of multisource

remote sensing data. I.E.E.E. Transactions on Geoscience and Remote Sensing, 28, 540-551

5. Bishop, C. M., 1995. Neural Networks for Pattern Recognition. Clarendon Press, Oxford Bowden, G.J., Dandy, G.C. and Maier, H.R., 2005. Input determination for neural network models in water resources applications. Part 1-background and methodology. Journal of Hydrology 301, p.75-92.

6. Chen, Q. and Mynett, A.E., 2003. Integration of data mining techniques and heuristic knowledge in fuzzy logic modeling of eutrophication in Taihu Lake. Ecological Modelling 162, p.55-67.

7. Clarici, E., 1995. Environmental Modelling Using Neural Networks, PhD Thesis, Imperial College.

8. Department of Environment Malaysia, DOE 1997. Malaysia environmental quality reports 1999. Kuala Lumpur: Ministry of Science, Technology and Environment.

9. Department of Environment Malaysia, DOE 1999. Malaysia environmental quality reports 1999. Kuala Lumpur: Ministry of Science, Technology and Environment.

10. DeSilets, L., Golden, B., Wang, Q., and Kumar, R., 1992. Predicting salinity in the Chesapeake Bay using backpropagation. Computer and Operations Research, 19, p.227-285

11. Elsner, J.B., and Tronis, A.A., 1992. Nonlinear prediction, chaos, and noise. Bull. Am. Meterol. Soc. 73(1), p.303-314.

12. Foody, G.M., McCulloch, M.B., and Yates, W.B., 1995. Classification of remotely sensed data by an artificial neural network: issues related to training data characteristics.

13. Photogrammetric Engineering and Remote Sensing.

14. Ha, H. and Stenstrom, M. K., 2003. Identification of land use with water quality data in stormwater using a neural network. Water Research, 37, p.4222-4230

15. Hafizan Juahir, Sharifuddin M. Zain, Zainol Mustafa, and Azme Khamis, 2001. Dissolved oxygen forecasting due to landuse activities using time series analysis at Sungai

16. Lang at, Hulu Lang at, Selangor. Ecological Environmental Modelling, Proceeding of the National Workshop, Universiti Sains Malaysia, 3-4 September, p.157-164.

17. Hafizan Juahir, Sharifuddin M. Zain, Mohd. Ekhwan Toriman, M. Nazari Jaafar and W.

18. Klaewtanong, 2003a. Performance of autoregressive integrated moving

average and neural network approaches for forecasting dissolved oxygen at Langat River Malaysia. Urban Ecosystem Studies In Malaysia: A study of change. Universal Publishers, p. 145-165.

19. Hafizan Juahir, Sharifuddin M. Zain, M. Nazari Jaafar, Zainal Majid and M. Ekhwan Toriman, 2003b. Land use temporal changes: application of GIS and statistical analysis on the impact of water quality at Langat River Basin, Malaysia. presented in 2nd Annual Asian Conference of Map Asia 2003, 17-19, Oct., PWTC Kuala Lumpur.

20. Hafizan Juahir, Sharifuddin M. Zain, M. Nazari Jaafar and Zainal Ahmad, 2004a. An Application of Second order backpropagation method in Modeling River Discharge at Sungai Langat, Malaysia. Water Environmental Planning: Towards integrated planning and management of water resources for environmental risks, IIUM, p.300-307.

21. Hafizan Juahir, Sharifuddin M. Zain, M. Ekhwan Toriman and Mazlin Mokhtar, 2004b. Application of Artificial Neural Network Model In the Predicting Water Quality Index. Jurnal Kejuruteraan Awam, 16 (2), p.42-55 Helena, B., Pardo, R., Vega, M., Barrado, E., Fernandez, J.M., Fernandez, L., 2000. Temporal evaluation of grounwater composition in an alluvial aquifer (Pisuerga river, Spain) by principal component analysis. Water Research 34, 807-816.

22. Hepner, G.F., Logan, T., Ritter, N., and Bryant, N., 1990. Artificial Neural Network classification using a minimal training set: comparison to conventional supervised classification. Photogrammetric Engineering and Remote Sensing, 56, 469-473

23. Jain, A. & Prasad Indurthy, S. K. V. (2003) Comparative Analysis of Event-based Rainfallrunoff Modeling Techniques-Deterministic, Statistical, and Artificial Neural Networks, Journal of Hydrologic Engineering, 8, p. 93-98.

24. Jiang, D., Zhang, Y., Hu, X., Zeng, Y., Tan, J. And Shao, D., 2004. Progress in developing an ANN model for air pollution index forecast. Atmospheric Environment 38, p.7055- 7064.

25. Kim, J.,-O., and Mueller, C.W., 1987. Introduction to factor analysis: what it is and how to do it. Quantitative Applications in the Social Sciences Series. Sage University Press, Newbury Park.

26. Kuo, J.-T., Hsieh, M.-H., Lung, W.-S. And She, N., 2007. Using artificial neural network for reservoir eutrophication prediction. Ecological Modelling, 200, p.171-177

27. Loska, K. and Wiechula, D., 2003. Application of principal component analysis for the estimation of source of heavy metal contamination in

surface sediments from the Rybnik Reservoir. Chemosphere 51, p.723-733.

28. Lowe, D. and Zapart, C., 1999. Point-Wise Confidence Interval Estimation by Neural Networks: A Comparative Study based on Automotive Engine Calibration. Neural Computing & Applications, Vol. 8, p.77-85.

29. Mohd. Ekhwan Toriman and Hafizan Juahir, 2003. Artificial Neural Network Modelling For Langat River Discharge: Implication For River Restoration. Pertandingan Minggu Penyelidikan dan Inovasi UKM, Pusat Pengurusan Penyelidikan, 3-5 Julai.

30. Muttil, N. and Chau, K.,-W., 2007. Machine-learning paradigms for selecting ecologically significant input variables. Engineering Application of Artificial Intelligence 20, p.735- 744.

31. Peddle, D.R., Foody, G.M., Zhang, A., Franklin, S.E., and Ledrew, E.F., 1994.

32. Multisource image classification II: An empirical comparison of evidential reasoning and neural network approaches. Canadian Journal of Remote Sensing, 12, 277-302.

33. Recknagel, F., Bobbin, J., Whigham, P., and Wilson, H., 2002. Comparative application of artificial neural networks and genetic algorithms for multivariate time-series modeling of algal blooms in freshwater lakes. Journal of Hydroinformatics 4(2), p.125- 134.

34. Rosnani Ibrahim, 2001. River Water quality Status In Malaysia. Proceedings National Conference On Sustainable River Basin Management In Malaysia, 13 & 14 November 2000, Kuala Lumpur, Malaysia.

35. Rumelhart, D. E., Hinton, E. and Williams, J., 1986. Learning internal representation by error propagation. Parallel Distributed Processing, 1, p. 318-362.

36. Ruslan Rainis, Kamarul Ismail and Hafizan Juahir, 2004. Modeling The Relationship Between River Water Quality Index (WQI) and Land Uses Using Artificial Neural Networks (ANN). Presented in JSPS Seminar, December 15-17, Kyoto, Japan.

37. Scardi, M., 2001. Advances in neural network modeling of phytoplankton primary production. Ecological Modelling 146, p.33-45.

38. Schalkoff, R., 1992. Pattern Recognition: Statistical, Structural and Neural Approaches. New York, Wiley Shao, R., Martin, E.B., Zhang, J. and Morris, A.J., 1997. Confidence bounds for neural network representations. Computers and Chemical Engineering, 21, p.1173-1178.

39. Silverman, D., and Dracup, J.A., 2000. Artificial neural networks and long-range precipitation in California. Journal of Applied Meteorology 31(1), p.57-66.

40. Simeonov, V., Stratis, J.A., Samara, C., Zachariadis, G., Voutsa, D., Anthemidis, A., Sofoniou, M. and Kouimtzis, Th., 2003. Assessment of the surface water quality in Northern Greece. Water Research 37, p.4119-4124.

41. Sousa, S.I.V., Martins, F.G., Alvim-Ferraz, M.C.M. and Pereira, M.C., 2007. Multiple linear regression and artificial neural networks based on principal components to predict ozone concentrations. Environmental Modelling & Software 22, p.97-103.

42. Stevens, J., 1986. Applied Multivariate Statistics for the Social Science. Hill Sdale: New Jersey, USA, p.515.

43. Tibshirani, R., 1999. A comparison of some error estimates for neural network models. Neural Computation, 8, p.152-163.

44. Townsend, N.W. and Tarassenko, L., 1999. Estimations of error bounds for neural-network function approximators. IEEE Trans Neural Netwoks, 10(2), p.217.

45. Wackerly, D.D., Mendenhall, W and Scheaffer, R.L., 1996. Mathematical Statistics with Applications. 5th. Ed., Duxbury Press: Belmont, USA.

46. Wunderlin, D.A., Diaz, M.P., Ame, M.V., Pesce, S.F., Hued, A.C., Bistoni, M.A., 2001. Pattern recognition techniques for the evaluation of spatial and temporal variations in water quality. A case study: Suquia river basin (Cordoba, Argentina). Water Research 35, 2881-2894.

47. Zarita Zainuddin, 2004. Modelling Nonlinear Relationship in Ecology and Biology using Neural Networks. In Koh Hock Lye and Yahya Abu Hassan, Ecological Environmental Modelling (ECOMOD 2001): Proceedings of the National Workshop, 3-4 September, USM, p.88-95

48. Zhang, J., Morris, A.J., Martin, A.J. and Kiparissides, C., 1998. Prediction of polymer quality in batch polymerization using robust neural networks. Chemical Engineering Journal. 69, p.135-143.

49. Zhang, Q. & Stanley, S. J.,1997. Forecasting raw-water quality parameters for North Saskatchewan River by neural network modeling. Water Resource, 31, p. 2340-2350.

Chapter 13

APPLICATION OF CHEMOMETRICS TO THE INTERPRETATION OF ANALYTICAL SEPARATIONS DATA

James J. Harynuk, A. Paulina de la Mata and Nikolai A. Sinkov

Department of Chemistry, University of Alberta Canada

INTRODUCTION

Interesting real-world samples are almost always present as mixtures containing the analyte(s) of interest and a matrix of components that are irrelevant to answering the analytical question at hand. Additionally, the compounds comprising the matrix are usually present in far greater abundance (both number and concentration) than the analytes of interest, making quantification or even detection of these analytes difficult if not impossible. When tasked with these types of samples, analysts turn to some form of separations technique such as gas or liquid chromatography (GC or LC) or capillary electrophoresis (CE) so that individual components in each sample may be quantified. More recently, more complex analytical questions are being probed, for example profiling blood or urine to identify a disease state or ascertaining the geographic origin of a food/beverage sample.

These tasks often go beyond the simple quantification of one or two analytes in a sample. For these and other similar questions, separations scientists are turning more often to chemometric tools as a means of visualizing and interpreting the rich data that they obtain from their separations systems. Here we present a brief overview of separations approaches, with a focus on the data that are derived from different methods and on phenomena in the separations approach that lead to challenges in data interpretation. This is followed by a discussion of approaches that exist for the chemometric interpretation of separations data, specific challenges that arise in the chemometric treatment of these data, and solutions that have been implemented to deal with these challenges.

Separations techniques

Chromatography is widely used for the separation, purification, and analysis of mixtures. In general, analytes contained in either a gaseous or liquid mobile phase are flowed past a stationary phase which is usually confined within a column. Depending on the chemistries of the analytes and the conditions of the separation (mobile/stationary phase compositions, temperature, etc.) different compounds will partition between the two phases to varying degrees. The separation arises due to this differential partitioning, with analytes which associate weakly with the stationary phase passing through the column more quickly than those with a greater affinity for the stationary phase (Miller, 2005; Cazes, 2010).

There are many types of chromatography, with the most common being liquid chromatography (LC) where analytes partition between a mobile liquid phase and an immobile stationary phase, and gas chromatography (GC) where the mobile phase is a gas and the stationary phase is a solid or more often a viscous, liquid-like polymer. There are numerous modes for LC separations, including for example reverse-phase (RPLC), normalphase (NPLC), ion (IC), size exclusion (SEC), and hydrophilic interaction (HILIC) to name a few. From a point of view of chemometric data interpretation and the discussion in this chapter, all of these LC separations generate data which are equivalent. In any chromatographic separation, the sample is delivered to the inlet of the column while the outlet is connected to a detector, which records a continuous signal.

The detector response rises and then falls to baseline based on the analyte flux passing through it, ideally generating one separate peak with an approximately Gaussian shape for each individual analyte. Assuming that the conditions for repeat analyses are not changed, the peak for a given analyte will appear at the same time in every analysis, with the peak area/height being proportional to the quantity of analyte present in a sample (Poole, 2003; Miller, 2005).

Another separations technique which is popular for some samples is capillary electrophoresis (CE). Here, an electric field applied across a fused silica capillary containing a buffer induces motion of the buffer and analytes in the sample. The CE separation is dependent on differential mobilities of analytes in the solution in the presence of the electric field. This difference in mobilities is based on the fact that different analytes have different charges and sizes in solution. While the separation mechanism of CE is fundamentally different from the chromatographic mechanism, the data are a series of peaks recorded as a function of time. Consequently, the same tools can be applied to data from a CE separation, and similar concerns exist for the interpretation of these data (Poole, 2003; Miller, 2005). For ease of readability, and because chemometrics are more often applied to chromatographic data than electrophoretic data, we will often refer to a chromatogram in this chapter. This could equally be an electropherogram; when considering the application of chemometric techniques to separations data whether the origin is electrophoretic or chromatographic is largely irrelevant.

When tasked with incredibly complex samples, analysts are now turning more and more frequently to so-called comprehensive multidimensional separations (e.g.: GC×GC, LC×LC, CE×CE) (Liu & Phillips, 1991; Erni & Frei, 1978; Michels et al., 2002). In these techniques, the mixture of compounds is sequentially separated by two different separation mechanisms. In the case of GC×GC, for example, a sample might be separated first on an apolar column, followed by a polar column. The exact workings of comprehensive multidimensional separations are beyond the scope of this work, and are discussed elsewhere (Górecki et al., 2004; Cortes et al., 2009; François et al., 2009; Kivilompolo et al., 2011; Li et al., 2011). However, these techniques are gaining in popularity, and are capable of separating exceedingly complex mixtures comprising thousands of individual compounds. Due to the vastly improved separation power of these techniques, the data are much more informationrich, and without some form of chemometric treatment it is essentially impossible to do more than scratch the surface of the information contained therein.

Separations data

The detector signal from a separations experiment, when plotted vs. time, yields a series of (ideally) Gaussian peaks, each representing one compound in the sample. Acquisition speed is one consideration for a chromatographic detector: it must be sufficient to faithfully record the profile of each compound as it passes through the detector. In order to obtain an accurate peak profile, the minimum number of acquisition points required across a peak is 10.

Thus, the required speed of the detector is intrinsically linked to the nature of the separation. In separations where the base width of the peaks are on the order of 5 s, a data rate of 2 Hz would be acceptable, but when peak widths are 100-200 ms, as in GC×GC, then detector rates on the order of 50-100 Hz are required for quantitative analysis. From a point of view of chemometric analysis of separations data, another important consideration is whether the detector is univariate or multivariate. Univariate detectors, such as the flame ionisation detector, or single-wavelength UV-visible spectrometer, record only one variable as a function of time, generating data which take the form of a vector of instrument response. Other detectors, typically mass spectrometers and multi-channel spectroscopic instruments, can be operated such that they record a multivariate response. Data from these instruments comprise an array of signal responses with each row representing a time when a response was recorded, and each column representing a variable that was recorded (e.g.: detector wavelength, ion mass-to-charge ratio).

To the chemometrician, it is immediately obvious that there are numerous advantages to collecting multivariate chromatographic data; however, it is worth noting that most of this advantage has been by and large ignored by chromatographers. Typically, only the profile of a single variable vs. time would be used to selectively quantify an analyte, or the detector response across all channels at a given time used to help identify a peak. One other aspect of raw separations data is the sheer number of variables measured for each sample. When a univariate detector is used for a 15 min separation, operating with an acquisition speed of 10 Hz, the data will be a vector of 9000 individual measurements per sample. If a multivariate detector is employed instead, for example a mass spectrometer operating over a 30-300 m/z mass range, this number increases to 2 439 000 individual variables arranged in a 9000 × 271 array per sample! In the case of GC×GC-MS analyses, which are typically 60 min in length but have a high-speed MS collecting data at rates of ~100 Hz, there are on the order of 100 million data points collected for each sample.

Challenges with chromatographic data

Variations in analytical separations data are, in principle, no different from those derived from any other instrument; being based on both chemical and non-chemical aspects of the analysis. All relevant information will be contained within the chemical variations and any chemometric approach to interpreting chromatographic data must be capable of identifying relevant chemical variation while minimizing the effects of irrelevant chemical and nonchemical variations. Sources of irrelevant chemical variation include matrix peaks, here defined as any chemical source of signal introduced with the sample, but having no bearing on the conclusions drawn from the data. Additionally, there is background signal which can for example derive from changes in mobile phase concentration which influence detector signals in LC or chemical "bleed" signatures from stationary phases as they degrade in GC. Non-chemical variations include, for example, baseline drift (for non-chemical reasons), retention time shifts (due to minor fluctuations in operating conditions), and electronic noise. These may easily interfere with the relevant chemical information, degrading model performance and the validity of results (de la Mata-Espinosa et al., 2011a). Figure 1 presents an overlay of several LC chromatograms of similar samples exemplifying the challenges of baseline drift and retention time shifts. One of the major challenges in handling chromatographic data using chemometric tools is appropriate pre-processing to remove as many non-chemical and irrelevant chemical variations as possible from the data set.

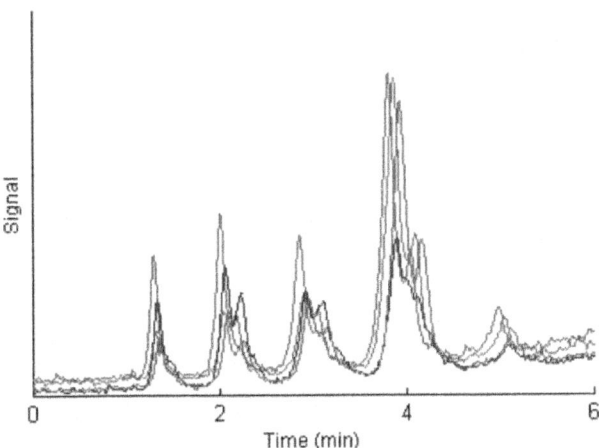

Figure. 1: LC chromatograms of edible oils showing a high degree of variation in baseline.

Initial efforts into the application of statistical and chemometric tools to chromatographic data were accomplished using data that were processed to provide a list of detected, integrated peak areas or heights (or the calibrated concentrations for known compounds). However, the trend in recent years has turned towards the direct chemometric interpretation of raw chromatographic signals (Watson et al., 2006; Johnson & Synovec, 2002). The reason for this trend is that many errors can occur during integration of raw signals (Asher et al., 2009; de la Mata-Espinosa et al., 2011b). By applying chemometric tools directly to the raw data, many of these errors can be avoided. Of course, when working with the raw data, other issues become more important, most notably retention time shifts and the population of available variables.

Baseline and noise

Baseline variations, such as noise and drift, are due to small changes in experimental conditions, for example changes in detector response due to the mobile phase gradient in LC separations or increased levels of stationary phase bleed at higher temperatures in temperature-programmed GC. Other sources of noise and drift could include changes in detector response as its components age, contamination of solvents or gases, and of course electronic noise (which is minimal in modern chromatographic systems).

Chemometric approaches to handling chromatographic data should incorporate baseline correction of some form. When raw chromatographic data are processed, the method of baseline correction and its importance are generally obvious to the analyst. In the case where integrated peak tables are used, this is often done automatically by the chromatographic software with little consideration by the analyst, even though the manner in which the baseline is calculated will significantly influence the determination of peak areas/heights.

Retention time shifts

In all separations, retention times of peaks can easily shift by a few seconds from one analysis to the next. This is not much of an issue with simple samples having only a few peaks which are then integrated prior to chemometric analysis. However, retention times of peaks are used for identifying the compounds. With complex separations, unstable retention times may result in unreliable peak identification, making comparisons from one run to the next impossible. When comparing raw data this is even more important as one must ensure that the peak for a given component is always registered in the exact same position in the data matrix so that the algorithms will recognize the signals correctly. The causes of retention time shifts depend on the separations technique being used.

In GC, peaks may shift due to degradation of the stationary phase, decreasing retention times over time; build-up of heavy matrix components which foul the column, effectively changing the chemistry of the stationary phase; minor gas leaks which alter the flow rate; or even matrix effects on the evaporation rate in the injector, affecting the rate of mass transfer to the column. In LC, peak shifts may be due to small fluctuations in mobile phase chemistry from one run to the next; temperature fluctuations which in turn affect solvent viscosity and solute diffusion coefficients, altering the kinetics as well as the thermodynamics of the separation; or degradation / fouling of the stationary phase of the column. CE is the technique most prone to drastic shifts in migration time, due to the instability of the electroosmotic flow in the capillary (Figureure 2). Electroosmotic flow depends on the applied voltage, the buffer concentration and composition, and is incredibly sensitive to the surface chemistry of the capillary.

The act of analyzing a sample by CE will often have a minor, possibly irreversible effect on the capillary surface, resulting in a change in the migration time of an analyte. Shifts in retention times are minimized by proper instrument maintenance, precise control of instrumental conditions or by using approaches such as retention time locking in GC to account for variations in instrument performance (Etxebarria et al., 2009; Mommers et al., 2011) and relative retention times in CE. Even with these approaches, some retention time shifting will occur and require more advanced alignment techniques for correction prior to chemometric analysis.

Incomplete separation

Another challenge with the interpretation of chromatographic data is incomplete separation of peaks. If two or more compounds have similar retention characteristics under a given set of separation conditions, they will not be completely resolved, as evidenced by the peak clusters in Figure 1. In these cases, apportioning the signal between the different compounds becomes a challenge, especially for univariate signals.

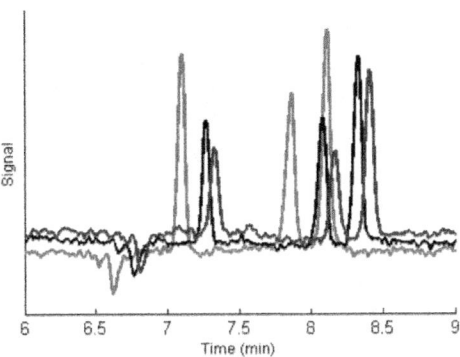

Figure. 2: CE of substituted benzenes showing extreme misalignment.

The general approach used for these cases is one of deconvolution: decomposing the analytical signal to determine the contribution of each coeluting compound, or to determine the contribution of the compound of interest, disregarding the remaining data.

Data overload

As shown in Section 1.2, raw chromatographic signals present an overabundance of data to the analyst. This poses several challenges. From a practical point of view, attempts to construct a chemometric model using the entirety of the data set could easily exceed the capabilities of the computer system being used. More fundamentally, if the raw data are considered, the number of variables measured for each sample will vastly outnumber the number of samples available in the data set. These overdetermined systems can defeat many chemometric techniques due, for example, to collinear variables. Finally, for most chromatograms, especially multidimensional ones, only a small fraction of the data points actually contain meaningful signal. Most of the signal is due to background noise or irrelevant matrix components. Consequently, the raw data must somehow be reduced in size prior to chemometric analysis. This is typically achieved via a feature selection approach, as discussed in Section 3.3.3.

PRE-PROCESSING STEPS FOR CHROMATOGRAPHIC DATA

Baseline correction

The aim of baseline correction is to separate the analyte signal of interest from signal which arises due to changes in mobile phase composition or stationary phase bleed and signal due to electronic noise. Several baseline correction methods have been proposed in literature, with the two most common approaches being to fit a curve to the data and subtract this value from the signal, and modeling the baseline to exclude it using factor models (Amigo et al., 2010).

Curve fitting is the classical approach used in virtually all commercial software packages provided by vendors of separations equipment. The algorithms used in this approach fit a polynomial function across segments of the chromatogram using regions where no analyte peaks elute to determine the coefficients of the polynomial and then interpolating the background signal for regions where peaks are eluting. The functions are usually first-order polynomials; however, higher-order polynomials or a series of connected first-order polynomials are also used in some situations. Having determined the equation of the background signal, the fitted line is then subtracted from the signal (Brereton, 2003; Gan et al., 2006; Kaczmarek et al., 2005; Zhang et al., 2010; Persson & Strang, 2003; Eilers, 2003). Correction of the baseline using curve fitting is demonstrated in Figure 3.

The approach of using models such as parallel factor analysis (PARAFAC) for background correction is analogous to the use of these approaches for deconvoluting coeluting peaks. As these models are more often used for this purpose than for simple background correction, they will be discussed in more detail in Section 3.3. These approaches often rely on having a multivariate signal and are applied to the chromatogram or more typically small selected regions where a single analyte elutes. The result of applying these deconvolution techniques for background correction is essentially the deconvolution of a single analyte peak, with the background noise making up the error matrix

(Amigo et al., 2010). These approaches are generally more powerful and likely result in better quality analytical data, but they are not widely used in separation science. The reason for this is likely historical as these tools have only recently become available to the separation sciences, while the classical curve fitting approach is well established, works with univariate detectors, and performs well in most practical situations.

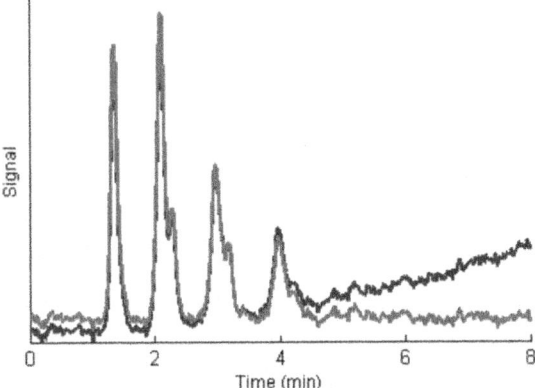

Figure. 3: An LC chromatogram before (blue) and after (red) baseline correction.

Alignment of separations data

The retention times of analytes in separations fluctuate from one analytical run to the next and, in order for chemometric techniques to be applied to separations data, these fluctuations must be corrected during pre-processing. This ensures that the signal from each analyte in each analysis is correctly registered within the data matrix to be processed. There are essentially two approaches to this problem: integrated peak tables, or mathematical warping and alignment of the raw signal.

Peak tables

Integrated peak tables are the simplest way to ensure that analytical separations data are properly aligned for chemometric processing. In order to use this approach, one must be able to reliably assign a unique identifier to each peak in each sample of the data set, and ensure that the same compound is identified with the same identifier in each sample. It should be noted that while the compound name is an obvious identifier, a series of labels such as Unknown x, where x is a numerical identifier would also be acceptable in the event that compound names were unknown, so long as compounds are matched correctly. Rather than identifying peaks by retention time, one could use relative retention times or retention indices in order to adjust for slight variations in the retention times of peaks. Algorithms for aligning peak tables exist and perform well, so long as some peaks can be easily and reliably matched across all chromatograms (Lavine et al., 2001). The challenges with this approach stem from its reliance on integrated peak tables. Thus, any integration errors due to poorly-resolved peaks or peaks that are missed due to falling outside of integration parameters in the software will impact any subsequent analysis

Raw signal alignment

Alignment of raw chromatographic signals prior to chemometric processing is more complex than the alignment of peak tables. In addition to the three more popular algorithms that will be presented below, there are several others that have been developed (Yao et al., 2007; Toppo et al., 2008; Eilers, 2004; Van Nederkassel et al., 2006). In deciding which approach to use, one of the first questions to be answered is if the analysis is to be qualitative or quantitative. This is because some alignment methods can distort peaks, affecting their quantification. Some of the more common algorithms include correlation optimized warping (COW) (Nielsen et al., 1998; Tomasi et al., 2004), correlation optimized shifting (coshift) (Van den Berg, 2005), and a piecewise peak-matching algorithm (Johnson et al., 2003). In instances where there are non-systematic peak shifts, COW is a popular algorithm. COW relies on stretching or compressing segments of a sample signal such that the correlation coefficient between it and a reference signal is maximized for each interval. Care must be taken with the selection of the input parameters to avoid significant changes in peak shapes as this approach to the warping of the chromatogram has been shown to affect peak areas, leading to poor quantitative conclusions (Nielsen et al., 1998; Tomasi et al., 2004). A fast and simple alignment algorithm is coshift.

This algorithm is useful when data only require a single left-right shift in retention time. The entire data matrix is shifted in one direction or the other by a set amount, maximizing the correlation between a target and the data matrix that required alignment. The single shifting value for the entire data matrix is a weakness, especially for chromatographic data where peaks can shift in different directions and to different extents in a single file. To handle this, an algorithm termed icoshift (interval-correlation-shifting) has been derived from coshift. Icoshift aligns each data matrix to a target by maximizing the cross-correlation between the sample and the target within a series of user-defined intervals (Savorani et al., 2010). The use of multiple intervals permits the alignment of separations data where shifts of different magnitudes and directions occur. These alignment algorithms have been used successfully for both one-dimensional data (de la Mata-Espinosa, 2011a; Liang, 2010; Laursen, 2010) and two-dimensional data, with some modifications (Zhang, 2008). It is important to note that the shifting of chromatograms using coshift or icoshift does not lead to distortions of peak shape, and consequently does not introduce errors into quantitative results.

The piecewise peak matching approach (Johnson et al., 2003) provides another avenue for chromatographic alignment. In this approach, peaks are identified in a target signal to which all other signals will be aligned. The algorithm then identifies peaks within the sample signals located within predetermined windows of the peaks in the target. Peaks within windows are deemed to come from the same compound, and matched. The chromatograms are aligned by stretching or compressing the regions between peak apexes. A variant of this algorithm can be used when MS data are available. In this case, the mass spectrum at the apex of each peak in the target signal is compared to the mass spectrum of each peak within a set window on the sample signal and peaks are matched if their spectra have a high enough match quality (Watson et al., 2006). A general scheme for peak alignment using this approach is described in Figureure 4. Depending on the number and relative positions of the peaks in chromatograms matched using this approach, peak shapes may be altered, possibly affecting quantitative results.

One of the biggest challenges for all alignment algorithms is that they depend on the data to be aligned being reasonably similar in terms of both matrix and analyte peaks. In some instances this will not be the case. In our laboratory, we have observed this when analyzing arson debris where the matrix and analytes form an incredibly complex and variable chromatogram from one sample to the next. A similar situation can be easily imagined when processing samples of biological origin. One solution to this issue is to add markers to every sample prior to the separation step in the analysis. These markers should be easily identifiable within the samples, even under conditions where they coelute with matrix components; should occur in multiple, evenly distributed locations along the chromatogram, and should not occur natively in the samples. One choice is a series of deuterated compounds which, with MS detection, are trivial to identify even in a complex mixture (Sinkov et al., 2011b). One additional benefit is that these compounds can act as internal standards if quantitative results are desired.

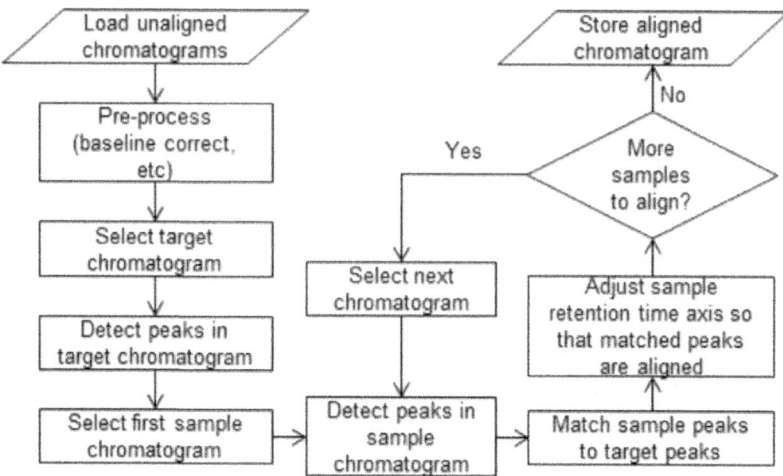

Figure. 4: Flowchart for target-based chromatographic alignment, adapted from (Johnson et al., 2003).

Deconvolution of overlapping peaks

The central issue in deconvolution is depicted in Figureure 5. The instrument response is represented as a black solid line which is the sum of the four dashed, coloured peaks. Ideally, the four signals should be individually quantified. This is a common problem for analytical separations, even those of relatively simple mixtures. Some of these issues may be solved by changing the experimental conditions or using characteristic features (wavelengths or ions) of the coeluting analytes and a multivariate detector to selectively detect and quantify them. However, in many cases this is insufficient and more advanced techniques must be used. The strategies used for deconvolution depend heavily on whether the detector signal is univariate or multivariate.

Deconvolution of univariate signals

In the case of univariate signals, one is typically limited to using univariate curve-fitting analyses where a number of Gaussian or modified Gaussian curves are determined such that the sum of these curves fits the experimentally observed cluster of peaks (Felinger, 1994). In these approaches, only a small window of chromatographic data (one peak cluster) should be processed at a time, and constraints such as fixed peak widths, shapes, unimodality, and non-negativity are often required to ensure the validity of the solution. To solve a univariate deconvolution problem, approaches such as evolving factor analysis (EFA) (Maeder, 1987) or multivariate curve resolution (MCR) (Tauler

& Barceló, 1993), among others (Vivó-Truyols et al., 2002; Sarkar et al., 1998; Kong et al. 2005) can be used. When these approaches are used with univariate data, the variables to be solved for are the number, positions, and abundances of each of the peaks that make up the signal.

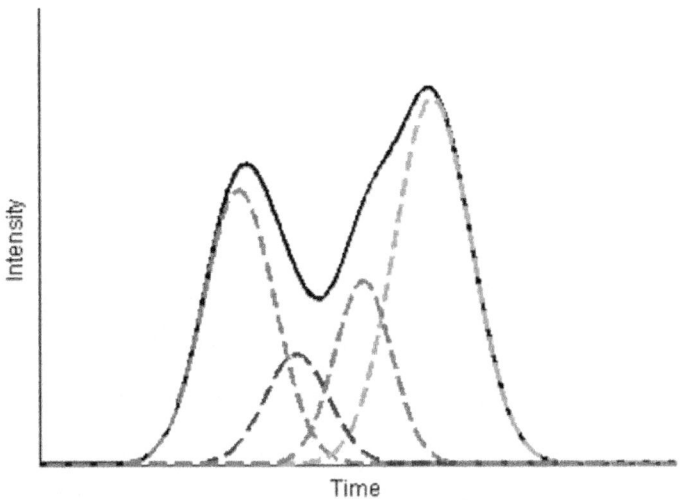

Figure. 5: Deconvolution of overlapping peaks. The black, solid trace represents the analytical signal observed at the detector, which is the sum of the four peaks represented by dashed lines.

Multivariate curve resolution is widely applicable to separations data and is one of the most common approaches (Franch-Lage et al., 2011; Marini et al., 2011, de la Mata-Espinosa et al., 2011a). The aim of this technique is to determine the number of components present in a sample and the contribution of each component to the sample. In performing MCR, the concentration and response profiles for each analyte are obtained, providing a qualitative and semi-quantitative overview of the components in an unresolved mixture without a priori knowledge of the mixture composition.

Deconvolution of multivariate signals

When multivariate detectors are used for separations, the additional dimension of information can be exploited to aid in deconvolution. MCR and EFA can also be used with multivariate data. In the case of MCR, the experimental matrix is decomposed into a matrix of concentration vs. time profiles (deconvoluted peaks) and pure spectral profiles of each compound. Knowledge of the number of components contributing to the signal in the region being deconvoluted is useful to guide the process and improve the results (de

Juan & Tauler, 2006), though strictly speaking it is not required. Parallel factor analysis (PARAFAC) (Harshman, 1970; Bro, 1997; Amigo et al., 2010) is a technique that is ideally suited for interpreting multivariate separations data. PARAFAC is a decomposition model for multivariate data which provides three matrices, A, B and C which contain the scores and loadings for each component. The residuals, E, and the number of factors, r, are also extracted. The PARAFAC decomposition finds the best

trilinear model that minimizes the sum squares of the residuals in the model through a procedure of alternating least squares.

The biggest advantage of using PARAFAC over other models is the uniqueness of the solution; PARAFAC is less flexible and uses fewer degrees of freedom, being a more restricted model. However, its unique solution reflects actual pure analyte profiles in both the time dimension and the spectral dimension. Thus, the results of PARAFAC analysis on a cluster of overlapping multivariate peaks provide both qualitative and quantitative data where the deconvoluted signals appear as analyte peaks. One restriction to the use of PARAFAC is that the data must be trilinear (Bro, 1997; Amigo et al., 2010). In the case of chromatographic techniques with a multivariate detector, the dimensions are retention time, detector signal, and samples. In the case of comprehensive multidimensional separations, such as GC×GC, PARAFAC considers retention in the two dimensions and the samples as the three dimensions.

Feature selection

High data acquisition rates combined with the length of time required for many separations results in a large number of data points collected for a given separation (see Section 1.2). In many situations, most of the data are collected when no analytes are eluting from the system, and represent background signal when only mobile phase is reaching the detector. In the case of spectroscopic and especially mass spectral detectors, at a given point in time, many of the recorded data in this dimension will not contain useful information, even when an analyte of interest is eluting. Furthermore, many components in the mixture can be completely irrelevant to analysis (Johnson & Synovec, 2002; Sinkov & Harynuk, 2011a). Consequently, only a small portion of separations data is potentially useful. It is also well known that any model will be heavily influenced by the specific variables that are included in its construction (Kjeldahl & Bro, 2010).

The inclusion of irrelevant data is detrimental to the model because the mathematics attempt to account for variations observed in these irrelevant variables. Consequently the model is forced to model noise, resulting in a decrease in its predictive ability. Worse yet, the model could fit the data well and provide a seemingly useful prediction, until crossvalidation shows otherwise. Finally, the inclusion of extraneous variables increases the demands on the computer system being employed, making model construction slower, or in some cases outright impossible. Thus, prior reduction of separations data to a manageable size is crucial. Figureure 6 depicts situations where either too few or too many variables were used to model a system.

One common manner to achieve data reduction is to use a table of integrated peaks instead of raw chromatographic data. This has the advantage of reducing the number of variables to those compounds included in the peak list, removing baseline noise and, if the analyst knows which exact peaks to use, removing signal from irrelevant compounds. Problems with this approach include the restriction to identified compounds, which may or may not include all of the information required for modeling, and integration errors that skew results. Finally, even with an error-free comprehensive peak table, the analyst must still perform feature selection since many peaks will undoubtedly be irrelevant to the analysis.

In the case of multivariate detection, it can be advantageous to monitor only one or a few channels (wavelengths, ions, etc.) as this will selectively detect only a portion of the analytes, allowing the analyst to avoid many interfering species while greatly reducing the size of the data. However, in these cases the analyst must know exactly what signals to use and runs the risk of missing important features of the data encoded in the channels that were ignored. Further, using this approach destroys much of the multivariate advantage that can be realized through using these more complex (and expensive) detection strategies. Objective feature selection techniques generally have two steps: variable ranking, and variable selection. Objective variable ranking techniques such as analysis of variance (ANOVA) (Johnson & Synovec, 2002), the discriminating variable test (DIVA) (Rajalahti et al., 2009a, 2009b), and informative vectors (Teofilo et al., 2009) have the distinct advantage that variables are ranked based on a mathematically calculable "perceived utility" and not on subjective analyst perception. In essence, the data are given the chance to inform the user of what is relevant and what is likely noise, providing an approach that can be generalized to any set of analytical data.

ANOVA is an effective method when the goal is to discriminate between classes of samples. ANOVA calculates the F ratio for each variable: the ratio of between-class variance to within-class variance. If the F ratio for a given variable is high, it is deemed to be more valuable for describing the difference between classes. Once the F ratio has been calculated for every data point in the chromatogram, the variables can be ranked in order of decreasing F ratio.

A chemometric model is then constructed using a fraction of variables having the highest F ratio. One significant advantage of ANOVA is that the algorithm can be written with memory conservation in mind and thus is easily applied to data sets with very large numbers of samples and variables (hundreds or thousands of samples, each containing millions of variables). Consequently, it can be easily applied to a set of GC-MS chromatograms across the entire chromatogram, something that is difficult for other feature ranking approaches.

DIVA is a feature ranking technique that aids feature selection prior to chemometric analysis (Rajalahti et al., 2009a, 2009b). This approach involves the creation of a PLS-DA model using all candidate variables. Projecting this PLS-DA model onto a new single LV yields what is termed a target projected (TP) model (Rajalahti et al., 2009a). From this, the ratio of explained variance to residual variance for each variable in the TP model provides its selectivity ratio, upon which variables are ranked (Rajalahti et al., 2009a, 2009b; Kvalheim, 1990; Kvalheim & Karstang, 1989). DIVA produces a ranking that is slightly different than that produced by ANOVA, though a direct comparison on chromatographic data has not yet been performed to our knowledge. Once variables have been ranked, those to be included in the model must be selected. This is generally achieved by constructing a model using a forward-selection or backwards elimination approach, in an attempt to maximize some metric of model quality. Model quality can be assessed based on several metrics such as mean correct classification rates (Rajalahti et al., 2009b) or the degree of separation between classes of samples in principal component (or latent variable) space, for example using either a Euclidian distance-based metric (Pierce et al., 2005) or a metric that accounts for size and shape of clusters (Sinkov & Harynuk, 2011a).

The one exception to the rank-and-select approach are genetic algorithms (Yoshida et al., 2001), though due to the sheer number of variables present in a typical separation, these are not often used on the raw separations data as arriving at the optimal number and combination of variables is computationally inefficient and uncertain. Sometimes, several feature selection methods are used for a given analysis. For example, an analyst might reduce chromatogram to a peak table, selecting a series of candidate variables of interest and then perform further variable ranking and optimization on the integrated peak table, especially in the case of multidimensional separations where hundreds, if not thousands of compounds can be resolved (Felkel et al., 2010).

Finally, cross-validation is extremely important, especially when processing raw separations data and using a feature ranking approach such as ANOVA. As discussed previously, raw separations data contain on the order of 10^5 to 10^6 data points for each sample. In these cases of overdetermined systems it is entirely possible that some combinations of variables containing only noise will, by random chance, indicate a difference between samples. When handling raw separations data, a good approach to avoid this problem is to break the data set into three separate sets: a training set to construct the model, an optimization set to optimize data processing parameters (such as alignment and feature selection), and finally a test set to determine if the optimized model has any meaning (Brereton, 2007). Of course this does require that one collect data for a large number of samples so that a representative population of samples exists for each of the three subsets of data.

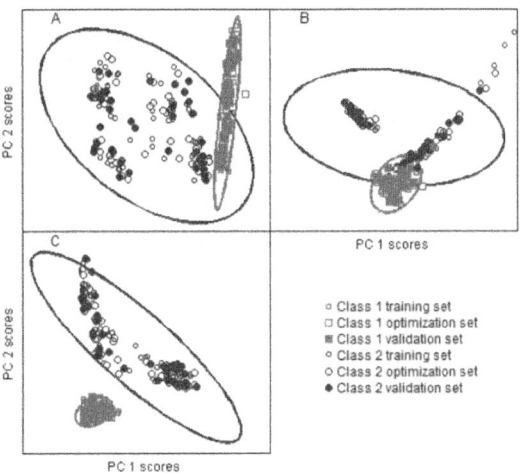

Figure. 6: Models constructed from the same data set using different numbers of top-ranked variables. (A) Too few variables; (B) Too many variables; (C) Optimal number of variables.

Applications and examples

After applying the appropriate pre-processing, different chemometric techniques can be applied according to the aim of the study. Pattern recognition is one of the chemometric methods most used in analytical chemistry and this is true for separations data. Pattern recognition can be generally divided into two classes: exploratory data analysis and unsupervised and supervised pattern recognition (Otto, 2007; Brereton, 2007). Exploratory data analysis aims to extract important information, detect outliers and identify relationships between samples and its use is recommended prior to the application of other chemometric techniques. Examples of the use of exploratory data analysis tools applied to separations data include principal component analyisis (PCA) (de la Mata-Espinosa et al., 2011a; Ruiz-Samblas et al., 2011) and factor analysis (Stanimirova et al.,

2011). Unsupervised pattern recognition techniques uncover patterns within a data set without a priori class assignment of samples. Here, the objective is to find patterns in the data which allow grouping of similar samples using, for example, cluster analysis which has been applied to separations data by Reid et al. (2007). When supervised pattern recognition is used, the classes of samples in a training set are known and used to calibrate a model, which is then used to predict class assignments of unknown samples. Some examples of which are linear discriminant analysis (LDA), and partial least squares-discriminant analysis (PLS-DA) (de la Mata-Espinosa et al., 2011b; Zorzetti et al., 2011; Sinkov et al., 2011b). In a study performed by Sinkov et al., two alignment techniques for chromatographic data were compared. The data comprised raw GC-MS chromatograms of simulated arson debris where some samples contained different types of gasoline weathered to different extents spiked into debris samples which themselves exhibited a high degree of variability in their chemical composition. The goal was to build a PLS-DA model that could correctly classify debris samples based on whether or not they contained gasoline (Figureure 7). As can be seen, the alignment algorithm used has a direct impact on the quality of the predictions. In Figureure 7A, there are multiple false positives, false negatives, and ambiguous samples. In Figureure 7B, all samples are classified correctly and there are no ambiguous samples.

Another example of applying chemometrics to separations data is depicted in Figureures 8 and 9. Here, interval PLS (iPLS) was applied to blends of oils in order to quantify the relative concentration of olive oil in the samples (de la Mata-Espinosa et al., 2011b). iPLS divides the data into a number of intervals and then calculates a PLS model for each interval. In this example, the two peak segments which presented the lower root mean square error of cross validation (RMSECV) were used for building the final PLS model. As mentioned in Section 3.3.2, PARAFAC is a chemometric tool for multidimensional data treatment. The scores and loadings obtained with PARAFAC can be used in two-way models for data exploration and quantitative analysis (Vosough et al., 2010). When small deviations in trilinearity exist within the data, usually due to relatively small shifts in retention time in the case of separations data, a modified version of PARAFAC called PARAFAC2 is recommended for use (Bro et al., 1999). Like PARAFAC, PARAFAC2 decomposes raw data into loading and score matrices but without the imposition of trilinearity as in PARAFAC. Even without this constraint, the PARAFAC2 model preserves the property of uniqueness that is so advantageous with PARAFAC. Thus, analyte profiles and concentrations can be estimated by PARAFAC2 even if chromatographic alignment is not perfect (Amigo et al., 2008; Skov et al., 2009).

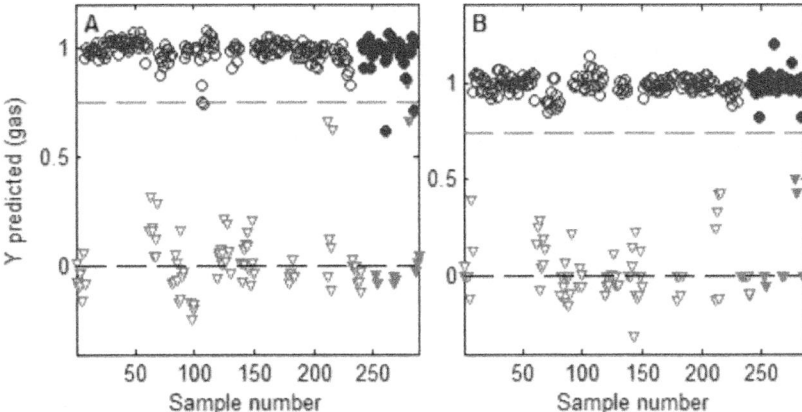

Figure. 7: PLS-DA Models for identifying gasoline in simulated arson debris derived from the same raw data, but aligned with different techniques. (A) Feature-based alignment; (B) Deuterated alkane ladder – based alignment. All other treatment and model construction algorithms were the same in both cases. Hollow markers indicate data in the training set while filled markers indicate data in the validation set. Circles represent debris containing gasoline while triangles represent gasoline-free debris. Reprinted from Sinkov et al., 2011b, with permission.

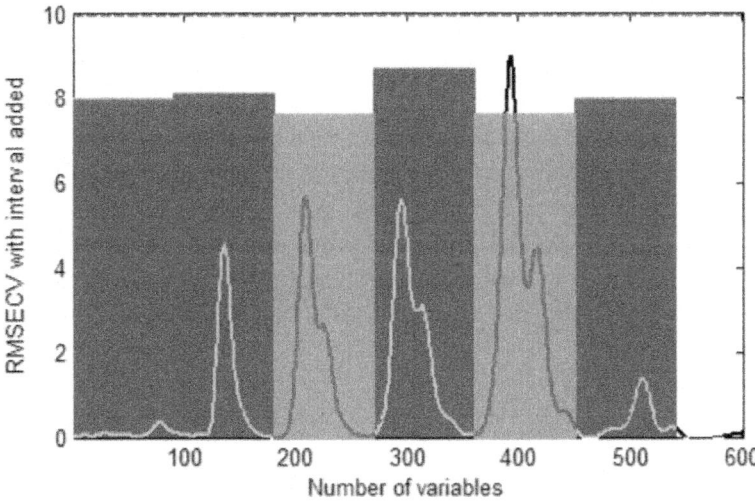

Figure. 8: Feature selection using iPLS. Segments in green showed lower RMSECV and were thus used to construct the final model. Reprinted from de la Mata-Espinosa et al., 2011b, with permission.

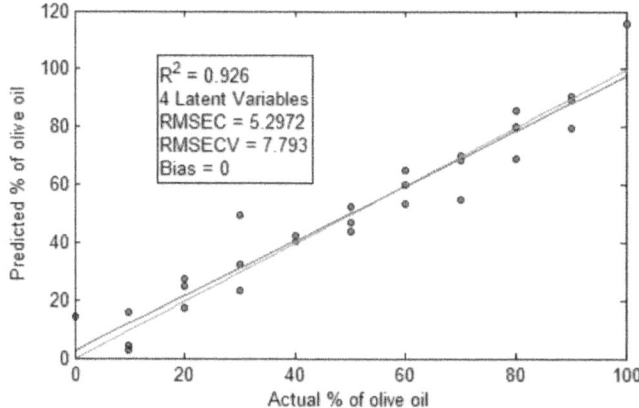

Figure. 9: Predicted vs. actual % olive oil using PLS model constructed based on results in Figureure 8. Reprinted from de la Mata-Espinosa et al., 2011b, with permission.

CONCLUSIONS

The analyst must choose from a plethora of methods for processing separations data, a potentially daunting task. It is our hope that this review will help chromatographers entertaining thoughts of applying chemometrics to their data understand what they must consider when choosing how to prepare their data. Likewise, it is hoped that we have informed chemometricians of some of the specific challenges associated with the processing of chromatographic data and the origins of those limitations. In the development of a chemometric model for the interpretation of separations data, there are numerous opportunities for missteps that will exclude key information from the model and/or generate meaningless results. However, when due care is taken there are also many opportunities to apply chemometric techniques to transform the rich data generated by these powerful analytical tools into valuable information effectively and efficiently.

REFERENCES

1.   Amigo, J.M.; Skov, T.; Bro, R.; Coello, J. & Maspoch, S. (2008). Solving GC-MS problems with PARAFAC2. Trends in Analytical Chemistry, Vol.27, No.8, (September 2008), pp. 714- 725, ISSN 0165-9936

2.   Amigo, J.M.; Skov, T. & Bro, R. (2010). ChroMATHography: solving chromatographic issues with mathematical models and intuitive graphics. Chemical Reviews, Vol.110, No.8, (May 2010), pp. 4582-4605, ISSN

1520-6890

3. Asher, B.J.; D'Angostino, L.A.; Way, J.D.; Wong, C.S. & Harynuk, J.J. (2009). Comparison of peak integration methods for the determination of enantiomeric fraction in environmental samples. Chemosphere, Vol.75, No.8, (May 2009), pp. 1042-1048, ISSN 0045-6535

4. Brereton, R.G. (2003). Chemometrics Data Analysis for the Laboratory and Chemical Plant, Wiley, ISBN 0-474-78977-8, UK

5. Brereton, R.G. (2007). Applied Chemometrics for Scientists, John Wiley & Sons Inc., ISBN 978-0-470-01686-2, Toronto, Canada

6. Bro, R. (1997). PARAFAC. Tutorial and applications. Chemometrics Intelligent Laboratory Systems, Vol.38, No.2, (October 1997), pp. 149-171, ISSN 0169-7439

7. Bro, R.; Andersson, C.A.& Kiers, H.A.L. (2009). PARAFAC-Part II. Modeling chromatographic data with retention times shifts. Journal of Chemometrics, Vol.13, No.3-4, (May-August 1999), pp. 295-309, ISSN 0886-9383

8. Casez, J. (2010). Encyclopaedia of Chromatography, (3rd ed.) CRC Press, ISBN 1-4200-8483, Florida, USA

9. Cortes, H.J.; Winniford, B.; Luong, J. & Pursch, M. (2009). Comprehensive two dimensional gas chromatography review. Journal of Separation Science, Vol.32, No.5-6, (March 2009), pp. 883-904, ISSN 1615-9306

10. de Juan, A. & Tauler, R. (2006). Multivariate Curve Resolution (MCR) from 2000: Progress in concepts and applications. Critical Reviews in Analytical Chemistry, Vol.36, No.3-4, (2006) pp. 163-176, ISSN 1040-8347

11. de la Mata-Espinosa, P.; Bosque-Sendra, J.M.; Bro, R. & Cuadros-Rodríguez, L. (2011a).

12. Discriminating olive and non-olive oils using HPLC-CAD and chemometrics.

13. Analytical and Bioanalytcial Chemistry, Vol.399, No.6, (February, 2011), pp. 2083-2092, ISSN 1618-2650

14. de la Mata-Espinosa, P.; Bosque-Sendra, J.M.; Bro, R. & Cuadros-Rodríguez, L. (2011b).

15. Olive oil quantification of edible vegetable oil blends using triacylglycerols chromatographic fingerprints and chemometric tools. Talanta, Vol.85, No.1, (July 2011), pp. 183-196, ISSN 0039-9140 Eilers, P.H.C. (2003). A perfect Smoother. Analytical Chemistry, Vol.75, No.14, (July 2003) pp.

16. 3631-3636, ISSN 0003-2700

17. Eilers, P.H.C. (2004). Parametric Time Warping. Analytical Chemistry, Vol.76, No.2, (January 2004), pp. 404-411, ISSN 0003-2700

18. Erni, F. & Frei, R.W. (1978). 2-Dimensional column liquid-chromatographic technique for resolution of complex mixtures. Journal of Chromatography, Vol.149, (February 1978), pp. 561-569 ISSN 0021-9673

19. Etxebarria, N.; Zuloaga, O.; Olivares, M.; Bartolomé. L.J. & Navarro, P. (2009). Retentiontime locked methods in gas chromatography. Journal of Chromatography A, Vol.1216, No.10, (March 2009), pp. 1624-1629 ISSN 0021-9673

20. Felinger, A. (1994). Deconvolution of overlapping skewed peaks. Analytical Chemistry, Vol.66, No.19, (October 1994), pp. 3066-3072, ISSN 0003-2700

21. Felkel, Y.; Dorr, N.; Glatz, F. & Varmuza, K. (2010). Determination of the total acid number (TAN) of used gas engine oils by IR and chemometrics applying a combined strategy for variable selection. Chemometrics and Intelligent Laboratory Systems, Vol. 01, No. 1, (March, 2010), pp. 14-22 ISSN 0169-7439

22. Franch-Lage, F.; Amigo, J.M.; Skibsted, E.; Maspoch, S. & Coello, J. (2011). Fast assessment of the surface distribution of API and excipients in tablets using NIR-hyperspectral imaging. International Journal of Pharmaceutics, Vol.441, No.1-2, (June 2011), pp. 27- 35, ISSN 0378-5173

23. François, I.; Sandra, K. & Sandra, P. (2009). Comprehensive liquid chromatography: undamental aspects and practical considerations—A review. Analytica Chimica Acta, Vol.641, No.1-2, (May 2009), pp. 14-31, ISSN 0003-2670

24. Gan, F.; Ruan, G. & Mo, J. (2006). Baseline correction by improved iterative polynomial fitting with automatic threshold. Chemometrics and Intelligent Laboratory Systems, Vol.82, No.1 (May 2006), pp. 59-65, ISSN 0169-7439

25. Górecki, T.; Harynuk, J. & Panić, O. (2004). The evolution of comprehensive twodimensional gas chromatography, Journal of Separation Science, Vol.27 (2004) pp.359-379, ISSN 1615-9306

26. Harshman, R.A. (1970). Foundations of the PARAFAC procedure: models and conditions for an 'exploratory' multimodal factor analysis. UCLA Working Papers Phonet. Vol 16, (1970), pp. 1-84

27. Johnson, K.J. & Synovec, R.E. (2002). Pattern recognition of jet fuels: comprehensive GC×GC with ANOVA-based feature selection and

principal component analysis.

28. Chemometrics and Intelligent Laboratory Systems, Vol.60, No.1-2, (January 2002), pp. 225-237, ISSN 0169-7439

29. Johnson, K.J.; Wright, B.W.; Jarman, K.H. & Synovec, R.E. (2003). High-speed peak matching algorithm for retention time alignment of gas chromatographic data for chemometric analysis. Journal of Chromatography A, Vol.996, No.1-2, (May 2003), pp. 141-155, ISSN 0021-9673

30. Kaczmarek, K.; Walczak, B.; de Jong, S. & Vandeginste, B.G.M. (2005). Baseline reduction in two dimensional gel electrophoresis images. Acta Chromatographica, Vol.15 (2005), pp. 82-96, ISSN 1233-2356

31. Kivilompolo, M.; Pol, J. & Hyotylainen, T. (2011). Comprehensive two-dimensional liquid chormatography (LC×LC): A review. LC GC Europe, Vol.24, No 5 (May 2011), pp.

32. 232-+, ISSN 1471-6577

33. Kjeldahl, K. & Bro, R. (2010). Some common misunderstandings in chemometrics. Journal of Chemometrics, Vol.24, No.7-8, (July-August, 2011), pp. 558-564, ISSN 0886-9383

34. Kong, K.; Ye, F.; Guo, L.; Tian, J. & Xu, G. (2005). Deconvolution of overlapped peaks based on the exponentially modified Gaussian model in comprehensive two-dimensional gas chromatography, Journal of Chromatography A, Vol.1086, No.1-2 (September 2005) pp. 160-164, ISSN 0021-9673

35. Kvalheim, O.M. & Karstang, T.V. (1989). Interpretation of latent-variable regression models.

36. Chemometrics and Intelligent Laboratory Systems, Vol.7, No.1-2, (December 1989), pp. 39-51, ISSN 0169-7439

37. Kvalheim, O.M. (1990). Latent-variable regression models with higher-order terms: An extension of response modelling by orthogonal design and multiple linear

38. regression. Chemometrics and Intelligent Laboratory Systems, Vol.8, No.1, (May 1990), pp. 59-67, ISSN 0169-7439

39. Lavine, B.K.; Brzozowski, D.; Moores, A.J.; Davidson, C.E. & Mayfield, H.T. (2001). Genetic algorithm for fuel spill identification. Analytica Chimica Acta, Vol.437, No.2, (June 2001), pp. 233-246, ISSN 0003-2670

40. Laursen, K.; Frederiksen, S.S.; Leuenhagen, C. & Bro, R. (2010). Chemometric quality contrl of chromatographic purity. Journal of Chromatography A, Vol.1217, No.42 (October 2010), pp. 6503-6510,

ISSN 0021-9673

41. Li, Y.H.; Wojcik, R & Dovichi, N.J. (2011). A replaceable microreactor for on-line protein digestion in a two dimensional capillary electrophoresis system with tandem mass spectrometry detection. Journal of Chromatography A, Vol.1218, No.15 (April 2011), pp. 2007-2011, ISSN 0021-9673

42. Liang, Y.; Xie, P. & Chau, F. (2010). Chromatographic fingerprinting and related chemometric techniques for quality control of traditional Chinese medicines.

43. Journal of Separation Science, Vol.33, No.3 (February 2010), pp. 410-421, ISSN 1615- 9314

44. Liu, Z. & Phillips, J.B. (1991). Comprehensive 2-dimensional gas-chromatography using a modulator interface. Journal of Chromatographic Science, Vol.29, No.6 (June 1991), pp. 227-231, ISSN 0021-9665

45. Maeder, M. (1987). Evolving factor analysis for the resolution of overlapping chromatographic peaks. Analytical Chemistry, Vol.59, No.3, (February 1987), pp 527- 530, ISSN 0003-2700

46. Marini, F.; D'Aloise, A.; Bucci, R.; Buiarelli, F.; Magri, A.L. & Magri, D. (2011), Fast analysis of 4 phenolic acids in olive oil by HPLC-DAD and chemometrics, Chemometrics and Intelligent laboratory systems, Vol.106, No.1, (March 2011), pp. 142-149, ISSN 0169- 7439

47. Michels, D.A.; Hu, S.; Schoenherr, R.M.; Eggertson, M.J. & Dovichi, N.J. (2002), Fully automated two-dimensional capillary electrophoresis for high sensitivity protein

48. analysis, Molecular & Cellular Proteomics, Vol.1, No.1, (January 2002), pp. 69-74, ISSN 1535-9476

49. Miller, J.M. (2005). Chromatography: concepts and contrasts, (2nd ed.) Wiley, ISBN 0471472077, Hoboken, USA

50. Mommers, J.; Knooren, J.; Mengerink, Y.; Wilbers, A.; Vreuls, R. & van der Wal, S. (2011). Retention time locking procedure for comprehensive two-dimensional gas

51. chromatography. Journal of Chromatography A, Vol.1218, No.21 (May, 2011), pp.3159-3165 ISSN 0021-9673

52. Nielsen, N-P.; Cartensen, J.M. & Smedsgaard, J. (1998). Aligning of single and multiple wavelength chromatographic profiles for chemometric data analysis using correlation optimised warping. Journal of Chromatography A, Vol.805, No.1-2 (May 1998), pp. 17-35 ISSN 0021-9673

53. Otto, M. (2007). Chemometrics, Wiley-VCH, ISBN 978-3-527-31418-8,

Weinheim, Germany Persson, P.O. & Strang, G. (2003). Smoothing by Savitzky-Golay and Legendre filters, In: Mathematical Systems Theory in Biology, Communications, Computation and Finance, Rosenthal J. Gilliam D.S., pp. 301-315, IMA Vol. Math. Appl., 134, Springer, ISBN 978-0387-40319-9, New York, USA

54. Pierce K.M.; Hope J.L.; Johnson K.J.; Wright B.W. & Synovec R.E. (2005). Classification of gasoline data obtained by gas chromatography using a piecewise alignment algorithm combined with feature selection and principal component analysis.

55. Journal of Chromatography A, Vol.1096, No.1-2, (November 2005), pp. 101-110, ISSN 0021-9673

56. Poole, C.F. (2003). The Essence of Chromatography, (1st ed.), Elsevier, ISBN 0444501983, Amsterdam, The Netherlands

57. Rajalahti, T.; Arneberg, R.; Berven, F.S.; Myhr, K.M.; Ulvik, R.J. & Kvalheim, O.M. (2009a).

58. Biomarker discovery in mass spectral profiles by means of selectivity ratio plot.

59. Chemometrics and Intelligent Laboratory Systems, Vol. 95, No. 1, (January 2009), pp. 35-48, ISSN 0169-7439

60. Rajalahti, T.; Arneberg, R.; Kroksveen, A.C.; Berle, M.; Myhr, K.M. & Kvalheim, O.M. (2009b). Discriminating Variable Test and Selectivity Ratio Plot: Quantitative Tools for Interpretation and Variable (Biomarker) Selection in Complex Spectral or Chromatographic Profiles. Analytical Chemistry, Vol. 81, No. 7, (April 2009), pp. 2581-2590, ISSN 0169-7439

61. Reid, R.G.; Durham, D.G.; Boyle S.P.; Low, A.S. & Wangboonskul, J. (2007). Differentiation of opium and poppy straw using capillary electrophoresis and pattern recognition techniques. Analytica Chimica Acta, Vol.605, No. 1, (December 2007), pp. 20-27, ISSN 0003-2670

62. Ruiz-Samblas, C.; Cuadros-Rodriguez, L.; Gonzalez-Casado, A.; Rodriguez Garcia, F.D.P; de a Mata-Espinosa, P.; Bosque-Sendra, J.M. (2011). Multivariate analysis of HT/GC-(IT)MS chromatographic profiles of triacylglycerols for classification of olive oil varieties, Analytical and Bionalytical Chemistry, Vol.399, No.6 (February 2011), pp. 2093-2103, ISSN 1618-2642

63. Sarkar, S.; Dutta, P.K. & Roy, N.C. (1998). A blind-deconvolution approach for chromatographic and spectroscopic peak restoration, IEEE transactions on instrumentation and measurement, Vol.47, No.4 (August 1998), pp. 941-947, ISSN 0018-9456

64. Savorani, F.; Tomasi, G. & Engelsen, S.B. (2010). Icoshift: A versatile tool for the rapid alignment of 1D NMR spectra. Journal of Magnetic Resonance, Vol.202, No.2, (February 2010), pp. 190-202 ISSN 1090-7807

65. Sinkov, N.A. & Harynuk, J.J. (2011a). Cluster resolution: A metric for automated, objective and optimized feature selection in chemometric modeling. Talanta, Vol.83, No.4, (January 2011), pp. 1079-1087, ISSN 0039-9140

66. Sinkov, N.A.; Johnston, B.M.; Sandercock, P.M.L. & Harynuk, J.J. (2011b). Automated optimization and construction of chemometric models based on highly variable raw chromatographic data. Analytica Chimica Acta, Vol.697, No.1-2, (July 2011), pp. 8-15, ISSN 1873-4324

67. Skov, T.; Hoggard, J.C.; Bro, R. & Synovec, R.E. (2009). Handling within run retention time shifts in two-dimensional chromatography data using shift correction and modeling. Journal of Chromatography A, Vol.1216, No.18, (May 2009), pp. 4020-4029, ISSN 0021-9673

68. Stanimirova, I.; Boucon, C. & Walczak, B. (2011). Relating gas chromatographic profiles to sensory measurements describing the end products of the Maillard reaction.

69. Talanta, Vol.83, No 4, (January 2011), pp. 1239-1246, ISSN 0039-9140 Tauler, R. & Barceló, D. (1993). Multivariate curve resolution applied to liquid chromatography-diode array detection. Trends in

70. Analytical Chemistry, Vol.12, No.8, (1993), pp. 319-327, ISSN 0165-9936 Teofilo, R.F.; Martins, J.P.A. & Ferreira, M.M.C. (2009). Sorting variables by using informative vectors as a strategy for feature selection in multivariate regression. Journal of Chemometrics, Vol.23, No.1-2, (January-February 2009), pp. 32-48, ISSN 0886-9383

71. Tomasi, G.; Van den Berg, F. & Andersson, C. (2004). Correlation optimized warping anddynamic time warping as preprocessing methods for chromatographic data. Journal of Chemometrics, Vol.18, No.5, (May 2004), pp. 231-241, ISSN 0886-9383

72. Toppo, S.; Roveri, A.; Vitale, M.P.; Zaccarin, M.; Serain, E.; Apostolidis, E.; Gion, M.,Mariorino, M. & Ursini, F. (2008). MPA: A multiple peak alignment algorithm toperform multiple comparisons of liquid-phase proteomic profiles. Proteomics, Vol.8,No.2, (January 2008), pp. 250-253 ISSN 1615-9861

73. Van den Berg, F.; Tomasi, G. & Viereck, N. (2005). Warping: investigation of NMR preprocessing and correction, In: Magnetic Resonance in Food Science:

74. The Multivariate Challenge, Engelsen, S.B., Belton, P.S., Jakobsen, H.J., pp. 131-138, Royal Society of Chemistry, ISBN 0854046488, Cambridge, UK

75. Van Nederkassel, A.M.; Dazykowski, M.; Eilers, P.H.C. & Vander Heyden, Y. (2006). A comparison of three algorithms for chromatograms alignment. Journal of Chromatography A, Vol.118, No.2 (June 2006), pp. 199-210 ISSN 0021-9673

76. Vivó-Truyols, G.; Torres-Lapasió, J.R.; Caballero R.D. & García-Alvarez-Coque, M.C. (2002).

77. Peak deconvolution in one-dimensional chromatography using a two-way data approach. Journal of Chromatography A, Vol.958, No.1-2, (June, 2002), pp. 35-49, ISSN 0021-9673

78. Vosough, M.; Bayat, M. & Salemi, A. (2010). Matrix-free analysis of aflatoxins in pistachio nuts using parallel factor modeling of liquid chromatography diode-array detection data. Analytica Chimica Acta, Vol.663, No.1, (March 2010), pp. 11-18. ISSN 0003-2670

79. Watson, N.E.; VanWingerden, M.M.; Pierce, K.M.; Wright, B.W. & Synovec, R.E. (2006).

80. Classification of high-speed gas chromatography-mass spectrometry data by principal component analysis coupled with piecewise alignment and feature selection. Journal of Chromatography A, Vol.1129, No.1, (September, 2006), pp. 111- 118, ISSN 0021-9673

81. Yao, W., Yin, X. & Hu Y. (2007). A new algorithm of piecewise automated beam search for peak alignment of chromatographic fingerprints. Journal of Chromatography A, Vol.1160, No.1-2, (August 2007), pp. 254-262. ISSN 0021-9673

82. Yoshida H.; Leardi R.; Funatsu K. & Varmuza K. (2001) Feature selection by genetic algorithms for mass spectral classifiers. Analytica Chimica Acta, 446, 1-2, (November 2001), pp. 485-494, ISSN 0003-2670

83. Zhang D.; Huang, X.; Regnier, F.E. & Zhang, M. (2008). Two-dimensional correlation optimized warping algorithm for aligning GC×GC-MS data. Analytical Chemistry, Vol.80, No.8 (April 2008), pp. 2664-2671, ISSN 0003-2700

84. Zhang, Z.M.; Chen, S. & Liang, Y.Z. (2010). Baseline correction using adaptive iteratively reweighted penalized least squares. Analyst, Vol.5 (February 2010), pp. 1138-1146, ISSN 0003-2654

85. Zorzetti, B.M.; Shaver, J.M. & Harynuk, J.J. (2011). Estimation of the age of a weathered mixture of volatile organic compounds. Analytica Chimica Acta, Vol.694, No.1-2, (May 2011), pp. 31-37, ISSN 0003-2670

CITATION

CHAPTER 1

Riccardo Guidetti, Roberto Beghi and Valentina Giovenzana (2012). Chemometrics in Food Technology, Chemometrics in Practical Applications, Dr. Kurt Varmuza (Ed.), ISBN: 978-953-51-0438-4, InTech, DOI: 10.5772/34148.

CHAPTER 2

Chibuike C. Udenigwe and Rotimi E. Aluko, Chemometric Analysis of the Amino Acid Requirements of Antioxidant Food Protein Hydrolysates, doi:10.3390/ijms12053148.

CHAPTER 3

Jun Jiang and Xiaobin Jia, Profiling of Fatty Acids Composition in Suet Oil Based on GC–EI-qMS and Chemometrics Analysis, doi:10.3390/ijms16022864.

CHAPTER 4

Bekzod Khakimov, Birthe Møller Jespersen and Søren Balling Engelsen, Comprehensive and Comparative Metabolomic Profiling of Wheat, Barley, Oat and Rye Using Gas Chromatography-Mass Spectrometry and Advanced Chemometrics, doi:10.3390/foods3040569.

CHAPTER 5

Yidan Bao, Wenwen Kong, Yong He, Fei Liu Tian Tian and Weijun Zhou, Quantitative Analysis of Total Amino Acid in Barley Leaves under Herbicide Stress Using Spectroscopic Technology and Chemometrics, doi:10.3390/s121013393.

CHAPTER 6

Nádia Reis, Adriana S. Franca, and Leandro S. Oliveira, "Concomitant Use of Fourier Transform Infrared Attenuated Total Reflectance Spectroscopy and Chemometrics for Quantification of Multiple Adulterants in Roasted and Ground Coffee," Journal of Spectroscopy, vol. 2016, Article ID 4974173, 7 pages, 2016. doi:10.1155/2016/4974173.

CHAPTER 7

Yulia B. Monakhova, Rolf Godelmann, Claudia Andlauer, Thomas Kuballa, and Dirk W. Lachenmeier, "Identification of Imitation Cheese and Imitation Ice Cream Based on Vegetable Fat Using NMR Spectroscopy and Chemometrics," International Journal of Food Science, vol. 2013, Article ID 367841, 9 pages, 2013. doi:10.1155/2013/367841.

CHAPTER 8

Cozzolino Daniel, The Role of Visible and Infrared Spectroscopy Combined with Chemometrics to Measure Phenolic Compounds in Grape and Wine Samples, doi:10.3390/molecules20010726.

CHAPTER 9

José Camacho (2012). Exploratory Data Analysis with Latent Subspace Models, Chemometrics in Practical Applications, Dr. Kurt Varmuza (Ed.), ISBN: 978-953-51-0438-4, InTech, DOI: 10.5772/32149

CHAPTER 10

Marcelo Maraschin, Shirley Kuhnen, Priscilla M.M. Lemos, Simone Kobe de Oliveira, Diego A. da Silva, Maíra M. Tomazzoli, Ana Carolina V. Souza, Rúbia Mara Pinto, Virgílio G. Uarrota, Ivanir Cella, Antônio G. Ferreira, Amélia R.S. Zeggio, Maria B.R. Veleirinho, Ivone Delgadillo and Flavia A. Vieira (2012). Metabolomics and Chemometrics as Tools for Chemo(bio)diversity Analysis - Maize Landraces and Propolis, Chemometrics in Practical Applications, Dr. Kurt Varmuza (Ed.), ISBN: 978-953-51-0438-4, InTech, DOI: 10.5772/32584.

CHAPTER 11

Hideyuki Shinzawa, Masakazu Nishida, Toshiyuki Tanaka, Kenzi Suzuki and Wataru Kanematsu (2012). PARAFAC Analysis for Temperature-Dependent NMR Spectra of Poly(Lactic Acid) Nanocomposite, Chemometrics in Practical Applications, Dr. Kurt Varmuza (Ed.), ISBN: 978-953-51-0438-4, InTech, DOI: 10.5772/33030.

CHAPTER 12

Rashid Atta Khan, Sharifuddin M. Zain, Hafizan Juahir, Mohd Kamil Yusoff and Tg Hanidza T.I. (2012). Using Principal Component Scores and Artificial Neural Networks in Predicting Water Quality Index, Chemometrics in Practical Applications, Dr. Kurt Varmuza (Ed.), ISBN: 978-953-51-0438-4, InTech, DOI: 10.5772/32577.

CHAPTER 13

James J. Harynuk, A. Paulina de la Mata and Nikolai A. Sinkov (2012). Application of Chemometrics to the Interpretation of Analytical Separations Data, Chemometrics in Practical Applications, Dr. Kurt Varmuza (Ed.), ISBN: 978-953-51-0438-4

INDEX

A

acid detergent fiber (ADF) 94
ANN (Artificial Neural Networks) 226
Antioxidant enzymatic 46
attenuated total reflectance (ATR 196
attenuated total reflectance (ATR) 140, 144
Attenuated Total Reflectance (ATR) 106

B

Back propagation (BP) 229
biochemical oxygen demand (BOD), 226
Box–Behnken design (BBD) 61, 62, 64

C

Capillary electrophoresis (CE) 243, 244
Cereal metabolites 78
chemical oxygen demand (COD), 226, 228
condensed tannins (CT) 141, 145
Cross-polarization magic-angle (CP-MAS) 208

D

differential scanning calorimetry (DSC) 218

Diffuse Reflectance Fourier Transform Infrared Spectroscopy (DRIFTS) 106
dummy scans (DS) 124

E

effective wavelengths (EWs) 93, 97
Eigendecomposition (ED 160
Employed technology 9
Exploratory Data Analysis (EDA) 153

F

Fatty acid (FA) 61
field-of-view (FOV) 95
Fourier transform (FT) 196
Fourier transform infrared vibrational spectroscopy (FTIR) 196

G

gas chromatography (GC) 120
Gas chromatography (GC) 244

H

hydrophilic interaction (HILIC 244

I

Important voices 1
Individual amino 45, 47, 49
infrared (IR) 196
Interim National Water Quality Standards for Malaysia (INWQS) 225
Internal standard (IS) 78

L

latent variables (LV) 109
latent variables (LVs) 154
Light source 5
limits of quantification (LOQ) 61
linear discriminant analysis (LDA), 260
liquid chromatography 243, 244, 264, 268, 269
low-density lipoprotein (LDL) 76, 88

M

Metabolite extraction protocol 86
modes for LC separations, including for example reverse-phase (RPLC), normalphase (NPLC), ion 244
Montmorillonite (MMT) 213
Multiple scatter correction (MSC) 109
multivariate curve resolution (MCR) 254

N

Namely, sets of temperature-dependent NMR 208
Near infrared (NIR) 94
neutral detergent fiber (NDF) 94
non-point source (NPS). 226
nuclear magnetic resonance (NMR) 119, 134, 135

O

orthogonal projection to latent structures (O-PLS) 143

P

Partial Least Square (PLS) 105

partial least squares (PLS) 45, 47, 48, 119
Partial least squares (PLS) 141
Phenolic compounds 137
point source (PS) 226
principal component analysis (PCA) 227
Principal component analysis (PCA) 207, 209
Principal Component Analysis (PCA) 83, 153
principal components analysis (PCA) 197, 200
Provides a tutorial on the fundamental concept of Parallel factor (PARAFAC) 207

Q

quantitative structure-activity relationship (QSAR) 47

R

ratio deviation in prediction (RPD) 143
Receiver gain (RG) 124
Region of near infrared (NIR) 2
Regression coefficients (RC) 96
Relatively unbiased 86
relative standard deviations (RSD) 125
root mean square errors of calibration (RMSEC) 144
root mean squares error of prediction (RMSEP) 93, 96, 97

S

size exclusion (SEC), 244
standard error in prediction (SEP) 141
standard normal variate (SNV) 95, 103
Standard Normal Variates (SNV) 109
suet oil (SO) 61, 64
sweep width (SW) 124

T

thermal desorption (TD) 120
total amino acid (TAA) 93, 100, 101
total ion chromatogram (TIC) 64

total ion current (TIC) 80
transmission electron microscope (TEM) 218

V

Variable Importance for the Projection (VIP) 50, 56

W

water quality index (WQI) 225